THE MORGAN KAUFMANN SERIES IN INTERACTIVE 3D TECHNOLOGY

Series Editor: David H. Eberly, Geometric Tools, Inc.

The game industry is a powerful and driving force in the evolution of computer technology. As the capabilities of personal computers, peripheral hardware, and game consoles have grown, so has the demand for quality information about the algorithms, tools, and descriptions needed to take advantage of this new technology. To satisfy this demand and establish a new level of professional reference for the game developer, we created the *Morgan Kaufmann Series in Interactive 3D Technology*. Books in the series are written for developers by leading industry professionals and academic researchers, and cover the state of the art in real-time 3D. The series emphasizes practical, working solutions and solid software-engineering principles. The goal is for the developer to be able to implement real systems from the fundamental ideas, whether it be for games or for other applications.

VISUALIZING QUATERNIONS

VISUALIZING QUATERNIONS

ANDREW J. HANSON

AMSTERDAM • BOSTON • HEIDELBERG
LONDON • NEW YORK • OXFORD
PARIS • SAN DIEGO • SAN FRANCISCO
SINGAPORE • SYDNEY • TOKYO

ELSEVIER   Morgan Kaufmann is an imprint of Elsevier

MORGAN KAUFMANN PUBLISHERS

Senior Editor	Tim Cox
Assistant Editor	Rick Camp
Editorial Assistant	Jessie Evans
Publishing Services Manager	Simon Crump
Senior Production Editor	Paul Gottehrer
Cover Design	Chen Design Associates
Text Design	Chen Design Associates
Composition	VTEX Typesetting Services
Technical Illustration	Dartmouth Publishing, Inc.
Interior printer	Transcontinental Printing
Cover printer	Transcontinental Printing

Morgan Kaufmann Publishers is an imprint of Elsevier.
500 Sansome Street, Suite 400, San Francisco, CA 94111

This book is printed on acid-free paper.

This material is based upon work supported by the National Science Foundation under Grant No. 0204112. Any opinions, findings, and conclusions or recommendations expressed in this material ate those of the author and do not necessarily reflect the views of the National Science Foundation.

Library of Congress Cataloging-in-Publication Data

ISBN 13: 978-0-12-088400-1
ISBN 10: 0-12-088400-3

For information on all Morgan Kaufmann publications, visit our Web site at www.mkp.com or www.books.elsevier.com

This book is moved to digital printing, with reprint corrections, 2016: reformatted with additional corrections, 2020.

To My Father
Alfred Olaf Hanson
1914–2005

ANDREW J. HANSON is a professor of computer science at Indiana University in Bloomington, Indiana, and has taught courses in computer graphics, computer vision, programming languages, and scientific visualization. He received a BA in chemistry and physics from Harvard College in 1966 and a PhD in theoretical physics from MIT in 1971. Before coming to Indiana University, he did research in theoretical physics at the Institute for Advanced Study, Cornell University, the Stanford Linear Accelerator Center, and the Lawrence-Berkeley Laboratory, and then in computer vision at the SRI Artificial Intelligence Center in Menlo Park, CA. He has published a wide variety of technical articles concerning problems in theoretical physics, machine vision, computer graphics, and scientific visualization methods. His current research interests include scientific visualization (with applications in mathematics, cosmology and astrophysics, special and general relativity, and string theory), optimal model selection, machine vision, computer graphics, perception, collaborative methods in virtual reality, and the design of interactive user interfaces for virtual reality and visualization applications.

Contents

* A dagger (†) denotes a section with advanced content that can be skipped on a first reading.

Foreword

My first experiences with Andy Hanson's work were with his visualizations of mathematical and geometric problems, such as his clever and engaging "Visualizing Fermat's Last Theorem" (1990).* It was clear to me that he was a talented geometer and an excellent communicator, so when the SIGGRAPH conference started to bundle courses with its technical program I encouraged Andy to develop an advanced course on geometry. This book is the result of that course. His course is still doing well at SIGGRAPH—the audience overflowed a good-sized room at the conference this year. Even some of Andy's students couldn't get in.

As a computer graphics instructor I have been interested in the way quaternions represent 3D rotations and I hoped Andy's book would help me learn more about that. It has indeed shown me how quaternions represent 3D rotation frames and how quaternion interpolation provides an elegant and powerful way to interpolate 3D rotations. But I have learned a great deal more than that. Andy puts quaternions into a historical and mathematical context and shows how unit quaternions can be understood through their projections into the closed 3D unit sphere $\mathbf{S}^3$. This enables him to take you from a discussion of twisted belts, rolling balls, and gimbal lock to an appreciation of how thinking in $\mathbf{S}^3$ helps you to understand quaternion curves and how to interpolate quaternions. This in turn helps you to interpolate rotations more effectively as fully described in the later chapters.

Andy has written the first section of this book to be accessible to anyone with a bit of mathematics background, such as a passing familiarity with complex variables, linear algebra, and analytic geometry. I think he has accomplished this, but even an experienced reader will want to pause occasionally in thought in this first part. My mathematics background is in noncommutative rings, so many of the ideas presented here were not totally new to me. Even so, Andy's fluency in moving between algebra and geometry and his intuition into the meaning of curves

---

* "Visualizing Fermat's Last Theorem," *SIGGRAPH Video Review*, issue 61, piece 4.

in quaternion space led me to resort to paper and pencil a few times to work out his conclusions. The result has been worth the effort. Quaternions are beautiful mathematics in their own right and lead to some mathematical ideas that tie geometry and algebra together in a most lovely way. I am looking forward to further exploring these ideas.

This book closes a circle for me since I first started doing computer graphics in order to explore mathematics (though at a rather elementary level for my undergraduate mathematics students). Now computer graphics has led me to a place where I can use my graphics interests to return to mathematical questions. I hope this book will lead you on a similar journey.

Steve Cunningham
August, 2005
Iowa City, Iowa

Preface

The purpose of this book is to examine both the properties and applications of quaternions, and, in particular, to explore visual representations that help to develop our intuition about quaternions and their exploitation. Of all the natural advantages of quaternions, several stand out clearly above all others:

- Normalized quaternions are simply Euclidean four-vectors of length one, and thus are points on a unit hypersphere (known to mathematicians as the *three-sphere*) embedded in four dimensions. Just as the ordinary unit sphere has two degrees of freedom, e.g., latitude and longitude, the unit hypersphere has three degrees of freedom.
- There is a relationship between quaternions and three-dimensional rotations that permits the three rotational degrees of freedom to be represented exactly by the three degrees of freedom of a normalized quaternion.
- Because quaternions relate three-dimensional coordinate frames to points on a unit hypersphere, it turns out that quaternions provide a meaningful and reliable global framework that we can use to measure the distance or similarity between two different three-dimensional coordinate frames.
- Finally, again because they are points on a hypersphere, quaternions can be used to define optimal methods for smooth interpolation among sampled sets of three-dimensional coordinate frames.

These features in fact characterize essential properties—an appealing geometric interpretation, the existence of meaningful similarity measures, and interpolatability of sampled data—that are advantageous in general for mathematical representations that are used to model physical actions and states. Part **I** of the book will focus on understanding these basic properties and how they can be useful to us.

For the reader who is looking for more novel ways to exploit the properties of quaternions, we turn in Part **II** of the book to a number applications and insights that go well beyond conventional quaternion methods. Among the more sophisticated applications of quaternions presented in the text, I emphasize particularly the exploitation of *quaternion manifolds*, which have appeared in scattered technical articles, but have no equivalent systematic treatment elsewhere. There are three basic types of quaternion manifolds whose properties can be briefly summarized as follows:

Quaternion Curves. Quaternion curves are familiar in the computer graphics literature as the means to implement smooth interpolations of three-dimensional coordinate frames; however, there are many other applications that lead to natural curves in quaternion space representing continuous sequences of orientation frames. In a typical application from elementary differential geometry, moving frames are attached to a curve, constrained so that a particular frame axis follows the curve's tangent. This leads to classical constructs such as the Frenet–Serret frame; however, there are other equally important, but less familiar, frames that complement the Frenet–Serret equations, and can be used when the latter fail. By studying the family of all possible frames using the quaternion representation, we are led to an elegant system of quaternion differential equations describing the evolution of curve framings in general; these equations, known since the 19th century, correspond in essence to the square root of the conventional equations. The dynamic quaternion frame equations are now commonly used to replace the more traditional Euler-angle gyroscope dynamics equations in modern approaches to physically-based modeling for computer graphics. The entire family of concepts is amenable to depiction as curves embedded in quaternion space, with a variety of possible visualizations, applications, and intuitive interpretations.

Quaternion Surfaces. Computer graphicists and others with some mathematical background may be familiar with the *Gauss map*, which is essentially a mesh on a unit sphere, each point of which is the normal direction of the corresponding vertex on a meshed surface; the Gauss map contains deep hints of the fundamental nature of Riemannian geometry. In graphical models, surfaces are almost *never* adequately described simply by their vertex positions and normals; typically, in order to apply a texture, for example, a complete *frame* is required. I have proposed the *quaternion Gauss map* [70] as a natural and fundamental approach to the surface frame problem. The properties of surface frame fields have no consistent representation in terms of Euler angles, but can be studied as a surface with meaningful metric properties using the quaternion Gauss map. Many fundamental properties of differential geometry, e.g., the absence of a global set of coordinate frames on an ordinary unit sphere, are inescapably exposed by the quaternion Gauss map and the associated quaternion surfaces. For those with mathematical inclinations, we present a novel quaternion extension of the Weingarten equations for the classical differential geometry of surfaces.

Quaternion Volumes. Since the unit quaternions that are the central focus of this book are themselves three-dimensional, and the entire quaternion space is a hypersphere, the only remaining spaces that can usefully be constructed from quaternions consist of bounded volumetric objects. Quaternion volumes have found a new and essential application in the construction of *orientation domains*; in particular, a human or robotic joint with three full degrees of freedom can have any state represented as a quaternion point. The collection of all such states is a volume, normally continuous, and the *permissible space* of states is delineated by the boundary of that volume, thereby giving an elegant means of clamping errorful orientation measurements or commands to the permitted domain. Quaternion volumes incorporate a natural method for determining consistent distances and optimal paths from an invalid state to the allowed orientation domain.

Finally, in Part **III** of the book, we briefly look at the issues involved in attempting to generalize the properties of quaternions to dimensions other than three.

Intended Audience: I assume initially that the reader has a passing familiarity with complex variables, linear algebra, and analytic geometry, and has tried to transform a vertex with a matrix a few times. Certain topics, such as complex variables, will be reviewed thoroughly since they will serve as a framework for producing analogies that will be used to study quaternions and their properties. As the text moves on to more complex concepts, and particularly in the later chapters, the mathematics becomes more challenging, though I have made substantial efforts to add intuitive remarks and self-explanatory notes so that most readers should still be able to enjoy and assimilate essential features of the material.

My intent is that the earlier chapters should be informative to almost anyone with the noted minimal background, and that these readers need not trouble themselves with the more technical chapters. However, those who can potentially benefit

from some of the more advanced mathematical concepts that arise in the study of
quaternions should find ample material to keep them occupied in the later chapters.

The book keeps several different levels of readers in mind, with signposts indicating when a more advanced background may be needed to benefit fully from a
particular section. I therefore hope that readers with any number of diverse back-
grounds and interests, plus enough intellectual curiosity to follow a given train of
thought through to its inevitable conclusions, can absorb the arguments in the sections that are relevant to their interests, find them as fascinating as I do, and perhaps
generate some of their own original insights by asking "why" one time too many!

The basic material presented here covers first the needs of those wishing to
deepen their intuitive understanding of the relationship between quaternions and
rotations, quaternion-based animation, and moving coordinate frames. The next
level of material deals with concepts of 3D curves and surfaces appearing extensively in computer graphics and scientific visualization applications. Finally, there is
a selection of topics addressing the needs of those familiar with advanced problems
and research, including an attempt to address some of the deeper theoretical un-
derpinnings of quaternions, such as the question of what aspects of quaternions are
and are not generalizable to other numbers systems and other dimensions. While
we do not focus extensively on practical implementations or attempt to provide a
"cookbook" for the graphics game developer, we do supply selected software examples to illustrate various points, and summarize the core body of code in an
appendix.

Illustrations: The illustrations form an essential part of the text, and my goal has
been to provide whatever visual cues are at my disposal to meet the objective of
allowing the reader to truly "visualize quaternions." There are three main styles of
graphics that complement the text: the first consists of simple line drawings, many
formatted originally in the xfig package, and largely redrawn by the publisher's
artists to correspond to my original drawings and sketches. The major portion of
the mathematically precise illustrations were obtained directly from their equations using Mathematica™; high-resolution 3D shaded models were produced by
the MeshView package [84,85], my own quaternion-friendly 3D-plus-4D visualization tool that accepts a wide variety of elementary modeling data. The actual
model files used by MeshView were themselves typically computed and created
using Mathematica.

Outline: The history of quaternions has been recounted many times, and in many
ways. Our introduction will pause only momentarily to dwell on the historical as-

pects and development of the field, referring the reader to a number of sources more concerned with this material than we are. The book is divided into several major thematic units. First we look into various ways in which rotations, and hence quaternions, enter into everyday experience, and then begin to study 2D and 3D rotations. We attempt to draw fruitful analogies between ordinary complex variables and quaternions, exploiting the surprising richness of two-dimensional rotations and following an inevitable path to the relationship of quaternions to three-dimensional orientation spaces. Next, we examine the quaternion description of moving frames of coordinate-axis triples on curves and surfaces, revealing many essential insights that are almost completely neglected in the standard literature. By introducing the quaternion Gauss map, we are able to do for frames what the Gauss map historically accomplished for surface normals, and then we explore the idea of characterizing properties of curves and surfaces by analyzing the quaternion fields of their moving frames. A family of vexing problems in modeling and geometry is solved by defining a method that constructs optimal moving frames. Finding optimal continuous relations among frames leads us to the historically-important study of quaternion applications to orientation splines and energy-constrained choices of smooth quaternion orientation paths. The application of quaternion volumes to delimit the orientation spaces of biological or mechanical joints completes the family of dimensions that can be examined as quaternion manifolds. A selection of particular techniques that helps exploit all these quaternion manifold applications includes overviews of spherical modeling primitives and spherical geometry. The final chapters of our treatment outline the larger mathematical framework of division algebras and Clifford Algebras, in which quaternions play an essential part, and thus shed light on the ways in which quaternion methods may (and may not) be extended to higher dimensions.

Musings. The presentation style of this book is a somewhat personal and idiosynchratic one. It has evolved over many years of writing about mathematics, and takes a form that makes it easy for me personally to comprehend and retain essential facts. It may not satisfy everyone. In particular, the mathematics is quite pragmatic, completely neglecting theorems and proofs; but for this I offer no apologies. My choices of notation and mathematical exposition are clearly rooted in my background as a classically trained theoretical physicist, not a mathematician or an engineer. However, the desire to make mathematical things visible has deep psychological roots, possibly artistic in nature, and certainly owes a debt to a year of painfully inadequate accomplishment but inspirational hands-on study of 3D sculpture squeezed into my education as an undergraduate science major. For me, visual representations have always formed a more essential part of any intuitive understanding of science than mathematical formulas. One must of course be wary that graphical representations cannot often substitute for a facility with the mathematics itself. Nevertheless, visual representations can certainly make the mathematics simpler to recall and reconstruct, aid in the recognition of essential features, and enhance the likelihood that correct intuitive insights can be generated and applied.

Notations indicating advanced or supplementary material. This book contains a wide variety of information. We have attempted to separate the basic material into the first set of Chapters in Part **I**, and reserve the later Chapters in Part **II** and Part **III** for more advanced topics and applications. The main text in Part **I** is intended to follow a fairly sequential train of thought, so that an "average" reader with strong computer graphics background should be able to follow the main sections. However, some material of an advanced nature is included in various places in Part **I** for completeness, and is marked with a dagger (†) to indicate that this material requires additional background knowledge and can be skipped if desired. In Parts **II** and **III**, the † symbol is generally omitted, since nearly every section requires some specialized knowledge or background. Extremely important points, relating concepts and ideas that convey major themes of the book, are singled out by placing the text inside a ⬚**box**. Finally, examples of elementary computer programs are given at critical points throughout the text, with a summary in an appendix.

URL for demonstration software. The software fragments listed in the tables can be found at *http://www.visualizingquaternions.com*, along with several demonstration systems that support the interactive visualization of quaternions and supplement many of the concepts presented in the text.

Acknowledgments

Many people contributed to the intellectual journey leading to the completion of this book. Steve Cunningham and Anselmo Lastra provided the essential initial encouragement to put together some of my thoughts and results on 4D geometry and quaternions in the form of Siggraph Conference tutorials. Claude Puech's hospitality for a year's sabbatical in 1995–1996 at iMagis in Grenoble (now the INRIA laboratory in Montbonnot, France) provided the time to start bringing together my initial thoughts on quaternion manifolds. Pascal Fua's generous hospitality during my sabbatical at EPFL in Lausanne, Switzerland, in 2002 was similarly invaluable. We are indebted to the National Science Foundation, grants NSF IRI-91-06389, CCR-0204112, and IIS-0430730, for supporting portions of this work. I would like to acknowledge Ken Brakke, Marie-Paule Cani, Mathieu Desbrun, George Drettakis, David Eberly, George Francis, Lorna Herda, James Kajiya, Tamara Munzner, Konrad Polthier, Peter Shirley, François Sillion, and Daniel Thalmann among the many colleagues who have helped along the way. Thanks are due to Sarah Rudenga for her able assistance in transcribing the later parts of the manuscript. The author particularly would like to thank Ji-Ping Sha for his wonderful patience in providing an endless stream of valuable mathematical explanations and insights when my own needed replenishing. Many essential concepts and software systems were developed in collaboration with my students, and I am extremely grateful to Pheng-Ann Heng, Hui Ma, Ying Feng, Robert Cross, Kostya Ishkov, Eric Wernert, Yinggang Li, and Philip Chi-Wing Fu for their many contributions. Sidharth Thakur in particular deserves special acknowledgment for his assistance in polishing and organizing the final versions of a number of the figures and demonstrations. Thanks are due to Lucy Battersby and Jill Cowden for their assistance with the photographic illustrations. Finally, Tim Cox, my editor at Morgan Kaufmann, and the referees he selected to review and comment on the drafts of the manuscript all contributed enormously to the ultimate quality of the organization and presentation of this fascinating material.

Elements of Quaternions

The first group of chapters in this book, beginning with Chapter 1 and ending with Chapter 14, sets forth the fundamental concepts of quaternions, methods of quaternion visualization, and what we consider to be the most basic examples of the ways that quaternions can be used to explain and manipulate the phenomena of 3D orientation.

Although we have attempted to make the chapters in Part I as basic as possible, there are still advanced technical issues of interest to selected readers that are noted from time to time. These paragraphs are denoted with a dagger (†) to indicate that they can be omitted at the discretion of the reader.

The Discovery of Quaternions

01

1.1   HAMILTON'S WALK

Quaternions arose historically from Sir William Rowan Hamilton's attempts in the midnineteenth century to generalize complex numbers in some way that would be applicable to three-dimensional (3D) space. Because complex numbers (which we will discuss in detail later) have two parts, one part that is an ordinary real number and one part that is "imaginary," Hamilton first conjectured that he needed one additional "imaginary" component. He struggled for years attempting to make sense of an unsuccessful algebraic system containing one real and two "imaginary" parts. In 1843, at the age of 38, Hamilton (see Figure 1.1) had a brilliant stroke of imagination, and invented in a single instant the idea of a three-part "imaginary" system that became the quaternion algebra. According to Hamilton, he was walking with his wife in Dublin on his way to a meeting of the Royal Irish Academy when the thought struck him. Concerning that moment, he later wrote to his son Archibald:

> On the 16th day of [October]—which happened to be a Monday, and a Council day of the Royal Irish Academy—I was walking in to attend and preside, and your mother was walking with me, along the Royal Canal, to which she had perhaps driven; and although

FIGURE 1.1 *Sir William Rowan Hamilton, 4 August 1805–2 September 1865.* (History of Mathematics web pages of the University of St. Andrews, Scotland.)

she talked with me now and then, yet an under-current of thought was going on in my mind, which gave at last a result, whereof it is not too much to say that I felt at once the importance. An electric circuit seemed to close; and a spark flashed forth, the herald (as I foresaw, immediately) of many long years to come of definitely directed thought and work, by myself if spared, and at all events on the part of others, if I should even be allowed to live long enough distinctly to communicate the discovery. Nor could I resist the impulse—unphilosophical as it may have been—to cut with a knife on a stone of Brougham Bridge, as we passed it, the fundamental formula with the symbols, i, j, k; namely,

$$i^2 = j^2 = k^2 = ijk = -1$$

which contains the Solution of the Problem. . .

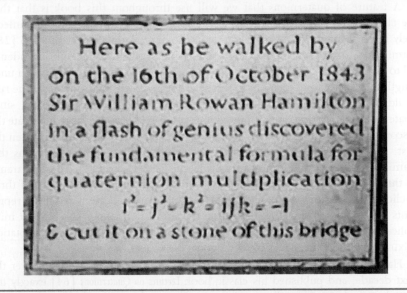

Here as he walked by
on the 16th of October 1843
Sir William Rowan Hamilton
in a flash of genius discovered
the fundamental formula for
quaternion multiplication
$$i^2 = j^2 = k^2 = ijk = -1$$
& cut it on a stone of this bridge

FIGURE 1.2 *The plaque on Broome Bridge in Dublin, Ireland, commemorating the legendary location where Hamilton conceived of the idea of quaternions. In fact, Hamilton and his wife were walking on the banks of the canal beneath the bridge, and the plaque is set in a wall there.* (Photograph courtesy University of Dublin.)

A curiosity is that there is no "Brougham Bridge" in Dublin. Hamilton apparently misspelled the name "Broome," which is indeed the name of the bridge on which he carved his equations, since the two spellings would have been pronounced the same.

We can see that Hamilton was so overwhelmed by his discovery that he feared he might collapse and die before he had a chance to tell anyone, and thus, for safety, carved the equations into the nearest wall, the side of a bridge arching over the canal along which he was walking! Although Hamilton's own carving soon disappeared with the weather, a plaque (Figure 1.2) was later placed on the very spot, and commemorates the event to this day.

A feature of quaternions that we will use throughout this book is that they are closely related to 3D rotations, a fact apparent to Hamilton almost immediately but first published by Hamilton's contemporary Arthur Cayley in 1845 [28]. Hamilton's quaternion multiplication rule, which contains three subrules identical to ordinary complex multiplication, expresses a deep connection between unit-length four-vectors and rotations in three Euclidean dimensions. Curiously, the rule could in principle have been discovered directly by seeking such a connection, since quaternion-like relations appear buried in rotation-related formulas that predate the discovery of quaternions by Hamilton. Rodrigues appears to actually have been the first, in 1840 [142], to write down an equivalent version of the equations that Hamilton scratched on "Brougham" Bridge in 1843 [5,17,112]. The appearance of three copies of the complex multiplication rule and the involvement of three Euclidean dimensions is not accidental. Complex multiplication faithfully represents the two-dimensional (2D) rotation transformation, and each complex multiplication subrule in the quaternion rule will be seen to correspond to a rotation mixing two of the three possible orthogonal 3D axes.

Hamilton proceeded to develop the features of quaternions in depth over the next decade, and published his classic book *Lectures on Quaternions* [64] exactly ten years later in 1853. The still more extensive *Elements of Quaternions* [65] was not published until 1866, shortly after Hamilton's death in 1865. The banner of quaternions was then picked up by Peter Tait, who had diplomatically awaited Hamilton's passing before publishing his own, perhaps more readable, opus, *Elementary Treatise on Quaternions*, [160] in 1867; Tait followed the *Treatise* with *Introduction to Quaternions* in 1873 [161], and produced a major revision of the *Treatise* in 1890 [162]. From there, the story continues in many directions that we shall not pursue here (see, for example, Altmann [4,5], Crowe [35], or van der Waerden [164]).

1.2   THEN CAME OCTONIONS

One of the fascinating questions that occurred to several people shortly after they learned of Hamilton's discovery was this: "If quaternions generalize complex numbers, can quaternions themselves be further generalized?" Although the technical answer to this question is complex, it was quickly found that, given certain compelling mathematical conditions, there is exactly one further generalization, the *octonions*.

The story of octonions is nearly as interesting as that of quaternions. Soon after Hamilton discovered quaternions, he sent a letter describing them to his friend

John T. Graves. Graves replied on 26 October, complimenting Hamilton on the boldness of the idea, but adding "There is still something in the system which gravels me. I have not yet any clear views as to the extent to which we are at liberty arbitrarily to create imaginaries, and to endow them with supernatural properties." He also asked: "If with your alchemy you can make three pounds of gold, why should you stop there?" [9].

On 26 December 1843, Graves wrote to Hamilton to tell him that he had successfully generalized quaternions to the "octaves," an 8-dimensional (8D) algebra, with which he was able to prove that the product of two sums of eight perfect squares is another sum of eight perfect squares. Unfortunately for Graves, Hamilton put off assisting him in publishing his results and octonions were discovered independently and published in 1845 by Arthur Cayley [29]. Despite the fact that both Graves and Hamilton published notes claiming Graves' priority over Cayley, the octonions were thereafter known widely as "Cayley numbers," and Graves' contribution is often overlooked. There are in fact several ways in which we can look for generalizations of quaternions. These are typically very advanced topics, but we have provided a brief treatment for the interested reader in Part III.

1.3   THE QUATERNION REVIVAL

Since the intense work on quaternions that occurred in the 19th century, many things have happened. One is that, although Hamilton lobbied tirelessly to get quaternions accepted as the standard notation for 3D vector operations such as the dot product and the cross product (see Chapter 3 or Appendix A), the notation was always perceived as awkward, and thus the alternative tensor notation set forward by Gibbs became the standard [35]. Even in this book, we will universally use Gibbs' notation $\mathbf{x} = (x, y, z)$ to express a 3D vector and to help "explain" the inner workings of quaternions, rather than the other way around. When the quantum mechanics of the electron was developed in the early 20th century, it became clear that quaternions were related to recently developed mathematical objects called spinors, and that electrons were related to both. Yet again, a quaternion-based notation was possible, but was discarded in favor of an alternate notation; the quaternion-equivalent notation that is standard in physics applications is based on 2×2 matrices, known as the Pauli matrices.

Both the physical theory of rotating elementary particles and essential elements of the mathematics of group theory incorporate quaternions, but the intense interest of Hamilton's era dimmed due to more transparent notational devices (an advantage never to be underestimated) and a lack of important practical applications. However, the theory of 3D rotations had one more trump card waiting in the wings of technology: the need for comprehensive methods of representing orientation frames and interpolating between them in the newly developed field of 3D computer graphics and animation. We will now, therefore, conclude our short history by devoting some space to the remarkable story of the revival of quaternions and their adaptation to the needs of the computer modeling and animation community.

Although the advantages of the quaternion forms for the basic equations of attitude control (clearly present in Cayley [28], Hamilton [64], and especially Tait [160]) had been noticed and exploited by the aeronautics and astronautics community [169,98,103], the technology did not penetrate the computer animation community until the landmark Siggraph 1985 paper of Ken Shoemake [149]. The importance of Shoemake's paper is that it took the concept of the orientation frame for moving 3D objects and cameras, which require precise orientation specification, exposed the deficiencies of the then-standard Euler-angle methods, and introduced quaternions to animators as a solution. This stimulated a wide variety of investigations and applications (e.g., [4,139,104,115]), ultimately creating a general awareness of quaternion exploitation as practiced by many different scientific communities, and potentially unifying the approaches so that perhaps the "quaternion wheel" will no longer be reinvented unnecessarily.

The primary tool introduced in Shoemake's original paper was the interpolation formula for a great circle connecting two points on an arbitrary-dimensional sphere. By analogy to the acronym "LERP" that might be used for ordinary linear interpolation, Shoemake coined the term "SLERP" for "spherical linear interpolation," a terminology that remains in common usage. Curiously, Shoemake himself did not present a derivation of the SLERP formula in his original paper, although there are several standard approaches, such as the Gram–Schmidt method, that we will present in later chapters.

Given the SLERP, the analogy to standard iterative procedures for constructing arbitrary-degree Euclidean splines from linear interpolation immediately suggests methods for constructing spherical splines. Spherical arcs from the SLERP can, if one analyzes the anchor points judiciously, be transformed to provide close analogs of the anchor-point and tangent-direction properties of the conventional families of Euclidean splines. It is reasonably straightforward to develop uniform quaternion spline families in an elegant practical form, and we will discuss quaternion splines thoroughly.

The computer graphics and modeling community now has general familiarity with quaternions and their uses, but we should note that there are still unresolved issues and controversies regarding the necessity or advantages of quaternion orientation representations. A number of industry-standard animation systems continue to omit quaternions entirely, requiring separate x-axis, y-axis, and z-axis interpolations for character and camera orientation control, and the author has more than once asked an animation team member of a graphics-effects-laden Hollywood movie whether quaternions were used to control orientations in their software and received a dismissive negative answer. The explanation is that good tools, or extremely skilled use of poor tools, can still produce good results, and that quaternion curve methods can easily suffer from their own anomalies unless skillfully applied. Without good tools and sophisticated designers, quaternion methods can produce results that are as poor as any other naively applied orientation spline method and that may show no clear superiority over a classic Euler-angle approach. Increasingly sophisticated quaternion interpolation tools, as well as awareness of their advantages, are necessary to generate an environment in which quaternions realize their potential for the representation and manipulation of orientation states. Creating and facilitating this awareness is one of the purposes of this book.

Folklore of Rotations

02

The rise of digital computing as an essential component of human existence has influenced everything from entertainment and the functionality of the family car to the conduct of warfare and the methods of scientific research. Both NASA astronauts and Hollywood film directors use computer systems to monitor and control the positions of objects in space. Computer systems perform operations inside the virtual mathematical world of computer models that would once have been done with sticks and clay, and this has altered forever the importance of the mathematics of rotations. What once might have been done simply by taking a physical object in hand is now done by a mathematical computer model. The traditional movie camera operator on a rolling miniature trolley car was replaced in *Star Wars* by elaborate computer-controlled platforms, and again in *Toy Story* by a computer model for each and every camera motion. The World War II pilot's controls that were mechanically connected to the flight surfaces have given way to computer-based controls for the Apollo Lunar Module, and a completely electronic fly-by-wire system for the F16 jet fighter.

Let us now look at some surprising examples of how orientation frames for 3D objects appear in our lives. We will gradually come to realize that the more complex are the tasks required of 3D orientation frames the more advantages we gain from discarding traditional representations of orientations and replacing them with quaternion-based methods.

We will examine three objects: an ordinary belt, a ball, and a gyroscope. We will see how each shows us a new property that is difficult, if not impossible, to explain properly in everyday language. Once we learn the language of quaternions a few chapters from now, we will be able to see these simple objects in a new way that will change our perceptions forever after.

2.1   THE BELT TRICK

Our first example is so simple it does not even require a computer to show us that something unexplainable by everyday logic is associated with *sequences of rotations*. The ancient "belt trick" parlor game is played with an ordinary leather belt or similar object as follows.

1   Two people begin by firmly holding opposite ends of the belt.

2   There is one rule: Each of the two people can move their end of the belt to any point in space they like, as long as *their end* of the belt keeps the *same orientation in space* they started with. Switching hands is fine, but the orientation cannot change. (Obviously, we also disallow any physical alteration of the belt.)

3   First, as shown in Figure 2.1, the players twist the belt by one full rotation (360 degrees). Following the keep-the-ends-pointing-the-same-way rule, the two players try every possible legal way to untwist the belt. It is discovered that they can *change the twist* from clockwise to counterclockwise (or vice versa) without violating the rule, but they never succeed at untwisting the belt.

4   Next, as shown in Figure 2.2, they twist the belt by *two full rotations* (720 degrees).
     It is discovered that under the same rules the belt can readily be returned to a flat, untwisted state!

Why?

There is no way to actually understand the results of the "belt trick" parlor game without quaternions or something equivalent to quaternions. We will show how to understand the deep reasons behind the belt trick in Chapter 12, after we have learned some basic quaternion visualization methods.

2.2   THE ROLLING BALL

Our next example is also so commonplace that many readers may have noticed it without even realizing that something unusual was happening. Again, no computer

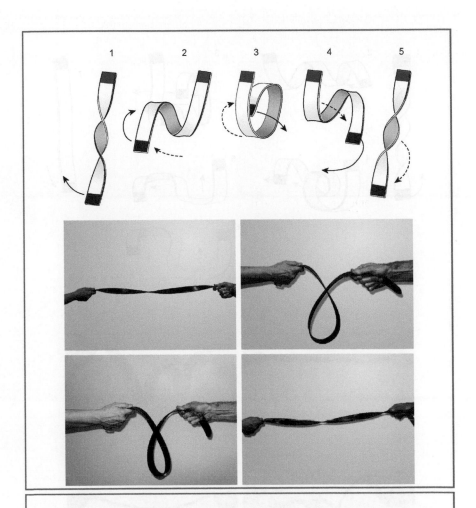

FIGURE 2.1 *The result of the belt-trick move for a 360-degree twist changes the orientation sequence from clockwise to counterclockwise, but cannot untwist the belt.*

is required—just a simple baseball, tennis ball, or even a beach ball. The "rolling ball" game, which again focuses on the result of a *sequence* of rotations, is played as follows.

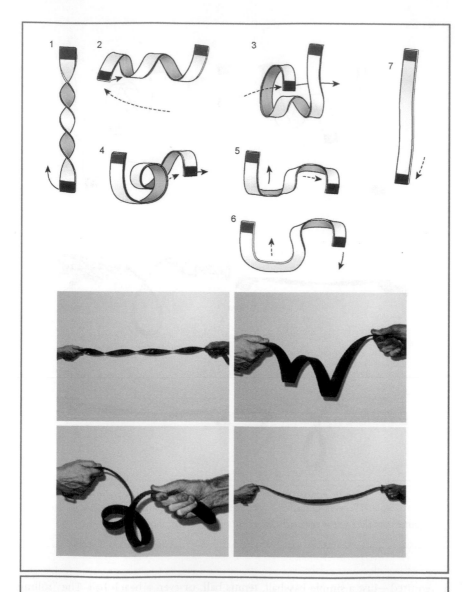

FIGURE 2.2 *A 720-degree twist can be continuously deformed to an untwisted belt.*

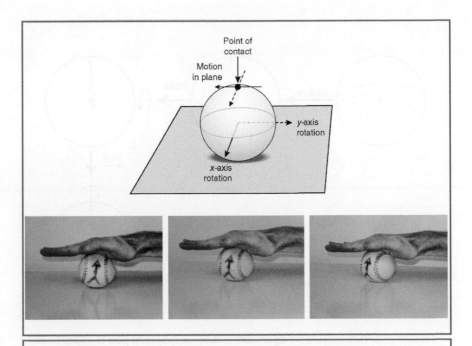

FIGURE 2.3 *Place a ball on a table, and put the palm of your hand on top of the ball. Making a circular rubbing motion with your hand parallel to the tabletop results in a spinning motion of the ball about the vertical axis, in the direction opposite your circular rubbing motion.*

1 Place a ball on a flat table.

2 Place your hand flat on top of the ball, with the palm horizontal and parallel to the table, touching the ball at only one point (the top of the ball). Moving the hand parallel to the tabletop will result in a rotation about an axis in the plane of the tabletop, as shown in Figure 2.3.

3 Without twisting your wrist in any way about the vertical axis, and keeping the palm absolutely flat and parallel to the table, make a circular rubbing motion, as though polishing the tabletop. Let only one point of the palm of your hand contact the ball, and let the ball roll freely.

4 Watch a point on the equator of the ball.

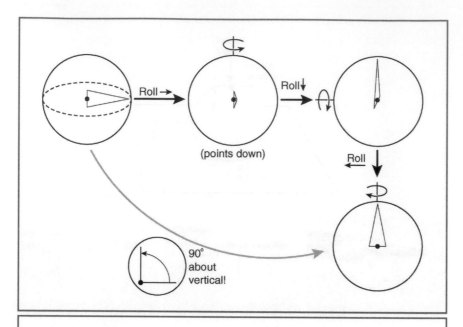

FIGURE 2.4 *Three successive 90-degree rotations about axes lying in the plane of the page result in a 90-degree motion about the axis perpendicular to the page, even though none of motions were around the perpendicular axis.*

5 If you rub in small clockwise circles, the point on the equator will rotate slowly counterclockwise. If you rub in small counterclockwise circles, the point on the equator will rotate slowly clockwise.

6 To obtain a more dramatic result, rotate the ball 90 degrees by moving your palm to the right, 90 degrees again by moving the palm toward your body, and 90 degrees once more by moving your palm to the left.

The result, illustrated in Figure 2.4, is a 90-degree counterclockwise twist about the vertical axis, yet *no motion* twisting the ball about the vertical axes was ever performed!

Why does this happen?

> The answer involves the deepest properties of group theory, and yet in Chapter 13 we will be able to make a simple quaternion picture that basically allows us to explain this phenomenon in practical terms. Quaternions permit us to make a representation of what is going on that is completely different from any other way of looking at the "rolling ball" game.

2.3   THE APOLLO 10 GIMBAL-LOCK INCIDENT

Quaternions are now used throughout the aerospace industry for attitude control of aircraft and spacecraft. The reader who is interested in further details may wish to consult the book *Quaternions and Rotation Sequences* by J. B. Kuipers [115] for an exhaustive list of applications. The physical devices used to sense and correct the orientation of spacecraft, however, have inescapable limitations that illustrate a family of problems that also occur in computer representations of orientations. A computer program can sidestep most such problems by properly exploiting quaternions, but conventional mechanical attitude control systems such as the Apollo Inertial Measurement Unit (IMU)—shown in Figure 2.5—cannot.

A vivid illustration is taken from the NASA logs and commentary [165] for 18 May 1969, describing the return of the Apollo 10 Lunar Module *Snoopy* to the Command Module *Charlie Brown* during the final "dry run" to test the Apollo systems in lunar orbit before the actual Apollo 11 lunar landing on 20 July 1969. As we join the event, the Lunar Module, manned by astronauts Stafford and Cernan, has finished its solo flight tests in lunar orbit and is returning to dock with the Command Module to start the trip back to Earth.

> After Stafford's camera failed, he and Cernan had little to do except look at the scenery until time to dump the descent stage. Stafford had the vehicle in the right attitude 10 minutes early. Cernan asked, "You ready?" Then he suddenly exclaimed, "Son of a bitch!" *Snoopy* seemed to be throwing a fit, lurching wildly about. He later said it was like flying an Immelmann turn in an aircraft, a combination of pitch and yaw. Stafford yelled that they were in gimbal lock—that the engine had swiveled over to a stop and stuck—and they almost were. He called out for Cernan to thrust forward. Stafford then hit the switch to get rid of the descent stage and realized they were 30 degrees off from their previous attitude. The lunar module continued its crazy gyrations across the lunar sky, and a warning light indicated that the inertial measuring unit really was about to reach its limits and go into gimbal lock. Stafford then took over in manual control, made a big pitch maneuver, and started working the attitude control switches. Snoopy finally calmed down.

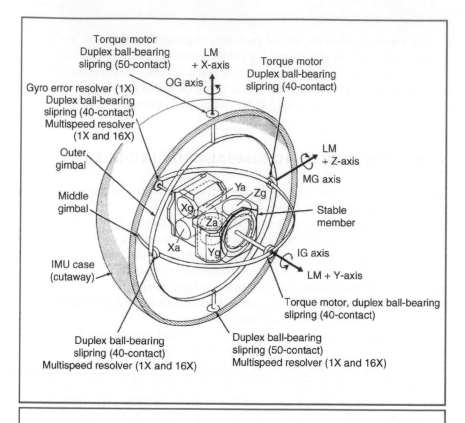

FIGURE 2.5 *Engineering drawing of the Apollo gimbal assembly.* (Images from NASA, Apollo Lunar Surface Journal at *www.hq.nasa.gov/office/pao/History/alsj/ lm_imu.gif.*)

In fact, there are a number of places in the Apollo development records and flight logs where it is mentioned that special maneuvers are required "to avoid gimbal lock." As shown in Figures 2.6 and 2.7, there are actual warning lights on the control panel and red-painted danger areas on the flight director attitude indicator that are intended to detect and prevent gimbal lock.

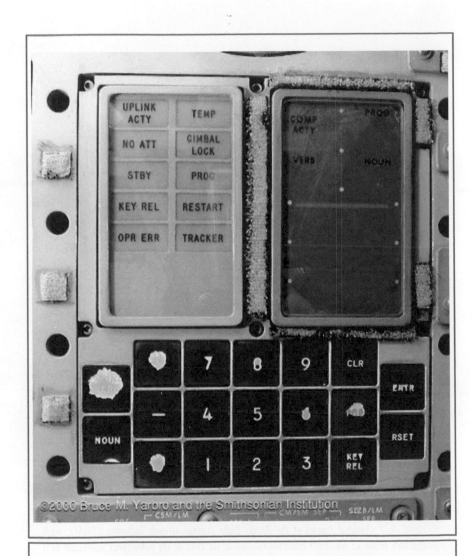

FIGURE 2.6 *The actual Apollo 13 guidance computer console with "Gimbal Lock" warning light.* (Source: *http://history.nasa.gov/afj/pics/dsky.jpg* from *http://history.nasa.gov/afj/compessay.htm.*)

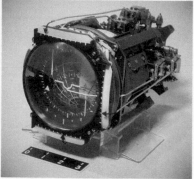

FIGURE 2.7 *Assorted images of the Apollo flight director attitude indicator (FDAI), with gimbal lock limits of spacecraft orientation marked as red disks at the poles (near yaw equal to 0 degrees or 180 degrees).* (From http://history.nasa.gov/ap08fj/09day3_green.htm, www.space1.com/Artifacts/Apollo_Artifacts/FLOWN_FDAI/flown_fdai.html, and www.space1.com/Artifacts/Apollo_Artifacts/FDAI/fdai.html.)

What is gimbal lock, and why did it become so visible in the Apollo missions? The basic context is that the Apollo IMU, shown in Figure 2.5, is a mechanical system with three apparent degrees of freedom. Figure 2.8 shows a schematic diagram of the IMU system. To understand the problem, we need to walk through the set of images shown in Figure 2.9, which are taken from a tutorial on gimbal lock compiled into the NASA Apollo 15 transcript [176]. The critical feature is a mechanical gyroscope system, which is designed to maintain constant orientation with respect to the outside world. The gyroscope system is attached to the spacecraft frame using three rotating rings of decreasing diameter. Initially, as shown in Figure 2.8, the rotation axes of the three rings are set up to be orthogonal, pointing along the current axes labeled 1, 2, and 3, respectively. The outer frame, representing the spacecraft structure, can rotate freely about axis 1 (Figure 2.9a) or about axis 3 (Figure 2.9b) and nothing worrisome happens.

However, if we rotate by 90 degrees about axis 2—as illustrated in Figure 2.9c— axes 1 and 3 line up, so that we have effectively *lost one of our degrees of freedom*: the ability to spin freely about the original axis 1 (Figure 2.9a). This would perhaps be acceptable if there were no gyroscope-based inertial guidance system involved. When we tried to rotate about the original axis 1, as shown in Figure 2.9d, we would note that the gray square fixed to the inner ring would be forced to change direction and would no longer point in the same direction as in all the other figures. Unfortunately, real gyroscopes do not behave this way. If you take a single gyroscope spinning along axis 3 in Figure 2.9d and forcibly rotate it around the vertical axis (original axis 1), the gyroscope will obey Newton's laws for applied torque (see Figure 2.10) and immediately start changing its direction until it points straight down, aligned with the axis of the forced rotation.

This spells disaster: If you have only a single gyroscope, the sensors attached to the gyroscope will be fooled into thinking the spacecraft has suddenly pitched by 90 degrees, and the control system will fire all the thrusters to (very quickly) realign the spacecraft structure with the gyroscope, possibly seriously damaging the vehicle in the process. (Note: Even this process can be described elegantly using quaternion versions of Newton's laws for gyroscopes, as we will see in Chapter 26.)

Reality is even worse: If you have multiple gyroscopes pointing in different directions, as in a typical real-world guidance system (see Figure 2.5), major incompatible forces are exerted and the system could actually self-destruct.

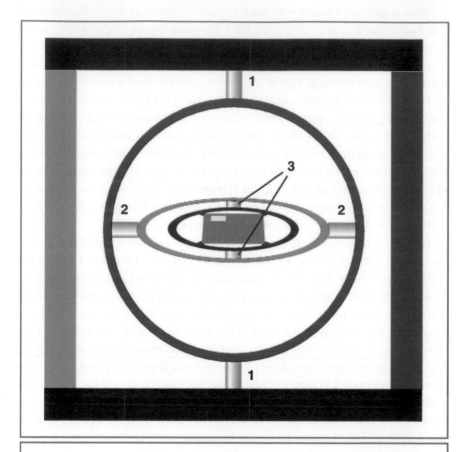

FIGURE 2.8 *A schematic representation of the gimbal system of a spacecraft guidance system. The outer rim represents the (presumably quite massive) structure of the vehicle. The cylinders connecting the rings to other structures represent nearly frictionless rotating bearings. The gray square in the middle is the inertial guidance system, which presumably is set to a particular orientation and is thereafter not supposed to change its direction. (Image from NASA Apollo 15 Flight Journal transcripts at https://history.nasa.gov/afj/ap15fj/15solo_ops3.html#gimballock, W. David Woods and Frank O'Brien.)*

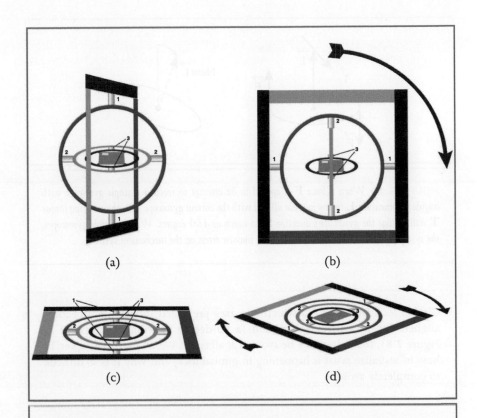

(a)

(b)

(c)

(d)

FIGURE 2.9 *How a spacecraft gets into gimbal lock. Starting from the initial configuration (Figure 2.8), we observe that each of the configurations a through c corresponds to rotations about a single axis. The 90-degree rotation about axis 2 in Figure 2.8 causes axes 1 and 3 to line up in c, losing a degree of freedom and potentially destroying the ability to further monitor the spacecraft's attitude. An external force causing rotation about the vertical axis in the bottom images will attempt to make the "locked" gyroscopes move with the spacecraft. A single gyroscope would simply flip in response to Newton's laws for torques. Multiple orthogonal gyroscopes such as those in the Lunar Module IMU would resist all torques and eventually cause disastrous damage to the spacecraft.* (Images from NASA Apollo 15 Flight Journal transcripts at https://history.nasa.gov/afj/ap15fj/15solo_ops3.html#gimballock, W. David Woods and Frank O'Brien.)

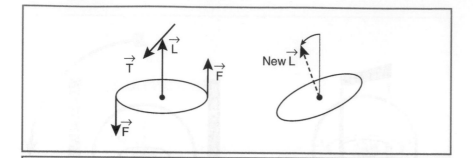

FIGURE 2.10 *When a force* **F** *is applied in an attempt to reorient a single gyroscope with angular momentum* **L** *to any axis not aligned with the current gyroscope axis, the resulting torque* **T** *will change the gyroscope's direction by as much as 180 degrees. With multiple gyroscopes, the system cannot follow the motion without massive stress on the mechanical system.*

Gimbal-lock situations, in which two axes presumed independent become aligned due to a particular motion (a 90-degree rotation about axis 2 in Figure 2.8), must therefore be avoided at all costs! We will see in Chapter 14 how to visualize what is happening in gimbal lock, and why it is so difficult to completely avoid it.

2.4   3D GAME DEVELOPER'S NIGHTMARE

Whereas mechanical devices such as the IMU (Figure 2.5) cannot escape the possibility of the gimbal-lock scenario (Figure 2.9), computer orientation systems—often designed with exactly the same features—can in principle do so using quaternions or equivalent techniques. In the "game developer's nightmare" scenario, there are no gyroscopes or torques but very similar (though hopefully less lethal) phenomena occur anyway. The following, for example, is a typical interchange from a newsgroup for a 3D modeling and animation system.

> Q. I do not understand the logic behind this behavior. What's worse is that if I try to drag the cube in random directions there does not seem to be any rhyme or reason to its behavior. It is completely mad. I've been struggling with this for hours and I give up.

A. What you are encountering is something called "gimbal lock" and is apparently not at
all simple to fix. I know that it is on the list of enhancements to be made to [the system],
but am not sure if this will be done in the current upgrade or a future one.

What has happened to user Q? He or she has most likely created a mathematical
model for the behavior of a simple cube that corresponds to a *tandem* continuous
rotation about all three axes in Figure 2.8 simultaneously. When the moving angle
of axis 2 passes through 90 degrees while the other two axes are also still applying
their own rotation angle changes, the results can be counterintuitive. In particular,
the object can twist backward and forward around the expected direct path between
the initial and target orientations, resulting in a very unphysical model of what
ought to be a physically satisfying motion. That is, the orientation sequence for
even a simple cube becomes poorly defined.

The "game developer's nightmare" is occurring for basically the same reason as
the Apollo 10 incident, but now the arena is the control of an object's orientation
within a computer—perhaps less threatening to life and limb, but just as frustrating
to the programmer! Again, we will see in Chapter 14, using quaternion representations for rotations, exactly why this must happen and how to picture what is going
on. Using a minimal arc-length quaternion path from the starting to the ending
orientation, as introduced in Chapter 10, eliminates the problem.

2.5   THE URBAN LEGEND OF THE UPSIDE-DOWN F16

Stories about the development of the high-performance F16 military fighter aircraft
provide yet another example of a situation in which a proper understanding of 3D
orientation issues is absolutely necessary. Although it is not clear whether quaternion technology could have avoided the reported situation, it is perfectly clear that
a lucid understanding of the nature of orientation computations, facilitated by the
study of quaternion orientations, would not have hurt!

The basic story is that in early testing it was discovered during simulated flight
that the F16 computer guidance system caused it *to turn upside down when it crossed the
equator*. The reason was apparently not gimbal lock but a glitch in the orientation
calculation that confused up and down when the *sign of the latitude changed*. You can
see that if "$|\sin \text{latitude}|$" (the absolute magnitude of the value) were used to de-
termine the "up" direction of the z axis instead of "$\sin \text{latitude}$" itself, the minute
the latitude became negative (as it does when crossing the equator) the default
"up" direction would be toward the North Pole, which would point the top of the

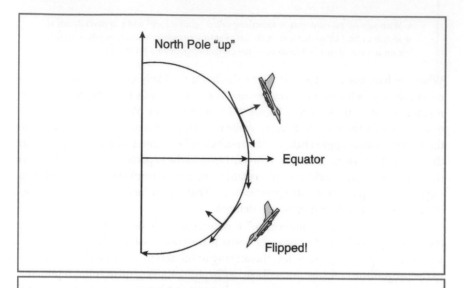

FIGURE 2.11 *One possible scenario for the legendary (simulated) F-16 flip. The "up" direction is assumed to be the North Pole, so that when the airplane crosses the equator this direction switches from being toward the sky to being toward the earth, causing the automatic guidance system to reorient the cockpit to point toward the earth!*

aircraft back in a northerly direction—*upside down* if you were over Brazil! Signs and continuity of rotations, which are handled more transparently by quaternions, *do* matter!

Figure 2.11 indicates one way this could happen. At each moment, the guidance system checks the current aircraft orientation with respect to the North Pole and defines the closest match between the current "up" and the North Pole. When the airplane flies south over the equator, "up" relative to the North Pole is in fact *toward* the earth rather than up toward the sky. The guidance system believes that somehow the plane just flipped over, and it is instructed to flip it back. Of course, the plane did not move, the guidance system did, and thus when the correction is applied the guidance system remains stable but the entire aircraft flips to align the cockpit as closely with "north" as it can.

There is an urban legend that this happened in real life, but we can find no verifiable report. However, the early versions of the Microsoft F-16 flight simulator game were rumored to have this same feature. Of the many alleged records of this problem, the following seemed more authentic than most.

F-16 Problems (from Usenet net.aviation) *http://catless.ncl.ac.uk/Risks/3.44.html*:
Bill Janssen janssen@mcc.com
Wed, 27 Aug 86 14:31:45 CDT

A friend of mine who works for General Dynamics here in Ft. Worth wrote some of the code for the F-16, and he is always telling me about some neato-whiz-bang bug/feature they keep finding in the F-16:

Since the F-16 is a fly-by-wire aircraft, the computer keeps the pilot from doing dumb things to himself. So if the pilot jerks hard over on the joystick, the computer will instruct the flight surfaces to make a nice and easy 4 or 5G flip. But the plane can withstand a much higher flip than that. So when they were "flying" the F-16 in simulation over the equator, the computer got confused and instantly flipped the plane over, killing the pilot [in simulation]. And since it can fly forever upside down, it would do so until it ran out of fuel.

Our point is that working with continuously changing orientations is a tricky business, and systems defined in terms of systems of angles—each considered a separately controlled activity—can lead to unexpected problems. Quaternions provide smooth continuity among orientation frames, and even when this particular functionality is not needed it is the case that learning to think about rotations from a quaternion point of view can improve one's intuitions and reduce the likelihood of errors such as the F-16 flip.

2.6 QUATERNIONS TO THE RESCUE

This concludes our introductory discussion of how rotation frames and some of their mysterious features appear in everyday life. All of the concepts involved in resolving the mysteries can be framed in terms of quaternions. We proceed in the next few sections to look at the basic properties of quaternions and how they can be exploited to achieve better insights and to find solutions to practical problems. In anticipation, we reiterate three fundamental features of quaternion technology we will eventually learn to use to help us avoid orientation disasters.

- *Shape*: Quaternions express a general 3D orientation frame as a point on a hypersphere, a well-understood topological space, yielding a geometric picture of related quaternion shapes possessing both elegance and clarity of notation.
- *Metric*: Quaternions provide a meaningful and reliable method of computing similarities or distances needed to compare orientation states.
- *Interpolatability*: Quaternions support clearly motivated optimal methods for smooth interpolation among intermediate orientation configurations.

Basic Notation

03

This chapter introduces a few basic conventions for vectors, matrices, and complex variables. If the reader finds that he or she is not familiar with the notation in this and the following chapters of Part I, a dense but fairly complete synopsis of the notation we use is presented in detail in Appendix A. For readers who are conversant with the conventions of 3D vector notation, matrices, and complex variables, this chapter and Appendix A should be elementary and can be skipped. Nevertheless, for anyone who might benefit from a quick summary, or who might be accustomed to notational conventions substantially different from those of the author, the material found here will be useful. In particular, the explanations of basic quaternion notation presented in Chapter 4 depend strongly on the notational conventions. Summaries of essential notation are provided in the sections that follow.

3.1 VECTORS

A vector $\mathbf{x}$ is a set of real numbers that we typically write in the form

$$\mathbf{x} = (x, y)$$

for 2D vectors, as

$$\mathbf{x} = (x, y, z)$$

31

for 3D vectors, and as

$$\mathbf{x} = (w, x, y, z)$$

for 4D vectors. Technically, we should treat this notation as a shorthand for a *column vector* because we are thinking in the back of our minds of multiplying these vectors on the left by rotation matrices to transform them to a new orientation.

3.2   LENGTH OF A VECTOR

The *length* of a Euclidean vector is computed from the Pythagorean theorem, generalized to higher dimensions. Several equivalent notations are common for the squared length of a Euclidean vector, e.g.,

$$\|\mathbf{u}\|^2 = \mathbf{u}^2 = \mathbf{u} \cdot \mathbf{u} = x^2 + y^2$$

in 2D,

$$\|\mathbf{u}\|^2 = \mathbf{u}^2 = \mathbf{u} \cdot \mathbf{u} = x^2 + y^2 + z^2$$

in 3D, and

$$\|\mathbf{u}\|^2 = \mathbf{u}^2 = \mathbf{u} \cdot \mathbf{u} = w^2 + x^2 + y^2 + z^2$$

in 4D. The *length* of the vector $\mathbf{u}$ is the square root of its squared length

$$\|\mathbf{u}\| = \sqrt{x^2 + y^2}$$

in 2D,

$$\|\mathbf{u}\| = \sqrt{x^2 + y^2 + z^2}$$

in 3D, and

$$\|\mathbf{u}\| = \sqrt{w^2 + x^2 + y^2 + z^2}$$

in 4D, respectively.

3.3   3D DOT PRODUCT

The inner product (or dot product) of two different vectors is closely related to the length, and is defined in 4D as

$$\mathbf{x}_1 \cdot \mathbf{x}_2 = x_1 x_2 + y_1 y_2 + z_1 z_2 + w_1 w_2,$$

where we drop terms in an obvious way to obtain the 2D and 3D cases.

3.4   3D CROSS PRODUCT

The cross product of two 3D vectors, used in an essential way in the definition of quaternion products, is given by

$$\mathbf{x}_1 \times \mathbf{x}_2 = (y_1 z_2 - y_2 z_1,\ z_1 x_2 - z_2 x_1,\ x_1 y_2 - x_2 y_1).$$

3.5   UNIT VECTORS

A unit vector $\hat{\mathbf{u}}$ is the vector that results when we divide a (nonzero) vector $\mathbf{u} = (x, y)$ or $\mathbf{u} = (x, y, z)$ by its Euclidean length, $\|\mathbf{u}\|$. That is,

$$\hat{\mathbf{u}} = \frac{\mathbf{u}}{\|\mathbf{u}\|}.$$

3.6   SPHERES

Spheres are described by the equation

$$\mathbf{u} \cdot \mathbf{u} = 1,$$

and are labeled by the dimension of the space that results if you cut out a bit of the sphere and flatten it. A circle then has dimension one, and a balloon dimension two (an exploded balloon can be flattened like a sheet of paper). Thus, a circle is the one-sphere $\mathbf{S}^1$ embedded in 2D space, a balloon is a two-sphere $\mathbf{S}^2$ embedded in 3D space, and the three-sphere used to describe quaternions is the hypersphere $\mathbf{S}^3$ embedded in 4D space.

3.7   MATRICES

Matrices are arrays of numbers, typically enclosed in square brackets. The multiplication of a 3×3 matrix times a column vector is written as follows.

$$\mathbf{x}' = \mathbf{R} \cdot \mathbf{x} = \begin{bmatrix} r_{11} & r_{12} & r_{13} \\ r_{21} & r_{22} & r_{23} \\ r_{31} & r_{32} & r_{33} \end{bmatrix} \begin{bmatrix} x \\ y \\ z \end{bmatrix} = \begin{bmatrix} x\,r_{11} + y\,r_{12} + z\,r_{13} \\ x\,r_{21} + y\,r_{22} + z\,r_{23} \\ x\,r_{31} + y\,r_{32} + z\,r_{33} \end{bmatrix} = \begin{bmatrix} x' \\ y' \\ z' \end{bmatrix}.$$

3.8   COMPLEX NUMBERS

Complex numbers can be written in Cartesian or in polar form as

$$z = x + iy$$
$$= r\cos\theta + ir\sin\theta$$
$$= re^{i\theta},$$

where i is the imaginary square root of -1, with $i^2 = -1$. Complex multiplication follows from the properties of i, so that, for example,

$$z_1 z_2 = (x_1 + iy_1)(x_2 + iy_2)$$
$$= (x_1 x_2 - y_1 y_2) + i(x_1 y_2 + x_2 y_1),$$
$$z_1 z_2 = r_1 e^{i\theta_1} r_2 e^{i\theta_2}$$
$$= r_1 r_2 e^{i(\theta_1 + \theta_2)}$$
$$= r_1 r_2 \cos(\theta_1 + \theta_2) + i r_1 r_2 \sin(\theta_1 + \theta_2).$$

We reiterate that further details of our notation conventions may be found in Appendix A.

What Are Quaternions?

04

If this book were a musical composition rather than a mathematical one, perhaps it would begin with a single melodic theme carried by a solo trumpet, followed by other instruments entering in harmony. It would then move to a full ensemble—echoing, developing, and exploring dozens of variations on the original theme, some barely recognizable—and finally reach a climactic crescendo, trailing with a lone trumpet reprising the original perfect melody. In this short chapter, we present that *mathematical melody*: the four quaternion equations from which every theme in the entire book is in some way derived.

Quaternions are four-vectors $q = (q_0, q_1, q_2, q_3) = (q_0, \mathbf{q})$ to which we assign the noncommutative multiplication rule

1. *Quaternion Multiplication*

$$p \star q = (p_0, p_1, p_2, p_3) \star (q_0, q_1, q_2, q_3)$$

$$= \begin{bmatrix} p_0 q_0 - p_1 q_1 - p_2 q_2 - p_3 q_3 \\ p_1 q_0 + p_0 q_1 + p_2 q_3 - p_3 q_2 \\ p_2 q_0 + p_0 q_2 + p_3 q_1 - p_1 q_3 \\ p_3 q_0 + p_0 q_3 + p_1 q_2 - p_2 q_1 \end{bmatrix}$$

$$= (p_0 q_0 - \mathbf{p} \cdot \mathbf{q}, \; p_0 \mathbf{q} + q_0 \mathbf{p} + \mathbf{p} \times \mathbf{q}), \qquad [4.1]$$

the inner product

$$p \cdot q = (p_0, p_1, p_2, p_3) \cdot (q_0, q_1, q_2, q_3)$$
$$= p_0 q_0 + p_1 q_1 + p_2 q_2 + p_3 q_3$$
$$= p_0 q_0 + \mathbf{p} \cdot \mathbf{q}, \qquad [4.2]$$

2. Quaternion Inner Product

and the conjugation rule

$$\bar{q} = (q_0, -q_1, -q_2, -q_3)$$
$$= (q_0, -\mathbf{q}), \qquad [4.3]$$

3. Quaternion Conjugation

which is constructed so that

$$q \star \bar{q} = (q \cdot q, \mathbf{0}).$$

We will generally restrict ourselves to *quaternions of unit length*, obeying the unit-length restriction

$$q \cdot q = (q_0)^2 + (q_1)^2 + (q_2)^2 + (q_3)^2 = (q_0)^2 + \mathbf{q} \cdot \mathbf{q} = 1. \qquad [4.4]$$

4. Unit-length Quaternions

Unit-length quaternions have three degrees of freedom rather than four. The modifier "unit length" will be assumed whenever we refer to a quaternion unless we explicitly specify otherwise. Equation 4.4 is the equation of the *hypersphere* (or the *three-sphere*), written formally as $\mathbf{S}^3$, embedded in a *4D Euclidean space* parameterized by the quaternion $q = (q_0, q_1, q_2, q_3)$. A representative set of computer programs implementing these basic features of quaternions is provided in Chapter 7, as well as in Appendix E.

> **Equations 4.1, 4.2, 4.3, and 4.4 form the foundation of all concepts presented in this book**.

This book focuses on visualizable representations of quaternions, and on their features, technology, folklore, and applications. We will heavily emphasize the representation and manipulation of orientation fields. We regard an orientation field simply as the assignment of a set of oriented orthonormal axes to each point of a topological structure (such as a curve) traced by a moving object (e.g., a movie camera). Quaternions provide a unique and powerful tool for characterizing the *relationships* among 3D orientation frames that the orthonormal axes themselves and traditional representations of them are unable to supply. We will also extensively study the geometry of spheres, which arise inescapably in every aspect of our study of unit quaternions. In that quaternion orientation-frame representations are specific to 3D Euclidean space, one might ask whether the methods of quaternions generalize in some way to allow the treatment of higher-dimensional analogs of the frame-representation problem. Although this is in itself an enormous subject, and far beyond the intended scope of this book, we introduce in the final chapters some material pertinent to this problem.

Our beginning chapters focus on the use of the simplified case of 2D rotations as a rich but algebraically simple proving ground in which we can see many of the key features of quaternion geometry in a very manageable context. The most elementary features of the relationship between 3D rotations and quaternions are then introduced as natural extensions of the 2D treatment. We will then be ready to start exploring the full implications of quaternions themselves.

† *Note on the multiplication rule:* The quaternion multiplication rule Equation 4.1, attributed to Hamilton, contains three subrules identical to ordinary complex multiplication, and expresses a deep connection between unit four-vectors and rotations in three Euclidean dimensions. This rule was in fact known earlier to Rodrigues [142] (see also Altmann [5]), though in a very different context and notation.

Road Map to Quaternion Visualization

05

Now that we have experienced a bit of the quaternion melody in Chapter 4, let us look forward along the road to develop a picture of the more elaborate themes to be explored.

5.1 THE COMPLEX NUMBER CONNECTION

The first thing we will do is to exploit the close relationship between complex numbers and quaternions. By building up a very simple framework for the understanding of rotations in 2D space using unit-length complex numbers, we will develop an intuitive picture that allows us to quickly perceive quaternion analogs. Thus, we can create a framework for 2D rotations based on complex numbers, and extend that to an already-recognizable framework for 3D rotations based on quaternions.

5.2 THE CORNERSTONES OF QUATERNION VISUALIZATION

Using the complex numbers and properties associated with 2D rotations as a starting point, we will study several basic conceptual structures that we will extend to the properties of 3D rotations and the effective exploitation of quaternions themselves. The four cornerstones of our edifice, presented schematically in Figure 5.1, are as follows.

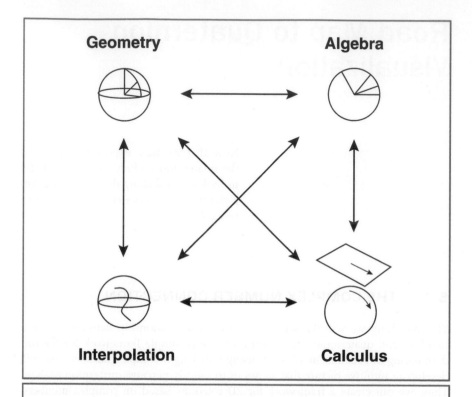

FIGURE 5.1 *The cornerstones of quaternion orientation visualization: geometry, algebra, calculus, and interpolation.*

- **Geometry:** Quaternions are unit four-vectors, and thus correspond geometrically to points at constant radius from an origin, which is the definition of a *sphere*. According to mathematical convention, spheres are named using the dimension of the space a tiny bug would perceive crawling around at the North Pole. Thus, a circle is a one-sphere, a balloon is a two-sphere, and quaternions describe a three-sphere. By studying the visualization of spheres in general, we will later be able to draw and see the geometry of quaternions.

- **Algebra:** Quaternion algebra has a geometric interpretation that includes complex numbers as a subalgebra, and the results of algebraic operations can be visualized using geometric methods.
- **Calculus, logarithms, and exponentials:** The relationships among quaternionic logarithms, their exponentials, and quaternion calculus provide other visualizable properties, which we will study starting from the polar form of a complex number.
- **Interpolation:** The interpolation from one quaternion to another has profound analogies to standard polynomial interpolation methods in Euclidean space. We will see that geodesic or "great circle" curves on spheres provide the starting point for a rich family of interpolation methods and their graphical depiction.

Starting from the nature of rotations in space, the geometry of quaternions, the algebraic properties of quaternion multiplication, quaternion calculus, and interpolatability, we will then proceed to successively more challenging developments of the main theme.

Fundamentals of Rotations

06 Quaternions are related in a fundamental way to 3D rotations, which can represent orientation frames and can act to produce changes in orientation frames. In this chapter, we begin by presenting the relationships among 2D rotation operations, 2D rotation matrices, and complex numbers. We then move on to 3D rotation matrices, examine an interesting idea that looks like the square root of a rotation, and finally relate all of this to 3D rotations and quaternions.

6.1   2D ROTATIONS

Rotations of 2D vectors are implemented by the action of 2D orthogonal matrices $\mathbf{R}_2$ with determinant one, and thus

$$
\begin{bmatrix} x' \\ y' \end{bmatrix} = \mathbf{R}_2(\theta) \cdot \begin{bmatrix} x \\ y \end{bmatrix} = \begin{bmatrix} \cos\theta & -\sin\theta \\ \sin\theta & \cos\theta \end{bmatrix} \begin{bmatrix} x \\ y \end{bmatrix}
$$

$$
= \begin{bmatrix} x\cos\theta - y\sin\theta \\ x\sin\theta + y\cos\theta \end{bmatrix}. \tag{6.1}
$$

43

The 2×2 matrix $\mathbf{R}_2$, written in the form

$$\mathbf{R}_2(\theta) = \begin{bmatrix} A & -B \\ B & A \end{bmatrix} = \begin{bmatrix} \cos\theta & -\sin\theta \\ \sin\theta & \cos\theta \end{bmatrix},$$

obeys by definition the constraints

$$\det \mathbf{R}_2 = A^2 + B^2 = +1$$

and

$$\mathbf{R}_2 \cdot \mathbf{R}_2^T = \begin{bmatrix} A^2 + B^2 & 0 \\ 0 & A^2 + B^2 \end{bmatrix} = \begin{bmatrix} 1 & 0 \\ 0 & 1 \end{bmatrix} \equiv I_2.$$

6.1.1 RELATION TO COMPLEX NUMBERS

We have already seen the algebra of complex numbers, $(x_1 + iy_1)(x_2 + iy_2) = (x_1 x_2 - y_1 y_2) + i(x_1 y_2 + x_2 y_1)$, in Chapter 3. The 2D rotations can be represented using unit-length complex numbers very simply as

$$z = x + iy,$$
$$z' = e^{i\theta}(x + iy)$$
$$= (x \cos\theta - y \sin\theta) + i(x \sin\theta + y \cos\theta), \tag{6.2}$$

which reproduces exactly the matrix results of Equation 6.1.

6.1.2 THE HALF–ANGLE FORM

With some foreknowledge of where we are going, we now examine an apparently trivial way of rewriting 2D rotations. What we shall see is that although the exact parallels to quaternions do not appear in the standard 2D form of the rotation matrix or its complex equivalents they begin to reveal themselves when we make a *half-angle transformation* on the 2D rotation framework. If we simply let

$$A = a^2 - b^2, \qquad B = 2ab,$$

then we can see that with $a = \cos(\theta/2)$ and $b = \sin(\theta/2)$ we recover the original matrix, in that

$$A = \cos^2(\theta/2) - \sin^2(\theta/2)$$

$$= \cos\theta,$$

$$B = 2\cos(\theta/2)\sin(\theta/2)$$

$$= \sin\theta,$$

and the rotation matrix is unchanged:

$$\mathbf{R}_2 = \begin{bmatrix} a^2 - b^2 & -2ab \\ 2ab & a^2 - b^2 \end{bmatrix} = \begin{bmatrix} \cos\theta & -\sin\theta \\ \sin\theta & \cos\theta \end{bmatrix}.$$

Note also that whereas $\det \mathbf{R}_2 = A^2 + B^2 = 1$, when we compute the determinant using the variables (a, b) we find the curious property

$$\det \mathbf{R}_2(a, b) = \left(a^2 - b^2\right)^2 + (2ab)^2 = \left(a^2 + b^2\right)^2 = 1. \tag{6.3}$$

We will see very soon that the pair (a, b) is interpretable as the simplest special case of a quaternion.

6.1.3 COMPLEX EXPONENTIAL VERSION

The complex version follows from the half-angle exponential

$$e^{i\theta/2} = \cos(\theta/2) + i\sin(\theta/2)$$

$$= a + ib,$$

where $e^{i\theta/2}e^{i\theta/2} = e^{i\theta}$. Thus, via Equation 6.2 we see that $e^{i\theta/2}$ is *literally* the square root of the original complex number representing 2D rotations.

6.2   QUATERNIONS AND 3D ROTATIONS

Like $\mathbf{R}_2$, the standard 3D rotation matrix $\mathbf{R}_3$ is also orthonormal:

$$\mathbf{R}_3 \cdot \mathbf{R}_3^T = \begin{bmatrix} 1 & 0 & 0 \\ 0 & 1 & 0 \\ 0 & 0 & 1 \end{bmatrix} \equiv I_3,$$

$$\det \mathbf{R}_3 = 1.$$

However, the following two new features appear in 3D.

- *Order dependence*: The product of two $\mathbf{R}_3$ matrices $\mathbf{S}$ and $\mathbf{T}$ may depend on the order in which the multiplication occurs. That is, except in special cases, we find that

$$\mathbf{S} \cdot \mathbf{T} \neq \mathbf{T} \cdot \mathbf{S}.$$

- *Single real eigenvector*: $\mathbf{R}_3$ has a single real eigenvector (Euler's theorem), and thus all 3D rotation matrices (and their products) can be written as follows in terms of one single final rotation matrix $\mathbf{R}_3(\theta, \hat{\mathbf{n}})$ that leaves a particular 3D direction fixed:

$$\mathbf{R}_3(\theta, \hat{\mathbf{n}}) \cdot \hat{\mathbf{n}} = \hat{\mathbf{n}}.$$

In addition to leaving $\hat{\mathbf{n}}$ fixed, $\mathbf{R}_3(\theta, \hat{\mathbf{n}})$ expresses all possible 3D rotations as a spinning by an angle θ about the direction $\hat{\mathbf{n}}$.

6.2.1   CONSTRUCTION

We now drop the subscript, write $\mathbf{R}_3 = \mathbf{R}$ for simplicity, and introduce the conventional set of three 3D rotation matrices that are simply rotations in the 2D plane of each pair of orthogonal axes.

$$\mathbf{R}_x = \begin{bmatrix} 1 & 0 & 0 \\ 0 & \cos\theta & -\sin\theta \\ 0 & \sin\theta & \cos\theta \end{bmatrix},$$

$$\mathbf{R}_y = \begin{bmatrix} \cos\theta & 0 & \sin\theta \\ 0 & 1 & 0 \\ -\sin\theta & 0 & \cos\theta \end{bmatrix},$$

$$\mathbf{R}_z = \begin{bmatrix} \cos\theta & -\sin\theta & 0 \\ \sin\theta & \cos\theta & 0 \\ 0 & 0 & 1 \end{bmatrix}.$$

These produce right-handed rotations fixing the basis vectors $\hat{\mathbf{x}} = (1, 0, 0)$, $\hat{\mathbf{y}} = (0, 1, 0)$, $\hat{\mathbf{z}} = (0, 0, 1)$, respectively. $\mathbf{R}_x$ rotates the yz plane about its origin, $\mathbf{R}_y$ the zx plane, and $\mathbf{R}_z$ the xy plane. We can explicitly construct $\mathbf{R}(\theta, \hat{\mathbf{n}})$ as follows.

1   Let $\hat{\mathbf{n}} = (\cos\alpha \sin\beta, \sin\alpha \sin\beta, \cos\beta)$, with $0 \leqslant \alpha < 2\pi$ and $0 \leqslant \beta \leqslant \pi$, denote the fixed axis of the eigenvector about which we wish to rotate, as shown in Figure 6.1.

2   Define $\hat{\mathbf{z}}$ as the column vector $(0, 0, 1)^T$, and note that

$$\hat{\mathbf{n}} = \mathbf{R}_z(\alpha) \cdot \mathbf{R}_y(\beta) \cdot \hat{\mathbf{z}}.$$

3   To construct the rotation matrix that spins about $\hat{\mathbf{n}}$, transform $\hat{\mathbf{n}}$ to $\hat{\mathbf{z}}$ by inverting the previously cited transformation, spin about $\hat{\mathbf{z}}$ by θ using the elementary matrix $\mathbf{R}_z(\theta)$, and tilt $\hat{\mathbf{z}}$ back to the direction $\hat{\mathbf{n}}$ (where it started):

$$\mathbf{R}(\theta, \hat{n}) = \mathbf{R}_z(\alpha) \cdot \mathbf{R}_y(\beta) \cdot \mathbf{R}_z(\theta) \cdot \mathbf{R}_y^T(\beta) \cdot \mathbf{R}_z^T(\alpha). \qquad (6.4)$$

Writing out all components of this product of matrices, and rewriting all appearances of α and β in terms of

$$\hat{\mathbf{n}} = (\cos\alpha \sin\beta, \sin\alpha \sin\beta, \cos\beta) = (n_1, n_2, n_3),$$

we find finally that the desired rotation matrix fixing the $\hat{\mathbf{n}}$ direction is

$$\mathbf{R}(\theta, \hat{\mathbf{n}}) = \begin{bmatrix} c + (n_1)^2(1-c) & n_1 n_2 (1-c) - s n_3 & n_1 n_3 (1-c) + s n_2 \\ n_2 n_1 (1-c) + s n_3 & c + (n_2)^2(1-c) & n_2 n_3 (1-c) - s n_1 \\ n_3 n_1 (1-c) - s n_2 & n_3 n_2 (1-c) + s n_1 & c + (n_3)^2(1-c) \end{bmatrix},$$

$$(6.5)$$

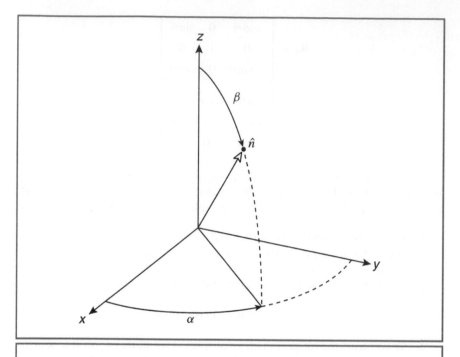

FIGURE 6.1 Polar coordinate conventions for construction of normal direction $\hat{n}$ on $\mathbf{S}^2$.

where $c = \cos\theta$, $s = \sin\theta$, and $\hat{\mathbf{n}} \cdot \hat{\mathbf{n}} = 1$ by construction. The manner in which the eigenvector $\hat{\mathbf{n}}$ emerges can be seen directly with the brief computation

$$\mathbf{R}(\theta, \hat{\mathbf{n}}) \cdot \hat{\mathbf{n}} = \begin{bmatrix} n_1 c + n_1 (n_1)^2 (1-c) + n_1 (n_2)^2 (1-c) + n_1 (n_3)^2 (1-c) \\ n_2 (n_1)^2 (1-c) + n_2 c + n_2 (n_2)^2 (1-c) + n_2 (n_3)^2 (1-c) \\ n_3 (n_1)^2 (1-c) + n_3 (n_2)^2 (1-c) + n_3 c + n_3 (n_3)^2 (1-c) \end{bmatrix}$$

$$= \begin{bmatrix} n_1 (c + 1 - c) \\ n_2 (c + 1 - c) \\ n_3 (c + 1 - c) \end{bmatrix}$$

$$= \hat{\mathbf{n}}.$$

6.2.2   QUATERNIONS AND HALF ANGLES

Finally, we return to our half-angle form of the 2D rotation and observe that in each 2D subplane there must be a 3D rotation that corresponds to the half-angle form. For example, with $n_1 = n_2 = 0$ we must have $n_3 = 1$, and thus

$$
\mathbf{R}_z = \begin{bmatrix} \cos\theta & -n_3\sin\theta & 0 \\ n_3\sin\theta & \cos\theta & 0 \\ 0 & 0 & 1 \end{bmatrix}
$$

$$
= \begin{bmatrix} a^2 - b^2 & -2ab & 0 \\ 2ab & a^2 - b^2 & 0 \\ 0 & 0 & 1 \end{bmatrix}, \tag{6.6}
$$

where $a = \cos\theta/2$, $b = \sin\theta/2$. It would seem highly probable that the general eigenvector form of the matrix can be expressed in terms of half angles as well, not just the Cartesian special cases. Thus, we are tempted to rewrite the entire matrix in terms of $\theta/2$ using $a = \cos(\theta/2)$ and $b = \sin(\theta/2)$ for notational convenience. The factor $(1 - c)$ in Equation 6.5 does not immediately take the same form as the expected 2D limit in Equation 6.6 until we have the insight that in fact $a^2 + b^2 = 1$. Then, motivated to make the substitution $1 - c = (a^2 + b^2) - (a^2 - b^2) = 2b^2$ we find the remarkable expression

$$
\mathbf{R}(\theta, \hat{\mathbf{n}}) = \begin{bmatrix} a^2 - b^2 + (n_1)^2(2b^2) & 2b^2 n_1 n_2 - 2abn_3 & 2b^2 n_3 n_1 + 2abn_2 \\ 2b^2 n_1 n_2 + 2abn_3 & a^2 - b^2 + (n_2)^2(2b^2) & 2b^2 n_2 n_3 - 2abn_1 \\ 2b^2 n_3 n_1 - 2abn_2 & 2b^2 n_2 n_3 + 2abn_1 & a^2 - b^2 + (n_3)^2(2b^2) \end{bmatrix}
$$

$$
= \begin{bmatrix} a^2 + b^2(n_1^2 - n_2^2 - n_3^2) & 2b^2 n_1 n_2 - 2abn_3 & 2b^2 n_3 n_1 + 2abn_2 \\ 2b^2 n_1 n_2 + 2abn_3 & a^2 + b^2(n_2^2 - n_3^2 - n_1^2) & 2b^2 n_2 n_3 - 2abn_1 \\ 2b^2 n_3 n_1 - 2abn_2 & 2b^2 n_2 n_3 + 2abn_1 & a^2 + b^2(n_3^2 - n_1^2 - n_2^2) \end{bmatrix}.
$$

Here we exploited the relation $(n_1)^2 + (n_2)^2 + (n_3)^2 = 1$ to make the expression even more symmetric by making the substitution

$$
-b^2 + 2b^2(n_1)^2 = -\big((n_1)^2 + (n_2)^2 + (n_3)^2\big)b^2 + 2b^2(n_1)^2
$$

$$
= b^2\big((n_1)^2 - (n_2)^2 - (n_3)^2\big)
$$

and its cyclic permutations.

This is a quadratic form that can now be written *solely* in terms of points on the hypersphere $\mathbf{S}^3$, and thus *in terms of quaternions!* To make this explicit, we choose the following parameterization for the set of quaternion variables on the hypersphere, an explicit choice that guarantees that $q \cdot q = 1$ (see Equation 4.4) is satisfied.

$$q_0 = a = \cos(\theta/2),$$

$$q_1 = n_1 b = n_1 \sin(\theta/2),$$

$$q_2 = n_2 b = n_2 \sin(\theta/2),$$

$$q_3 = n_3 b = n_3 \sin(\theta/2). \tag{6.7}$$

With this substitution, we find the final result

$$\mathbf{R}(q) = \begin{bmatrix} q_0^2 + q_1^2 - q_2^2 - q_3^2 & 2q_1q_2 - 2q_0q_3 & 2q_1q_3 + 2q_0q_2 \\ 2q_1q_2 + 2q_0q_3 & q_0^2 - q_1^2 + q_2^2 - q_3^2 & 2q_2q_3 - 2q_0q_1 \\ 2q_1q_3 - 2q_0q_2 & 2q_2q_3 + 2q_0q_1 & q_0^2 - q_1^2 - q_2^2 + q_3^2 \end{bmatrix}.$$

$$\tag{6.8}$$

Equations 6.5 and 6.8 are seen to be identical, $\mathbf{R}(\theta, \hat{\mathbf{n}}) \equiv \mathbf{R}(q)$, when we substitute Equation 6.7 into 6.8. In summary:

The quaternion equation 6.7, $q(\theta, \hat{\mathbf{n}}) = (\cos\frac{\theta}{2}, \hat{\mathbf{n}}\sin\frac{\theta}{2})$, when substituted into Equation 6.8, produces the standard matrix $\mathbf{R}(\theta, \hat{\mathbf{n}})$ of Equation 6.5 for a rotation by θ in the plane perpendicular to $\hat{\mathbf{n}}$, where $\hat{\mathbf{n}} \cdot \hat{\mathbf{n}} = 1$ is a unit three-vector lying on an ordinary sphere (the two-sphere $\mathbf{S}^2$) and θ is an angle obeying $0 \leqslant \theta < 4\pi$ rather than $0 \leqslant \theta < 2\pi$. This extension of the range of θ allows the values of q to reach all points of the hypersphere $\mathbf{S}^3$.

Relation of Quaternions to 3D Rotations

---

† We now can verify various properties of Equation 6.8 that are important in linear algebra. For example, whereas the determinant of $\mathbf{R}_2$ in Equation 6.3 is a square of a sum of squares, for $\mathbf{R}_3$ it is a cube because there is one additional power in a determinant of a matrix that is one dimension larger:

$$\det \mathbf{R}_3 = (q \cdot q)^3 = 1.$$

In addition, although it is complicated to verify, each row (or column) of Equation 6.8 has unit magnitude, as in, for example,

$$\left((q_0)^2 + (q_1)^2 - (q_2)^2 - (q_3)^2\right)^2 + (2q_1q_2 - 2q_0q_3)^2 + (2q_1q_3 + 2q_0q_2)^2$$
$$= (q \cdot q)^2 = 1,$$

where we have used $q \cdot q = 1$. In addition, each row (or column) is orthogonal to its neighbor, as in, for example,

$$\sum_{i=1}^{3} \mathbf{R}_{i1} \mathbf{R}_{i2} = 0,$$

and so on. These properties (each column or row has unit length and is orthogonal to its neighbors) define $\mathbf{R}$ as an orthogonal matrix.

6.2.3 DOUBLE VALUES

Because the quaternion values appear only *quadratically* in the rotation matrix Equation 6.8, the matrix obeys

$$\mathbf{R}(q) = \mathbf{R}(-q),$$

and thus:

Quaternions Double Rotations

> The quaternions q and $-q$ generate the same 3D rotation matrix. Therefore, quaternions realize a double covering (2:1 mapping) of ordinary 3D rotations.

6.3   RECOVERING θ AND $\hat{n}$

Given an arbitrary rotation matrix $\mathbf{M}$, one may recover the Euler-theorem eigenvector $\hat{n}$ and the value of θ using an efficient direct calculation, avoiding the complexity of a general linear-algebra approach. The steps are as follows.

1 *Axis computation:* As long as $\sin\theta$ is nonzero and large enough to avoid numerical errors, $\hat{n}$ may be computed by subtracting the transpose and using

Equation 6.5 to find

$$\mathbf{R} - \mathbf{R}^T = \begin{bmatrix} 0 & -2n_3 \sin\theta & +2n_2 \sin\theta \\ +2n_3 \sin\theta & 0 & -2n_1 \sin\theta \\ -2n_2 \sin\theta & +2n_1 \sin\theta & 0 \end{bmatrix}$$

$$= \begin{bmatrix} 0 & -c & b \\ c & 0 & -a \\ -b & a & 0 \end{bmatrix}.$$

Numerically, we simply compute $d = \sqrt{a^2 + b^2 + c^2}$ and normalize to yield the result

$$\hat{\mathbf{n}} = (a/d, b/d, c/d).$$

2   *Angle computation:* Again, we return to Equation 6.5 and next examine its trace to find that

$$t = \text{Trace}\,\mathbf{R} = 1 + 2\cos\theta.$$

Thus, the calculation is completed by taking

$$\cos\theta = \frac{1}{2}(t - 1),$$

$$\sin\theta = +\sqrt{1 - \cos^2\theta}.$$

Because $\mathbf{R}(\theta, \hat{\mathbf{n}}) = \mathbf{R}(-\theta, -\hat{\mathbf{n}})$, the sign of $\sin\theta$ may be taken as positive in the square root without loss of generality.

† There are some cases for which this basic approach requires further (sophisticated) modifications. For example, if $\sin\theta \leqslant 0$ we must be more careful. Details of the fully rigorous method are given in Chapter 16.

6.4   EULER ANGLES AND QUATERNIONS

There are many common ways to parameterize the standard 3×3 rotation matrices. So far, we have focused on the standard axis-angle version $\mathbf{R}(\theta, \hat{\mathbf{n}})$ of Equation 6.5

and its corresponding quaternion

$$q(\theta, \hat{\mathbf{n}}) = \big(\cos(\theta/2), \hat{\mathbf{n}} \sin(\theta/2)\big),$$

where the eigenvector of $\mathbf{R}$ may itself be parameterized as the $\mathbf{S}^2$ point

$$\hat{\mathbf{n}} = (\cos\alpha \sin\beta, \sin\alpha \sin\beta, \cos\beta).$$

This form has no degenerate variable states (due essentially to the explicit appearance of the eigenvector) and hence avoids gimbal-lock anomalies.

Many applications, however, treat rotations as *sequences* of axis-angle rotations, and this is one source of the gimbal-lock anomalies we will work through in Chapter 14. Triple sequences of axis-angle rotations are usually called Euler-angle representations, and we can easily relate these representations to corresponding quaternion parameterizations. The following are the two most common forms.

- XYZ: Because this sequence of rotations involves all three independent Cartesian rotation axes, it is sometimes mistakenly thought to be singularity-free. The explicit matrix is typically written as

$$\mathbf{R}_{xyz}(\alpha, \beta, \gamma) = \mathbf{R}(\alpha, \hat{\mathbf{x}}) \cdot \mathbf{R}(\beta, \hat{\mathbf{y}}) \cdot \mathbf{R}(\gamma, \hat{\mathbf{z}})$$

$$= \begin{bmatrix} \cos\beta \cos\gamma & -\cos\beta \sin\gamma & \sin\beta \\ \sin\alpha \sin\beta \cos\gamma + \cos\alpha \sin\gamma & \cos\alpha \cos\gamma - \sin\alpha \sin\beta \sin\gamma & -\sin\alpha \cos\beta \\ -\cos\alpha \sin\beta \cos\gamma + \sin\alpha \sin\gamma & \sin\alpha \cos\gamma + \cos\alpha \sin\beta \sin\gamma & \cos\alpha \cos\beta \end{bmatrix}.$$

(6.9)

The corresponding quaternion is given by

$$q_0 = \cos\frac{\alpha}{2} \cos\frac{\beta}{2} \cos\frac{\gamma}{2} - \sin\frac{\alpha}{2} \sin\frac{\beta}{2} \sin\frac{\gamma}{2},$$

$$q_1 = \sin\frac{\alpha}{2} \cos\frac{\beta}{2} \cos\frac{\gamma}{2} + \cos\frac{\alpha}{2} \sin\frac{\beta}{2} \sin\frac{\gamma}{2},$$

$$q_2 = \cos\frac{\alpha}{2} \sin\frac{\beta}{2} \cos\frac{\gamma}{2} - \sin\frac{\alpha}{2} \cos\frac{\beta}{2} \sin\frac{\gamma}{2},$$

$$q_3 = \cos\frac{\alpha}{2} \cos\frac{\beta}{2} \sin\frac{\gamma}{2} + \sin\frac{\alpha}{2} \sin\frac{\beta}{2} \cos\frac{\gamma}{2}.$$

(6.10)

One can see from the quaternion coordinates that at $\beta = \pm\pi/2$, this is a function only of $\alpha + \gamma$ or $\alpha - \gamma$, respectively,

$$q(\alpha, \pm\pi/2, \gamma) = \frac{1}{\sqrt{2}}\left(\cos\frac{\alpha \pm \gamma}{2}, \sin\frac{\alpha \pm \gamma}{2}, \pm\cos\frac{\alpha \pm \gamma}{2}, \pm\sin\frac{\alpha \pm \gamma}{2}\right).$$

- ZYZ: The following sequence corresponds to a coordinate system used traditionally in classical and quantum physics to parameterize the motion of a spinning physical object.

$$\mathbf{R}_{zyz}(\alpha, \beta, \gamma) = \mathbf{R}(\alpha, \hat{\mathbf{z}}) \cdot \mathbf{R}(\beta, \hat{\mathbf{y}}) \cdot \mathbf{R}(\gamma, \hat{\mathbf{z}}) = \mathbf{R}(\gamma, \hat{\mathbf{c}}) \cdot \mathbf{R}(\beta, \hat{\mathbf{b}}) \cdot \mathbf{R}(\alpha, \hat{\mathbf{z}})$$

$$= \begin{bmatrix} \cos\alpha\cos\beta\cos\gamma - \sin\alpha\sin\gamma & -\sin\alpha\cos\gamma - \cos\alpha\cos\beta\sin\gamma & \cos\alpha\sin\beta \\ \sin\alpha\cos\beta\cos\gamma + \cos\alpha\sin\gamma & \cos\alpha\cos\gamma - \sin\alpha\cos\beta\sin\gamma & \sin\alpha\sin\beta \\ -\sin\beta\cos\gamma & \sin\beta\sin\gamma & \cos\beta \end{bmatrix},$$

(6.11)

where $\hat{\mathbf{c}} = (\cos\alpha\sin\beta, \sin\alpha\sin\beta, \cos\beta)$ and $\hat{\mathbf{b}} = (-\sin\alpha, \cos\alpha, 0)$. The corresponding quaternion is given by

$$q_0 = \cos\frac{\beta}{2}\cos\frac{1}{2}(\alpha + \gamma),$$

$$q_1 = \sin\frac{\beta}{2}\sin\frac{1}{2}(\gamma - \alpha),$$

$$q_2 = \sin\frac{\beta}{2}\cos\frac{1}{2}(\gamma - \alpha),$$

$$q_3 = \cos\frac{\beta}{2}\sin\frac{1}{2}(\alpha + \gamma).$$

(6.12)

One can see that at $\beta = 0$, this is a function only of $(\alpha + \gamma)$, and that at $\beta = \pi$ this is a function only of $(\alpha - \gamma)$.

6.5   † OPTIONAL REMARKS

6.5.1   † CONNECTIONS TO GROUP THEORY

The fact that quaternion multiplication preserves membership in $\mathbf{S}^3$ leads to the observation that unit quaternions are a realization of the actual *group manifold* of the group $\mathbf{SU}(2)$, whereas ordinary 3D rotations belong to the group $\mathbf{SO}(3)$. Because the quaternion is a point q on the three-sphere $\mathbf{S}^3$, and because ordinary rotations identify each pair $(q, -q)$ as the same point in their own manifold, ordinary 3D rotation matrices correspond not to the geometry of $\mathbf{S}^3$ but to $\mathbf{S}^3/\mathbf{Z}_2$, which is $\mathbf{RP}^3$, the real three-dimensional projective space. (For an extensive treatment of the properties of this space, we refer the reader to Weeks [167].)

6.5.2 † "PURE" QUATERNION DERIVATION

Many treatments of quaternion rotations begin from a quaternion with a vanishing q_0 component and derive Equation 6.8 directly by bracketing the "pure quaternion" $v = (0, \mathbf{v})$ between a pair of unspecified conjugate quaternions. In fact, when we carry out the computation as follows, we do find exactly the set of matrix components in Equation 6.8.

$$q \star v \star \bar{q} = q \star (0, \mathbf{v}) \star \bar{q} = \big(0, \mathbf{R}(q) \cdot \mathbf{v}\big). \qquad [6.13]$$

However, the interpretation of $v = (0, \mathbf{v})$ as a 3D vector in the usual sense in this derivation is a subject of some controversy, eloquently discussed by Altmann [5]. It can be argued that, although Hamilton himself believed strongly in this interpretation, in fact $(0, \mathbf{v})$ only *coincidentally* obeys the same transformation law as a pure vector. The alternative viewpoint is that, to be precise, $v = (0, \mathbf{v})$ should be considered instead as *a rotation by* $\theta = \pi$. Whichever viewpoint one takes, Equation 6.13 does in fact serve as a convenient alternative derivation of Equation 6.8.

6.5.3 † QUATERNION EXPONENTIAL VERSION

In parallel with the complex exponential, an exponential approach using quaternions can be found. We will explore this in detail in a subsequent section. What we will see is that we can write the following exponential and expand it in a power series to find a quaternion expression corresponding exactly in apparent form to the complex expression for 2D rotations.

$$e^{\mathbf{i}\cdot\hat{\mathbf{n}}\theta/2} = \cos(\theta/2) + \mathbf{i} \cdot \hat{\mathbf{n}} \sin(\theta/2)$$
$$= q_0 + \mathbf{i} \cdot \mathbf{q}$$
$$= q_0 + iq_1 + jq_2 + kq_3,$$

where the three components of $\mathbf{i} = (i, j, k)$ are the quaternion "imaginaries" of Hamilton's original notation (Chapter 1), obeying $i^2 = j^2 = k^2 = ijk = -1$. As a consequence, for each component (e.g., just the i component with $q_2 = q_3 = 0$) we recover precisely $e^{i\theta/2}$, which is literally the square root of the complex representation of this 2D subset of the 3D rotations. More details and related approaches are presented in Chapter 15.

6.6   CONCLUSION

We now know all of the basic properties of the relationship between complex variables and 2D rotations, and the remarkable ways in which they extend to the more complicated structures relating quaternions to 3D rotations. From this basis, we will build an elaborate framework for the analysis of the orientation properties of many different structures, ranging from curves in space to the properties of the human shoulder.

Visualizing Algebraic Structure

In this chapter we continue to lay the foundation for quaternion visualization methods, examining first the geometric interpretation of the algebra of complex numbers and then extending that intuition into the quaternion domain. As we will see, the quaternion algebra itself has a geometric interpretation that includes complex numbers as a subalgebra, and the results of algebraic operations can be visualized using geometric methods.

7.1   ALGEBRA OF COMPLEX NUMBERS

In our context, an algebra is a rule for the combination of sets of numbers. We encounter one particular algebra frequently in ordinary scientific computation: the algebra of complex numbers. The quaternion algebra is one of only two possible generalizations of complex numbers. (The octonion algebra is the other. See Chapter 30 for more technical details.)

The quaternion algebra has been expressed in a wide variety of equivalent forms, each of which has aspects that parallel those of the multiplication of ordinary complex numbers. Thus, we will work toward the possibility of visualizing the quaternion algebra by looking first at the properties of complex numbers.

7.1.1 COMPLEX NUMBERS

We can represent a complex number z in the following alternative ways.

$$z[\text{ complex Cartesian }] = x + iy,$$

$$z[\text{ complex polar }] = re^{i\theta} = r\cos\theta + ir\sin\theta,$$

$$z[\ 2 \times 2 \text{ matrix }] = \begin{bmatrix} x & -y \\ y & x \end{bmatrix},$$

$$z[\ 2 \times 1 \text{ matrix }] = \begin{bmatrix} x \\ y \end{bmatrix}. \tag{7.1}$$

The algebra of complex multiplication follows from each of these forms, either by explicitly using $i^2 = -1$ or from matrix multiplication. Matrix multiplication itself can be used to represent the complex algebra in two alternate forms: one in which the complex number acted upon is itself a matrix, and one in which it is a column vector. Thus, for example,

$$z_1 z_2 = (x_1 + iy_1)(x_2 + iy_2)$$

$$= (x_1 x_2 - y_1 y_2) + i(x_1 y_2 + x_2 y_1),$$

$$z_1 z_2 = r_1 e^{i\theta_1} r_2 e^{i\theta_2}$$

$$= r_1 r_2 e^{i(\theta_1 + \theta_2)},$$

$$z_1 z_2 = \begin{bmatrix} x_1 & -y_1 \\ y_1 & x_1 \end{bmatrix} \cdot \begin{bmatrix} x_2 & -y_2 \\ y_2 & x_2 \end{bmatrix}$$

$$= \begin{bmatrix} x_1 x_2 - y_1 y_2 & -(x_1 y_2 + x_2 y_1) \\ x_1 y_2 + x_2 y_1 & x_1 x_2 - y_1 y_2 \end{bmatrix},$$

$$z_1 z_2 = \begin{bmatrix} x_1 & -y_1 \\ y_1 & x_1 \end{bmatrix} \begin{bmatrix} x_2 \\ y_2 \end{bmatrix}$$

$$= \begin{bmatrix} x_1 x_2 - y_1 y_2 \\ x_1 y_2 + x_2 y_1 \end{bmatrix}.$$

7.1.2 ABSTRACT VIEW OF COMPLEX MULTIPLICATION

This leads us to the useful concept of defining the algebra abstractly, independently of any particular representation using "imaginary" numbers or matrices. We may simply define complex multiplication to be the algebra mapping two pairs of numbers into a new single pair of numbers using a rule. Starting by reexpressing z in Equation 7.1 as the pair of real numbers (x, y), we can then straightforwardly write the abstract algebra of complex numbers as the following quadratic map from reals to reals, where we now use $\star$ to remind us that this is no longer ordinary multiplication:

$$(x_1, y_1) \star (x_2, y_2) = (x_1 x_2 - y_1 y_2, x_1 y_2 + x_2 y_1). \tag{7.2}$$

Table 7.1 realizes Equation 7.2 as an elementary computer program. We remark that the algebra might also be expressed in polar coordinates, as in, for example,

$$(r_1, \theta_1) \star (r_2, \theta_2) = (r_1 r_2, \theta_1 + \theta_2).$$

We see that algebras may have more than one equivalent abstract form. The polar form is related to a logarithmic map, and follows from

$$\log z = \log r + i\theta + 2n\pi i,$$

with $\log(z_1 \star z_2) = \log z_1 + \log z_2$, where the phase ambiguity of the exponential of this logarithm is specified by the integer n.

The *complex conjugation* operation, defined as

$$\bar{z} = x - iy, \tag{7.3}$$

allows us to write the *modulus* or magnitude $|z|$ of a complex number in terms of the complex product of a number with its conjugate,

$$|z|^2 = z\bar{z} = \left(x^2 + y^2, 0\right). \tag{7.4}$$

An essential property of the algebra is the existence of an *identity element*. We can see that multiplying the pair $(1, 0)$ by any element gives back the same element:

$$(x, y) \star (1, 0) = (x \times 1 - y \times 0, x \times 0 + 1 \times y)$$
$$= (x, y).$$

```
void
ComplexProduct(double x1, double y1, double x2, double y2,
               double *x, double *y)
{
  *x = x1*x2 - y1*y2;
  *y = x1*y2 + x2*y1;
}

void
ComplexSum(double x1, double y1, double x2, double y2,
           double *x, double *y)
{
  *x = x1 + x2;
  *y = y1 + y2;
}

void
ComplexConjugate(double x1, double y1, double *x, double *y)
{
  *x = x1;
  *y = -y1;
}

double
ComplexModulus(double x1, double y1)
{
 return(sqrt(x1*x1 + y1*y1));
}
```

TABLE 7.1   *Elementary C code implementing the complex operations for multiplication, addition, complex conjugation, and modulus. Identical results can be obtained using C++ classes and alternate methods of passing the results. To ensure completely elementary code, we return multiple values as results only through pointers such as* `double *x`.

Thus, $(1, 0)$ is the (unique) identity element of the algebra.

Alongside the identity property, we also have the property that *every nonzero complex number can be divided into another complex number*. Division is defined as multiplication by the inverse,

$$z^{-1} = \frac{\bar{z}}{|z|^2},$$

which obviously fails for the excluded case when $z = 0$, and otherwise satisfies $z \star z^{-1} = (1, 0)$ as required.

7.1.3 RESTRICTION TO UNIT–LENGTH CASE

Finally, we add the requirement that our pairs all be of unit length, and thus points in $\mathbf{S}^1$. Given the unit-length conditions

$$\left\| (x_1, y_1) \right\|^2 = (x_1)^2 + (y_1)^2 = 1,$$

$$\left\| (x_2, y_2) \right\|^2 = (x_2)^2 + (y_2)^2 = 1,$$

we can verify by explicit computation that the product has unit length as well!

$$\left\| (x_1, y_1) \star (x_2, y_2) \right\|^2 = (x_1)^2(x_2)^2 + (y_1)^2(x_2)^2 + (x_1)^2(y_2)^2 + (y_1)^2(y_2)^2$$

$$= \left\| (x_1, y_1) \right\|^2 \left\| (x_2, y_2) \right\|^2$$

$$= 1.$$

Thus, we find a fundamental and far-reaching property of unit-length complex numbers (number pairs obeying the algebra of Equation 7.2):

Complex Unit Circle

> The algebraic form of complex multiplication preserves membership in the unit circle.

† This property makes the unit-length complex numbers into a group—the unitary group $\mathbf{U}(1)$.

We may now *visualize* the algebra directly by watching how complex multiplication transports points around a circle, as illustrated in Figure 7.1. The entire essence of the unit-circle-preserving picture can now be seen alternatively as a consequence

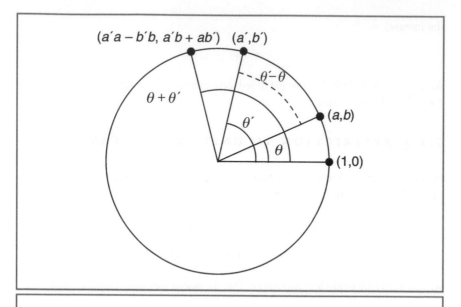

FIGURE 7.1 *Visualizing the complex algebra for unit-length complex numbers. Points on the circle produce new points on the same circle when the complex algebra is applied. The point* $(1,0)$ *has the properties of the identity; when multiplied by* (a, b), *it produces* (a, b) *itself.*

of the rules for trigonometric functions of sums: if $(a, b) = (\cos\theta, \sin\theta)$ and $(a', b') = (\cos\theta', \sin\theta')$, then

$$(a, b) \star (a', b') = (\cos\theta\cos\theta' - \sin\theta\sin\theta', \sin\theta\cos\theta' + \sin\theta'\cos\theta')$$
$$= \big(\cos(\theta + \theta'), \sin(\theta + \theta')\big),$$

which is exactly the rigid 2D rotation shown in Figure 7.1. We note that the *scalar product* has a somewhat different interpretation, in that it singles out the *relative* angle between the two directions on the circle:

$$(a, b) \cdot (a', b') = \cos\theta\cos\theta' + \sin\theta\sin\theta'$$
$$= \cos(\theta - \theta').$$

7.2   QUATERNION ALGEBRA

7.2.1   THE MULTIPLICATION RULE

For completeness in this context, we first review the formula for the quaternion product of two quaternions p and q, which may be written as

$$p \star q = (p_0 q_0 - \mathbf{p} \cdot \mathbf{q}, \; p_0 \mathbf{q} + q_0 \mathbf{p} + \mathbf{p} \times \mathbf{q}),$$

or more explicitly in component form as

$$p \star q = \begin{bmatrix} [p \star q]_0 \\ [p \star q]_1 \\ [p \star q]_2 \\ [p \star q]_3 \end{bmatrix} = \begin{bmatrix} p_0 q_0 - p_1 q_1 - p_2 q_2 - p_3 q_3 \\ p_1 q_0 + p_0 q_1 + p_2 q_3 - p_3 q_2 \\ p_2 q_0 + p_0 q_2 + p_3 q_1 - p_1 q_3 \\ p_3 q_0 + p_0 q_3 + p_1 q_2 - p_2 q_1 \end{bmatrix}. \tag{7.5}$$

The only noncommutative part is the ordinary 3D Euclidean cross-product term, and our signs are chosen so that this has a positive coefficient. (See Table 7.2 for an implementation of the basic quaternion operations as an elementary computer program.)

Just as we did with complex numbers, we can represent the quaternion product in two different ways using matrix multiplication itself: one in which the quaternion acted upon is a represented as column vector, and a second in which the quaternion itself is represented a matrix:

$$p \star q = \mathbf{P} \cdot q$$

$$= \begin{bmatrix} p_0 & -p_1 & -p_2 & -p_3 \\ p_1 & p_0 & -p_3 & p_2 \\ p_2 & p_3 & p_0 & -p_1 \\ p_3 & -p_2 & p_1 & p_0 \end{bmatrix} \begin{bmatrix} q_0 \\ q_1 \\ q_2 \\ q_3 \end{bmatrix},$$

```
double MIN_NORM = 1.0e-7;

void
QuaternionProduct
(double p0, double p1, double p2, double p3,
 double q0, double q1, double q2, double q3,
 double *Q0, double *Q1, double *Q2, double *Q3)
{ *Q0 = p0*q0 - p1*q1 - p2*q2 - p3*q3;
  *Q1 = p1*q0 + p0*q1 + p2*q3 - p3*q2;
  *Q2 = p2*q0 + p0*q2 + p3*q1 - p1*q3;
  *Q3 = p3*q0 + p0*q3 + p1*q2 - p2*q1;
}

double
QuaternionDot
(double p0, double p1, double p2, double p3,
 double q0, double q1, double q2, double q3,
{return(p0*q0 + p1*q1 + p2*q2 + p3*q3); }

void
QuaternionConjugate
(double q0, double q1, double q2, double q3,
 double *Q0, double *Q1, double *Q2, double *Q3)
{ *Q0 =   q0;
  *Q1 = -q1;
  *Q2 = -q2;
  *Q3 = -q3;
}

void
NormalizeQuaternion(
    double *q0, double *q1, double *q2, double *q3)
    {double denom;
    denom =sqrt((*q0)*(*q0) + (*q1)*(*q1) + (*q2)*(*q2)
                                          + (*q3)*(*q3));
 if(denom > MIN_NORM) { *q0 = (*q0)/denom;
                        *q1 = (*q1)/denom;
                        *q2 = (*q2)/denom;
                        *q3 = (*q3)/denom; }
}
```

TABLE 7.2   *Elementary C code implementing the quaternion operations of Equations 4.1 through 4.3, and forcing unit magnitude as required by Equation 4.4. In this straight C-coding method, we return multiple values as results only through pointers such as* double *Q0.

$$p \star q = \mathbf{P} \cdot \mathbf{Q}$$

$$= \begin{bmatrix} p_0 & -p_1 & -p_2 & -p_3 \\ p_1 & p_0 & -p_3 & p_2 \\ p_2 & p_3 & p_0 & -p_1 \\ p_3 & -p_2 & p_1 & p_0 \end{bmatrix} \cdot \begin{bmatrix} q_0 & -q_1 & -q_2 & -q_3 \\ q_1 & q_0 & -q_3 & q_2 \\ q_2 & q_3 & q_0 & -q_1 \\ q_3 & -q_2 & q_1 & q_0 \end{bmatrix}.$$

The matrix forms show that the quaternion algebra is also equivalent to multiplication by an orthogonal matrix in 4D Euclidean space (a matrix whose transpose is the same as its inverse). One can see that $\mathbf{P}$ is an orthogonal matrix by checking explicitly that for unit quaternions $\mathbf{P}^T \cdot \mathbf{P} = I_4$.

† Because $\mathbf{P}$ has only three free parameters, and because 4D orthogonal matrices (which are 4D *rotations*) have six free parameters, $\mathbf{P}$ does not itself include all 4D rotations.

7.2.2 SCALAR PRODUCT

For completeness, we note also the behavior of the *scalar product* of quaternions, which is determined by the *relative angle* ϕ between the two directions on the hypersphere. If we let $p = (\cos(\theta/2), \hat{\mathbf{n}}\sin(\theta/2))$, and $q = (\cos(\theta'/2), \hat{\mathbf{n}}'\sin(\theta'/2))$,

$$p \cdot q = \cos(\theta/2)\cos(\theta'/2) + \hat{\mathbf{n}} \cdot \hat{\mathbf{n}}' \sin(\theta/2)\sin(\theta'/2)$$

$$= \cos\phi.$$

7.2.3 MODULUS OF THE QUATERNION PRODUCT

One of the very special properties of the quaternion algebra is that the modulus of a product is the product of the moduli. We can check this fact explicitly as follows:

$$|p|^2 |q|^2 \overset{?}{=} |p \star q|^2$$

$$\big((p_0)^2 + \mathbf{p}^2\big)\big((q_0)^2 + \mathbf{q}^2\big) = (p_0 q_0 - \mathbf{p} \cdot \mathbf{q})^2$$

$$+ (p_0\mathbf{q} + q_0\mathbf{p} + \mathbf{p} \times \mathbf{q}) \cdot (p_0\mathbf{q} + q_0\mathbf{p} + \mathbf{p} \times \mathbf{q})$$

$$\begin{aligned}
&= (p_0)^2(q_0)^2 - 2p_0q_0\mathbf{p}\cdot\mathbf{q} + (\mathbf{p}\cdot\mathbf{q})^2 \\
&\quad + (p_0\mathbf{q} + q_0\mathbf{p})^2 + \mathbf{p}^2\mathbf{q}^2 - (\mathbf{p}\cdot\mathbf{q})^2 \\
&= (p_0)^2(q_0)^2 + (q_0)^2\mathbf{p}^2 + (p_0)^2\mathbf{q}^2 + \mathbf{p}^2\mathbf{q}^2 \\
&= \big((p_0)^2 + \mathbf{p}^2\big)\big((q_0)^2 + \mathbf{q}^2\big).
\end{aligned} \tag{7.6}$$

We note that the cancellations of the cross dependences of p and q that make the expression on the right-hand side factor into separable components are quite nontrivial.

† Quaternions, like complex numbers, are one of the four division algebras (see Chapter 30) that preserve the Euclidean norm, which is the deeper significance of Equation 7.6.

7.2.4 PRESERVATION OF THE UNIT QUATERNIONS

If $p \cdot p = 1$ and $q \cdot q = 1$, we see that

$$(p \star q) \cdot (p \star q) = q^T \mathbf{P}^T \mathbf{P} q = q^T q = q \cdot q = 1,$$

and thus:

> Quaternion multiplication preserves membership in the space $\mathbf{S}^3$ of unit quaternions.

Quaternion Algebra Preserves the Sphere

† This property makes the unit-length quaternions into a *group*—the special unitary group $\mathbf{SU}(2)$ (also known as $\mathbf{Spin}(3)$)—which is the double cover of the group of ordinary 3D rigid rotations, the special orthogonal group $\mathbf{SO}(3)$.

We may now *visualize* the action of the quaternion algebra schematically as shown in Figure 7.2, which symbolizes the motion of a unit quaternion acted upon by any other unit quaternion. This creates a new unit quaternion, i.e., a new point restricted to the unit hypersphere just like both the original quaternion q and the quaternion p that acts on it. Thus, the quaternion multiplication algebra itself has a geometric interpretation, and the systematic changes resulting from the algebra can be visualized using our geometric methods.

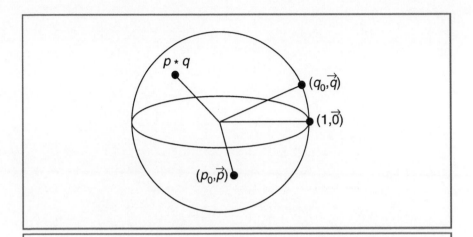

FIGURE 7.2 *Visualizing the algebra of unit-length quaternions. A quaternion point q on the hypersphere is transformed to new point $p \star q$ on the same hypersphere when the quaternion algebra is used to multiply p and q. The point $(1, \mathbf{0})$ has the properties of the identity; when multiplied by $(q_0, \mathbf{q})$, it produces $(q_0, \mathbf{q})$ itself.*

Visualizing Spheres

In this chapter we begin the task of exploring the techniques that can help us visualize quaternions and their properties. Because a quaternion is nothing more than a point on a generalized sphere, the first thing to do is to look at the properties of spheres, starting with simple ones, and work our way up to hyperspheres and quaternions. Additional material extending the treatment in this chapter is presented in Chapter 17 and in Appendix I.

To clearly understand our options for making graphical visualizations of quaternions, we need to begin by examining ways in which points on spheres can be viewed in reduced dimensions. Although the full import will not be apparent until later chapters, we will discover in this section that 3D graphics is luckily just sufficient to make a usable interactive graphics system for looking at individual quaternion points, quaternion curves, and even quaternion surfaces, volumes, and sets of streamlines.

A sphere can, in general, be described as the set of points lying at a constant radius from the origin. The local space in which a bug living on a sphere would move is one dimension lower than the Euclidean space used to describe the constant-radius equation. A sphere is parameterized using the dimension in which a bug would live, and thus for example a circle on a piece of paper in 2D space is a one-sphere (or $\mathbf{S}^1$) and a balloon in physical 3D space is a two-sphere (or $\mathbf{S}^2$), which denotes the surface of the standard sphere of everyday language. Quaternion geometry is the geometry of the hypersphere, typically referred to as the three-sphere (or $\mathbf{S}^3$). Thus, we can work up to understanding quaternion geometry by studying spherical geometry in the following sequence of examples:

Description	Equation	Embedding Dimension
Circle ($\mathbf{S}^1$)	$q \cdot q = (q_0)^2 + (q_1)^2 = 1$	$\mathbb{R}^2$
Sphere ($\mathbf{S}^2$)	$q \cdot q = (q_0)^2 + (q_1)^2 + (q_2)^2 = 1$	$\mathbb{R}^3$
Hypersphere ($\mathbf{S}^3$)	$q \cdot q = (q_0)^2 + (q_1)^2 + (q_2)^2 + (q_3)^2 = 1$	$\mathbb{R}^4$

In each of these cases, we can use a different version of the same visualization method to perceive the structure of the sphere in its local dimension, a dimension one lower than the embedding dimension.

> The basic "trick" to seeing a sphere is based on the fact that if we have any unit vector, the sphere describing its degrees of freedom can be made quantitatively visible in the next-lower dimension.

The Sphere Trick

8.1   2D: VISUALIZING AN EDGE-ON CIRCLE

Our first example of the sphere trick follows from examining a circle, a "ring" in 2D, as shown in Figure 8.1. We shall argue that the horizontal projection (the projection perpendicular to q_0) is sufficient to determine the entire unit vector up to the sign of q_0, and that a two-mode depiction of this horizontal line resolves that ambiguity nicely. The fundamental algebraic equation of the one-sphere $\mathbf{S}^1$ is

$$(q_0)^2 + (q_1)^2 = 1. \tag{8.1}$$

Our goal is to describe points on the resulting circle using a single 1D line segment. Although our main point is to note that the circle is a curve whose points solve Equation 8.1, Figure 8.1 shows us another important concept; namely, that the *interior* of the circle, with $(q_0)^2 + (q_1)^2 < 1$, is also a significant component. Depending on how one wants to think of it, one can describe the interior of a circle (and we will generalize this to any sphere) either as an *empty space* enclosed by the curve or as a space that is "filled up" and thus has dimensions of area. In any event, the points interior to the curve but never part of the curve are an important feature to keep in mind.

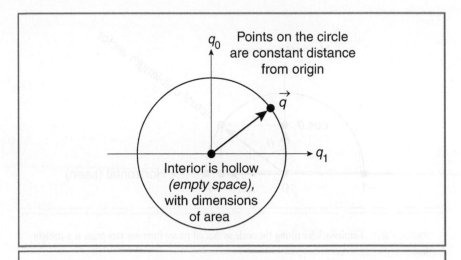

FIGURE 8.1 *The simplest sphere is a circle—a curve of dimension one consisting of the points at a constant distance from the origin and enclosing an empty space (a disk with the dimensions of area).*

Equation 8.1 has only one free parameter, even though we draw it as a locus of points (q_0, q_1) in 2D. We next examine various ways of choosing the single variable parameterizing the points of $\mathbf{S}^1$.

8.1.1 TRIGONOMETRIC FUNCTION METHOD

Knowing that the trigonometric functions cosine and sine satisfy Equation 8.1, we could choose to solve the equations as

$$q_0 = \cos\theta,$$

$$q_1 = \sin\theta \qquad\qquad (8.2)$$

and create a preliminary image (Figure 8.2) showing what would be seen and unseen if we knew only the projection onto the horizontal axis. In this form, the two values of $q_0 = \pm\sqrt{1 - \sin^2\theta}$ that solve the equation for each $-1 < q_1 < +1$ emerge automatically from the periodicity of the trigonometric functions.

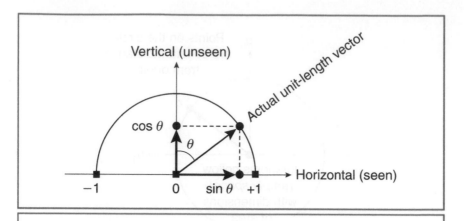

FIGURE 8.2 *Framework for tilting the circle so that all we see from our viewpoint is a straight line.*

8.1.2 COMPLEX VARIABLE METHOD

We recall from our earlier chapter on 2D rotations and complex variables that we can think of a complex variable with unit length as an exponential obeying Euler's famous identity,

$$e^{i\theta} = \cos\theta + i\sin\theta.$$

Thus, our circle (as described, for example, by Equation 8.2) can be written as a single unit-magnitude complex variable

$$z = q_0 + iq_1$$
$$= e^{i\theta}$$
$$= \cos\theta + i\sin\theta.$$

Recalling the definition of the complex conjugate $\bar{z} = q_0 - iq_1$, we verify that $|z|^2 = z\bar{z} = 1$.

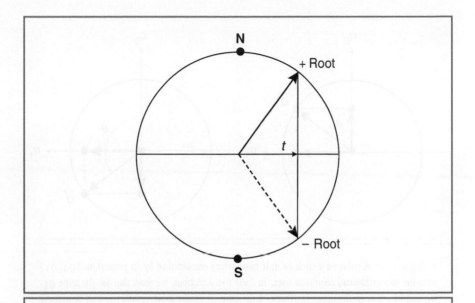

FIGURE 8.3 *The same horizontal projection is shared by the north vector* $(+\sqrt{1-t^2}, t)$ *and the south vector* $(-\sqrt{1-t^2}, t)$.

8.1.3 SQUARE ROOT METHOD

However, we are looking for a way to represent the circle that *does not depend on a 2D view*. Thus, we examine the alternate parameterization given by the two separate, single-valued roots of Equation 8.1,

$$q_1 = t,$$

$$q_0 = \pm\sqrt{1-t^2}, \tag{8.3}$$

as illustrated by the ambiguous projection shown in Figure 8.3. Taking $q_1 = t$, we thus produce an alternate interpretation of the ambiguity of q_0 due to the fact that a quadratic equation has two roots: one in the northern hemisphere and one in the southern hemisphere. Considered *only* as an intrinsic 1D mathematical object, we can imagine going from the full view in Figure 8.4 to an alternative

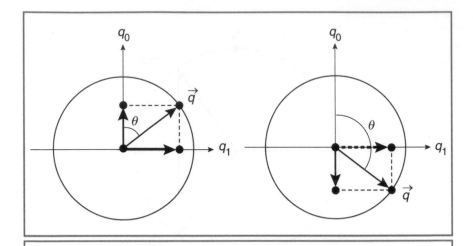

FIGURE 8.4 *A point on a circle (a unit two-vector) characterized by its projections* (q_0, q_1)
onto the two orthogonal coordinate axes. In these two depictions, we show that for the same q_1,
q_0 *can be either in the upper (northern) half-space, or in the lower (southern) half-space. We
distinguish these by drawing the horizontal projection with either a solid or dashed line. Thus, the
horizontal projection and its drawing mode (e.g., solid or dashed) together determine the entire
unit vector* **q** *without ambiguity.*

two-part view (shown in Figure 8.5). We see only two copies of $-1 \leqslant q_1 \leqslant 1$,
corresponding to the plus and minus signs in Equation 8.3, and the equator (the
two points $q_1 = \pm 1$). These two points, which are the degenerate two-point sphere
$\mathbf{S}^0$ satisfying $x^2 = 1$, effectively sew the two pieces together to enclose the invisible
2D area (seen "sideways" in Figures 8.3 and 8.4). We must now *imagine* this unseen
area as lying between the North and South Poles, and this is what is represented in
Figure 8.6.

8.2   THE SQUARE ROOT METHOD

Let us now focus on the two-part square root representation (Equation 8.3) of the
circle. We can see that every value of q_1 implies the value of q_0 up a sign. Knowing the
position of the unit-length mark and the position of q_1, we already know what the

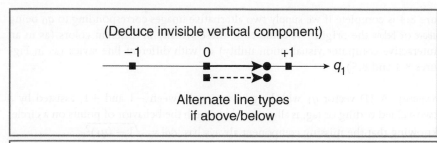

FIGURE 8.5 *Result of transforming the circle to an edge-on projection, leaving only a line segment with length between -1 and $+1$.*

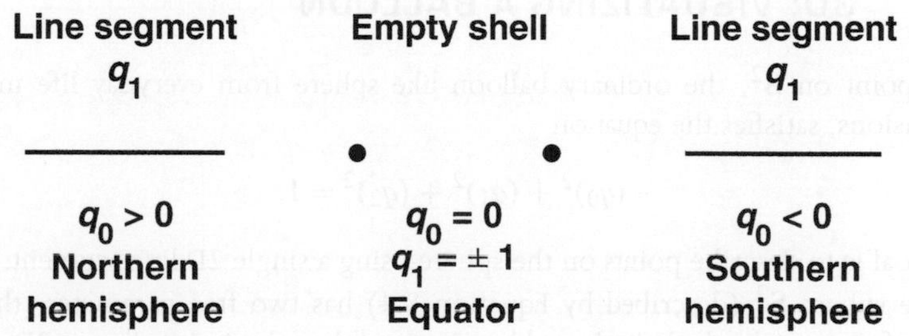

Line segment   **Empty shell**   **Line segment**
        q_1 q_1

————————   ●   ●   ————————

      $q_0 > 0$ $q_0 = 0$ $q_0 < 0$
    Northern $q_1 = \pm 1$ **Southern**
   hemisphere   **Equator**   **hemisphere**

FIGURE 8.6 *The simple circle $\mathbf{S}^1$ visualized as two line segments, the northern and southern hemispheres, together with the equatorial point pair (this is actually $\mathbf{S}^0$) that forms the border exactly between the two, at $q_0 = 0$, or equivalently at $|q_1| = 1$.*

size of q_0 must be. We illustrated this in Figure 8.2, showing also the trigonometric parameterization of the positive root of Equation 8.3 for additional clarity.

From Figure 8.2, we can now imagine tilting the circle until we are looking perpendicular to the q_0 axis, finally doing away with q_0 altogether. Figure 8.5 shows what happens if we display just q_1 and use our knowledge of the fact that $|q_0| = \sqrt{1 - (q_1)^2}$ to deduce the missing information. The picture implied by Figure 8.4 is complete if we supply two alternative images corresponding to q_0 being *above* or *below* the origin. We could code these visually in different colors (as in an interactive computer visualization utility) or with different line styles (as in Figures 8.4 and 8.5).

Summary: A 1D vector q_1 with length varying between -1 and $+1$, assisted by a two-valued coding or tag, is all we need to watch the behavior of points on a circle, knowing that the missing component always has $|q_0| = \sqrt{1 - (q_1)^2}$.

There is one other remarkable fact we should note, and need to remember for the next case. From Figure 8.1, we see that there is a completely empty *interior region* within the circle. The region, obeying the constraint $(q_0)^2 + (q_1)^2 < 1$, *exists* and is potentially important. It *cannot be seen* in Figures 8.5 and 8.6 but must be held in mind.

8.3   3D: VISUALIZING A BALLOON

Each point on $\mathbf{S}^2$, the ordinary balloon-like sphere from everyday life in three dimensions, satisfies the equation

$$(q_0)^2 + (q_1)^2 + (q_2)^2 = 1. \tag{8.4}$$

Our goal is to describe points on the sphere using a single 2D line segment.

The sphere $\mathbf{S}^2$ (described by Equation 8.4) has two free parameters that describe, for example, the latitude and longitude of the spherical surface as 3D points. $\mathbf{S}^2$ can be parameterized by two variables in several ways, but it cannot be described in terms of complex variables as we did for $\mathbf{S}^1$. As in the case of the circle $\mathbf{S}^1$, there is an "interior" of $\mathbf{S}^2$, the set of points for which $(q_0)^2 + (q_1)^2 + (q_2)^2 < 1$ that can either be thought of as the space hollowed out inside $\mathbf{S}^2$ or more appropriately as a solid ball filling the portion of the 3D space enclosed by $\mathbf{S}^2$. In mathematical language this is referred to as a three-ball, denoted by B^3. $\mathbf{S}^2$ is the *boundary* of this object—the skin you would peel off the surface of the solid ball.

8.3.1   TRIGONOMETRIC FUNCTION METHOD

Knowing the polar form of a 3D unit vector, we can easily find trigonometric functions satisfying Equation 8.4.

$$q_0 = \cos\theta,$$

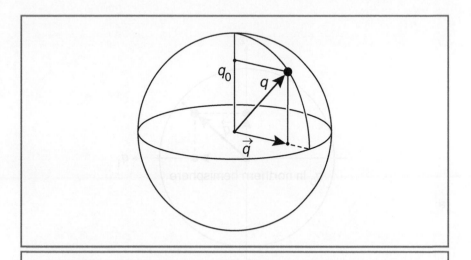

FIGURE 8.7 *Projection of the basic properties of a point on* $\mathbf{S}^2$ *as seen from a 3D perspective.*

$$q_1 = \sin\theta\cos\phi,$$
$$q_2 = \sin\theta\sin\phi.$$

This polar form produces graphic representation shown in Figure 8.7, as well as the projection to 2D shown in Figure 8.8. The 2D projected vectors are modulated simply by $\sin\theta$, and thus the value of $q_0 = \cos\theta$ is known up to a sign and we can, for example, draw the vector as a solid line for the northern hemisphere and as a dashed line for the southern.

8.3.2 SQUARE ROOT METHOD

As before, we can easily write down an alternate parameterization given by the two single-valued roots of Equation 8.4.

$$q_0 = \pm\sqrt{1 - (t_1)^2 - (t_2)^2}, \tag{8.5}$$

$$q_1 = t_1, \tag{8.6}$$

$$q_2 = t_2. \tag{8.7}$$

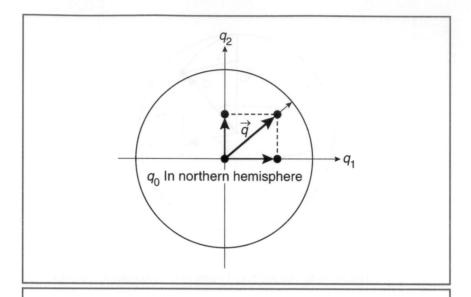

FIGURE 8.8 *Pure 2D view of the basic properties of $\mathbf{S}^2$. Exploiting the constraints on q_0 allows the deduction of the entire unit three-vector from the projected 2D component and the which-polar-region sign marking.*

We thus produce yet another formulation of the ambiguity of q_0 due to the fact that a quadratic equation has two roots. Note that $q_0 = 0$ at the equator, where

$$(q_1)^2 + (q_2)^2 = 1.$$

During a continuous motion from the northern hemisphere to the southern hemisphere, the projected vector approaches the equator from the north (where $q_0 > 0$), touches it at $q_0 = 0$, and then reverses apparent direction in the projection as it crosses into the southern hemisphere (where $q_0 < 0$).

In summary, a 2D vector $\mathbf{q} = (q_1, q_2)$ with length varying between 0 and $+1$, assisted by a two-valued coding or tag, is all we need to watch the behavior of points on a sphere, knowing that the missing component always has $|q_0| = \sqrt{1 - (q_1)^2 - (q_2)^2}$. An interior volume with $q \cdot q < 1$ lies hidden, sandwiched between the disk denoting the projection of the northern hemisphere to the plane

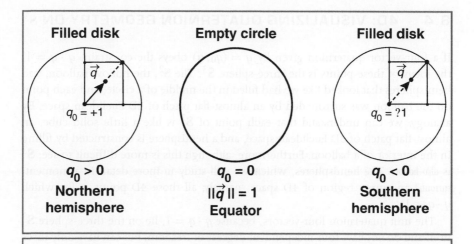

FIGURE 8.9 *The standard sphere* $\mathbf{S}^2$ *visualized as two filled 2D discs, the northern and southern hemispheres, together with the equatorial circle (i.e.,* $\mathbf{S}^1$*) that forms the border exactly between the two, at* $q_0 = 0$*, or equivalently at* $\|\mathbf{q}\| = 1$*.*

and the disk denoting the projection of the southern hemisphere to the plane. The equator of the sphere at $q_0 = 0$ is the circular outer boundary of both disks that joins the northern hemisphere continuously to the southern hemisphere, which we see merged "on top" of each other.

We picture this situation explicitly in Figure 8.9. Again, considered only as an intrinsic 2D mathematical object we can imagine an alternative two-part view of $\mathbf{S}^2$, as shown in Figure 8.9. We see only two copies of the filled disk—with $(q_1)^2 + (q_2)^2 \leqslant 1$, corresponding to the plus and minus signs in Equation 8.7—and the equator, which is now the circle $q_0 = 0$ or $\|\mathbf{q}\|^2 = (q_1)^2 + (q_2)^2 = 1$. This $\mathbf{S}^1$ circular curve once again serves to sew together the two filled-in disks surrounding the North and South polar regions to enclose the invisible 3D volume (seen "sideways" in Figure 8.7). We must now *imagine* this unseen volume as lying between the North and South Poles.

8.4   4D: VISUALIZING QUATERNION GEOMETRY ON $\mathbf{S}^3$

If a four-vector quaternion given by $q = (q_0, \mathbf{q})$ obeys the constraint $q \cdot q = 1$, the locus of these points is the three-sphere $\mathbf{S}^3$. The $\mathbf{S}^2$, the hollow balloon, had hemispheres that looked like we had filled in the middle of a circle, and each point on the balloon was surrounded by an almost-flat patch of 2D Euclidean space. By analogy, we can understand that each point of $\mathbf{S}^3$ is like a little solid cube, an almost-flat patch of 3D Euclidean space, and a hemisphere is constructed by filling in the interior of a balloon. Furthermore, although this is more difficult to see, $\mathbf{S}^3$ is *also hollow*. The hemispheres, which we will study in more detail in a moment, surround an empty region of 4D space; namely, all those 4D points q for which $q \cdot q < 1$.

The unit quaternion four-vectors, because $q \cdot q = 1$, lie on the three-sphere $\mathbf{S}^3$ itself and do not have four independent degrees of freedom, but can have only three independent components. Just as we did for each of the simpler spheres, we will argue that if we display just three independent components, which we write as $\mathbf{q}$, we can in principle infer the value of $q_0 = \pm\sqrt{1 - \mathbf{q} \cdot \mathbf{q}}$ using the sphere projection trick.

By the sphere projection trick, q_0 is essentially redundant information and can be inferred from the projection and the hemisphere flag. Figure 8.10 shows a *solid* sphere with a 3D vector $\mathbf{q}$ having length *less than or equal to* unity representing the entire quaternion. At the center ($\mathbf{q} = 0$) we have $q_0 = +1$, and this point— $q = (+1, 0, 0, 0)$—is the *identity quaternion*. Moving out to $\|\mathbf{q}\| = 1$, we touch the $\mathbf{S}^3$ *equator*, which is itself a two-sphere ($\mathbf{S}^2$) surface—the surface at which $q_0 = 0$ and which has no ambiguity. Continuing on a smooth path, we pass through the $q_0 = 0$ equator and "down" into the southern hemisphere, where $q_0 < 0$. When the southern hemisphere flagged-vector $\mathbf{q}$ approaches the origin ($\mathbf{q} = 0$) for the second time, we have arrived at $q_0 = -1$, which characterizes the South Pole.

Quaternion space, as shown in Figure 8.11, thus consists of the following components to be kept in mind during the visualization:

- *Northern hemisphere*: The northern hemisphere is a *solid ball* of unit radius with the identity quaternion $q = (1, 0, 0, 0)$ appearing at the origin, namely, $\mathbf{q} = 0$.
- *Southern hemisphere*: The solid ball of unit radius with $q_0 < 0$ and the conjugate identity quaternion $q = (-1, 0, 0, 0)$, the South Pole, at the origin where $\mathbf{q} = 0$.

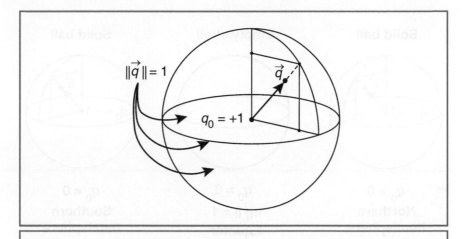

FIGURE 8.10 *A quaternion value displayed as a three-vector* $\mathbf{q}$—*a vector from the origin (the location of the identity quaternion* $(1, 0, 0, 0)$ *or* $q_0 = 1$) *to the point* $\mathbf{q} = (q_1, q_2, q_3)$ *inside a solid ball. The equator* $\|\mathbf{q}\| = 1$ *is the surface of the ordinary sphere* $\mathbf{S}^2$ *enclosing the solid ball.*

- Equator, a two-sphere: The equator is the $\mathbf{S}^2$ outer skin of both solid balls, respectively representing the northern hemisphere and the southern hemisphere. The equator is the set of points where $q_0 = 0$ and $\|\mathbf{q}\| = 1$, and thus lies exactly halfway between the poles $q_0 = +1$ and $q_0 = -1$, splitting the entire $\mathbf{S}^3$ right down the middle into two identical parts. One can think of the equator as the "stitching" where the two identical halves (the solid balls) are sewn together to make the entire three-sphere.
- Four-ball interior: One must imagine an enclosed 4D volume sandwiched between the two solid balls denoting the northern hemisphere and the southern hemisphere (shown in Figure 8.11). This is the interior four-ball with $\|q\| \leqslant 1$, whose boundary is in fact the entire $\mathbf{S}^3$ at $\|q\| = 1$.

One could in principle plot any three quaternion variables or their appropriate linear combinations and infer the unseen coordinate as usual using $q_{\text{unseen}} = \pm\sqrt{1 - (q_{\text{seen}})^2}$. One could also implement arbitrary 4D rigid transformations,

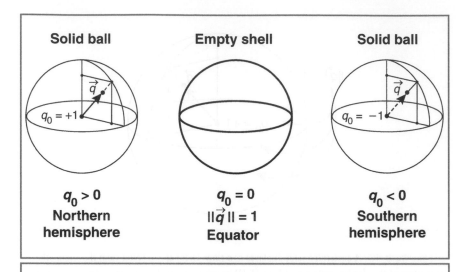

FIGURE 8.11 *The quaternion* $\mathbf{S}^3$ *visualized as two solid 3D balls, the northern and southern hemispheres, together with the equatorial* $\mathbf{S}^2$ *that forms the border exactly between the two, at* $q_0 = 0$, *or equivalently at* $\|\mathbf{q}\| = 1$.

and project to 3D using the top three rows of an arbitrary interactively adjustable 4×4 orthogonal rotation matrix in four Euclidean dimensions.

8.4.1 SEEING THE PARAMETERS OF A SINGLE QUATERNION

Any (unit) quaternion is a point on $\mathbf{S}^3$ and therefore described by three parameters incorporated in the standard parameterization

$$q(\theta, \hat{\mathbf{n}}) = \left(\cos\frac{\theta}{2}, \hat{\mathbf{n}}\sin\frac{\theta}{2} \right), \tag{8.8}$$

where $0 \leqslant \theta < 4\pi$ and where the eigenvector of the rotation matrix (unchanged by the rotation) is a point on the two-sphere $\mathbf{S}^2$ representable as

$$\hat{\mathbf{n}} = (\cos\alpha\sin\beta, \sin\alpha\sin\beta, \cos\beta),$$

with

$$0 \leqslant \alpha < 2\pi \quad \text{and} \quad 0 \leqslant \beta \leqslant \pi.$$

An informative visualization of quaternions can be constructed by examining their properties carefully. If we simply make a 3D display of the vector part of the quaternion, $\hat{\mathbf{n}} \sin \frac{\theta}{2}$, we see that the scalar element of the quaternion is redundant because for each θ,

$$q_0 = \cos \frac{\theta}{2} = \pm \left(1 - \left\| \hat{\mathbf{n}} \sin \frac{\theta}{2} \right\|^2 \right)^{1/2}. \tag{8.9}$$

That is, q_0 is just the implicitly known height of the 4D unit vector in the unseen projection direction, as illustrated in Figure 8.12a. In Figure 8.12b, we schematize the mental model of metric distance required to complete the interpretation of the visualization. If we imagine dividing the arc of the semicircle shown in Figure 8.12a into equal angular segments, the arc lengths are all the same distance apart in spherical coordinates. Projected onto the $\mathbf{q}$ plane, however, the projected spacing is nonuniformly scaled by a factor of $\sin \theta$. Thus, to keep our vision of distance consistent we imagine the space to be like 3D graph paper with concentric spheres drawn at equal distances in the special scale space. Such a 3D graph paper representation would look like that shown in Figure 8.12b. For small 3D radii, distances are essentially Euclidean near the 3D origin; they are magnified as the radius approaches unity to make the marked spheres equidistant in conceptual space.

If we assume the positive root is always taken for q_0, we effectively restrict ourselves to a single hemisphere of $\mathbf{S}^3$ and eliminate the twofold redundancy in the correspondence between quaternions and the rotation group. Alternatively, despite the fact that quaternions with both signs of q_0 map to the same point in this projection, we can indicate the simultaneous presence of both hemispheres using graphical cues. One possible method is to use saturated colors in the "front" hemisphere and faded colors (suggesting distance) for objects in the "back" hemisphere.

8.4.2 HEMISPHERES IN $\mathbf{S}^3$

To clarify the terminology, we note that a projected hemisphere for $\mathbf{S}^2$ is a filled disk (a two-ball) in the plane, and the full surface of the sphere consists of two such disks joined at the outer circular boundary curve. For $\mathbf{S}^3$, we use the word *hemisphere* to indicate a filled solid two-sphere (technically a three-ball) and imagine the full volume of the three-sphere to consist of two such spherical solids joined on the skin (a two-sphere) of the surface enclosing both.

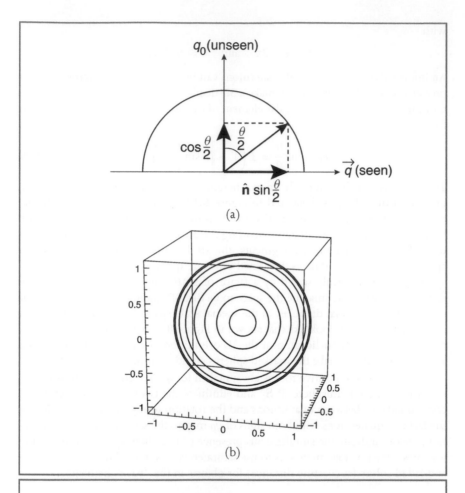

(a)

(b)

FIGURE 8.12 (a) Illustration of how the $q_0 = \cos(\theta/2)$ part of a quaternion is "known" if we have a 3D image of the vector part $\mathbf{q} = \hat{\mathbf{n}} \sin \frac{\theta}{2}$ of the quaternion. (b) Schematic representation of the concentric-sphere uniform distance scales needed to form a mental model of the metric distances in quaternion space between two points in the parallel 3D projection. Distances are roughly Euclidean near the origin ($\mathbf{q} \approx 0$) and equal-length lines appear increasingly compressed as the radius approaches the $\mathbf{S}^2$ equator at $\theta = \pi$, or equivalently $\|\mathbf{q}\| = \sin(\theta/2) = 1$, or $q_0 = 0$.

The family of possible values of Equation 8.8 projects to a double-valued line (actually an edge-on projection of a circle), which is a directed diameter of the unit two-sphere in the direction of $\hat{\mathbf{n}}$. In a polar projection, this circle becomes a line to infinity through the origin.

Any particular 3D rotation is represented twice, because the quaternion circle is parameterized by $0 \leqslant \theta < 4\pi$. A simple parallel projection thus produces two solid balls on top of each other in the 3D projection—one the analog of the North Pole disk of a two-sphere projected parallel from 3D to a screen, and the other the analog of the South Pole disk of a two-sphere.

Visualizing Logarithms and Exponentials

09 The calculus of quaternions involves studying infinitesimal forms and rates of change. These infinitesimal transformations of quaternions are closely related to their logarithms and exponentials, and we will begin by exploring the relationships among quaternionic logarithms, their exponentials, and quaternion calculus. As is now our custom, we will begin to study these objects and their visualizable properties starting from complex numbers, from which we will see that the polar form of a complex number has particularly useful properties that we will be able to exploit in assisting our intuitions.

9.1 COMPLEX NUMBERS

Infinitesimal transformations of quaternions are closely related to their logarithms, and the relationship between quaternionic logarithms and their exponentials provides another visualization viewpoint. We introduce the subject here by using, once again, unit-length complex numbers. The quaternion formulas will be very similar, with the important exception that the order of multiplication is not arbitrary.

The logarithm is of course simply the object that when exponentiated produces the argument of the logarithm. In the case of unit-length complex numbers,

$$\log\left(e^{i\theta}\right) = i\theta,$$

up to an arbitrary additive factor of $2n\pi i$, since

$$e^{i(\theta + 2n\pi)} \equiv e^{i\theta}$$

for an integer n. However, the real essence of this mathematical tool, which ultimately allows us to work with the order-dependent quaternion algebra instead of just the order-independent complex algebra, is the relation among the logarithm, infinitesimal quantities, and the power series expansion for the exponential. (The power series ultimately gives us a way of handling order dependence.) To begin, the formula for the exponential series is

$$e^t = \sum_{n=0}^{\infty} \frac{t^n}{n!},$$

where by definition

$$\log e^t = t.$$

We are interested in unit-length complex numbers and their derivatives. One way of studying them is to replace t with $it\theta$, converting the exponential into a unit-length complex number due to Euler's formula. This is represented as

$$e^{it\theta} = \sum_{n=0}^{\infty} \frac{(it\theta)^n}{n!}$$

$$= \cos(t\theta) + i\sin(t\theta),$$

which can be proven in a number of elegant ways, e.g., from the power series for the trigonometric functions.

However, the most striking properties of this expression are found by exploiting the power series to investigate small changes in t, leading to derivative formulas and a sense of the nature of calculus for unit-length complex numbers—ultimately leading to quaternions. For example, if we take a unit-length complex number $e^{i\theta}$ to a power we find

$$\left(e^{i\theta}\right)^t = e^{it\theta}.$$

Since $de^t/dt = e^t$, the derivative with respect to t is

$$\frac{de^{it\theta}}{dt} = i\theta e^{it\theta}.$$

Similarly, because $d\log x = dx/x$ we have

$$\frac{d\log(e^{it\theta})}{dt} = i\theta.$$

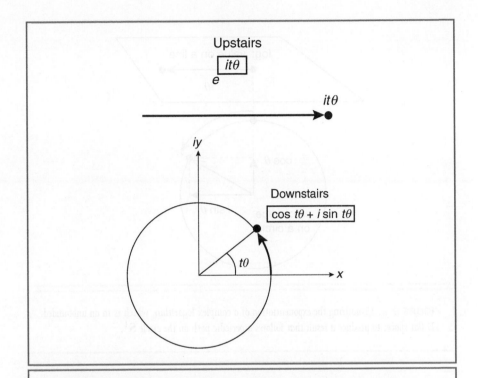

FIGURE 9.1 *A straight, purely imaginary, line "upstairs" in the argument of the exponential function becomes a periodic circle "downstairs" in the complex plane.*

What is more interesting for us is this: Imagine a picture (see Figure 9.1) in which we think of the logarithm as living "upstairs" in the exponential and the result of the entire power series as living "downstairs" on the bottom line of the equation. Then we can start to see the following.

- *The exponential is linear upstairs:* The logarithm is a pure imaginary nonperiodic number whose magnitude has no bound and whose imaginary part therefore looks like the real line.
- *The exponential is curved, possibly periodic downstairs:* There is a direct correspondence between the flat-space appearance of the logarithm values and the image

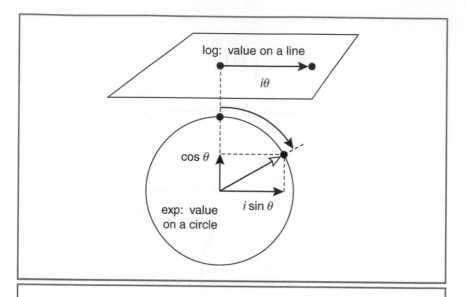

FIGURE 9.2 *Visualizing the exponentiation of a complex logarithm, which is in an unbounded 1D flat space, to produce a result that follows a periodic path on the circle* $\mathbf{S}^1$.

as each point is mapped to $\mathbf{S}^1$ by the exponential function. In particular, because $i(\theta + 2n\pi)$ produces the same point on the circle as $i\theta$ for any integer n, the same point on the circle may correspond to an infinite number of distinct points in the logarithm.

In summary:

> The exponential map of a pure imaginary number generates a circle. The logarithm of a point on the circle is not unique but many-valued.

The Exponential Map

We may therefore *visualize* the logarithms and their correspondence to points on the circle $\mathbf{S}^1$ directly by watching the map from a point on a straight line to a possibly periodically repeated point on a circle, as illustrated in Figure 9.2.

† *Mathematical note:* "Upstairs" is essentially where the Lie algebra of a group lives, and "downstairs" is where we see the Lie group itself. For example, if we look at the coefficient of i in the exponential form of a unit-length imaginary number it can take any value on the real line. Once we look at the value of the exponential, however, it must lie on a unit circle in the complex plane, and this circle is precisely the topological group $\mathbf{U}(1)$.

9.2   QUATERNIONS

Quaternion logarithms follow almost exact parallels to unit-modulus complex numbers. In fact, for any quaternion with $\hat{\mathbf{n}}$ contained completely in one Cartesian axis, for example, $\hat{\mathbf{n}} = (1, 0, 0)$, we recover a unit-modulus complex number. Thus, knowing that

$$e^{i\frac{\theta}{2}} = \cos\frac{\theta}{2} + i\sin\frac{\theta}{2},$$

and knowing that the quaternion $p = (a, b, 0, 0)$ obeys exactly the same algebra as complex numbers,

$$p_1 \star p_2 = (a_1 a_2 - b_1 b_2, a_1 b_2 + a_2 b_1),$$

we can deduce that

$$p = \left(\cos\frac{\theta}{2}, \sin\frac{\theta}{2}, 0, 0\right) = e^{\log p}$$

implies

$$\log p = \left(0, \frac{\theta}{2}, 0, 0\right).$$

We are quickly led to attempt to expand the power series of a pure-vector quaternion of the form $\log p = (0, n_1\frac{\theta}{2}, n_2\frac{\theta}{2}, n_3\frac{\theta}{2})$. Substituting into the exponential power series and using the quaternion algebra of Equation 7.5, we find by direct computation that

$$e^{(0, \hat{\mathbf{n}}\frac{\theta}{2})} = \left(\cos\frac{\theta}{2}, \hat{\mathbf{n}}\sin\frac{\theta}{2}\right).$$

Therefore, because by definition $p = \exp(\log p)$,

$$\log\left(\cos\frac{\theta}{2}, \hat{\mathbf{n}}\sin\frac{\theta}{2}\right) = \left(0, \hat{\mathbf{n}}\frac{\theta}{2}\right). \tag{9.1}$$

Standard operations such as taking derivatives proceed in the usual way. Because $\log q = (0, \hat{\mathbf{n}}\frac{\theta}{2})$ appears in every term of the exponential power series

$$q^{at} = \exp(at \log q) = \sum_{k=0}^{\infty} \frac{1}{k!}(at)^k \prod_{j=0}^{k} \star \left(0, \hat{\mathbf{n}}\frac{\theta}{2}\right),$$

where the products are quaternion products, we can pull out a factor of $\log q$ at either the beginning or the end and obtain the same result. Using the power series to compute the derivative terms, we may resume with the factor of $\log q$ at either end, yielding the derivative formula

$$\frac{dq^{at}}{dt} = a(\log q)q^{at} = aq^{at}(\log q).$$

Polar (logarithmic) version of the multiplication rule: Just as the complex-number multiplication rule exhibits some simplification in polar form, we can use quaternion logarithms to seek a similar quaternion expression. Just as we could rewrite the complex multiplication algebra suggestively as $(r_1, \theta_1) \star (r_2, \theta_2) = (r_1 r_2, \theta_1 + \theta_2)$, we can do the same for quaternions in a polar form in which the "angle" is a pure quaternion vector:

$$\mathbf{p} \star \mathbf{q} = \left(r_1, \hat{\mathbf{n}}_1 \frac{\theta_1}{2}\right) \star \left(r_2, \hat{\mathbf{n}}_2 \frac{\theta_2}{2}\right) = \left(r_1 r_2, \hat{\mathbf{m}} \frac{\theta_{12}}{2}\right).$$

Here,

$$\hat{\mathbf{m}} = \frac{p_0 \mathbf{q} + q_0 \mathbf{p} + \mathbf{p} \times \mathbf{q}}{\|p_0 \mathbf{q} + q_0 \mathbf{p} + \mathbf{p} \times \mathbf{q}\|},$$

and $\cos(\theta_{12}/2) = p_0 q_0 - \mathbf{p} \cdot \mathbf{q}$. Noncommutativity makes this form much less useful than the complex case, but because $\|(r, \theta \hat{\mathbf{n}})\| \equiv r$ and $(r_1)(r_2) = (r_1 r_2)$ it does show that the norm-preserving algebra holds directly.

Visualizing Interpolation Methods

10

In this chapter we complete our set of fundamental visualization methods by studying interpolation in the context of spheres, and eventually in the context of quaternion points. The interpolation from one quaternion to another has profound analogies with standard polynomial interpolation methods in Euclidean space. We will see that geodesic curves on spheres provide the starting point for a rich family of interpolation methods and their graphical depiction.

10.1 BASICS OF INTERPOLATION

We will begin with the most fundamental object—the interpolation that creates a great circle on a sphere of any dimension. This interpolation is in fact slightly nontrivial to derive even for $\mathbf{S}^1$. The derivation we present is the classic method used throughout the mathematical and group theory literature but less often seen in the computer graphics literature.

10.1.1 INTERPOLATION ISSUES

In Euclidean space, linear interpolation (sometimes abbreviated LERP) takes the form

$$\mathbf{x}(t) = \mathbf{x}_0 + t(\mathbf{x}_1 - \mathbf{x}_0) = (1-t)\mathbf{x}_0 + t\mathbf{x}_1,$$

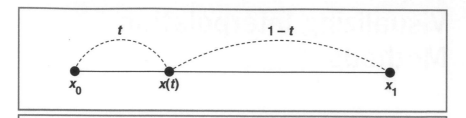

FIGURE 10.1 *The 1D linear interpolation producing points on a straight line.*

where—with $\mathbf{x}(0) \equiv \mathbf{x}_0$ and $\mathbf{x}(1) \equiv \mathbf{x}_1$—taking $0 \leqslant t \leqslant 1$ restricts the parametric curve to the straight line segment between $\mathbf{x}_0$ and $\mathbf{x}_1$. The LERP is shown in Figure 10.1. However, it would be silly to apply the LERP to points $\mathbf{q}$ on a sphere, because even if $\|\mathbf{q}_0\| = \|\mathbf{q}_1\| = 1$ the linearly interpolated point

$$\mathbf{p}(t) = (1 - t)\mathbf{q}_0 + t\mathbf{q}_1$$

will not have the desired properties. We can see easily from Figure 10.2 and the direct computation

$$\mathbf{p}(t) \cdot \mathbf{p}(t) = 1 - 2t + 2t^2 + 2t(1 - t)\cos\phi,$$

where $\mathbf{q}_0 \cdot \mathbf{q}_1 = \cos\phi$, that $\|\mathbf{p}(t)\|$ can be anywhere from 0 to 1 and thus will not describe a point on the same circle as $\mathbf{q}_0$ and $\mathbf{q}_1$. Figure 10.3, in contrast, shows the more desirable circular arc interpolation.

Note: Of course it is possible to *renormalize* a linearly interpolated $\mathbf{p}(t)$ at every point to give it unit length and therefore force it to lie on the sphere. However, this neglects the fact that one of the goals of a linear interpolator is the enforcement of constant velocity in the parameter t. Therefore, the correct analog for the spherical linear interpolator is the enforcement of *constant angular velocity*. Renormalizing $\mathbf{p}(t)$ cannot achieve this goal. Examining the projection of equally spaced portions of the straight line from $\mathbf{q}_0$ to $\mathbf{q}_1$ in Figure 10.4 to the circle, we see that the resulting circular arc segments are never equally spaced. The subtle but important differences between the angular velocity properties of the normalized LERP and the spherical linear interpolation (SLERP) are illustrated explicitly in Figures 10.4 and 10.5.

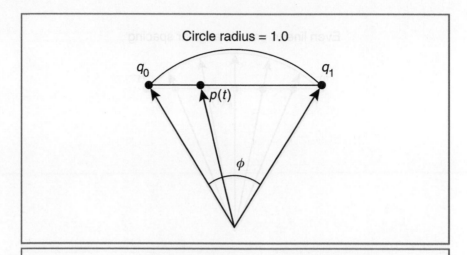

FIGURE 10.2 *A linear interpolant between the two points* **q**$_0$ *and* **q**$_1$ *on a circle in any dimension is not on the circle in general.*

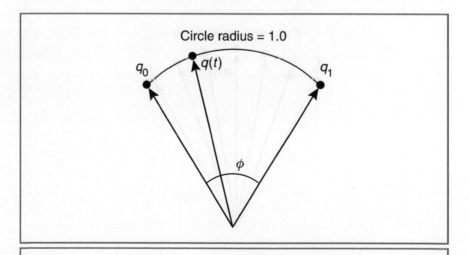

FIGURE 10.3 *Spherical, length-preserving interpolant of a path on the circle* **S**^{1}.

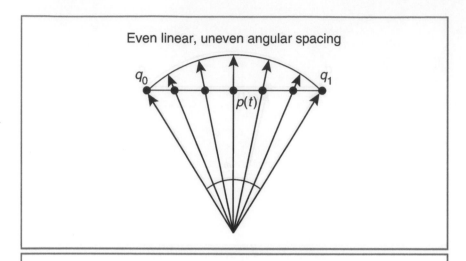

FIGURE 10.4 *A linear interpolant between the two points* $\mathbf{q}_0$ *and* $\mathbf{q}_1$ *on a circle in any dimension is not on the circle in general.*

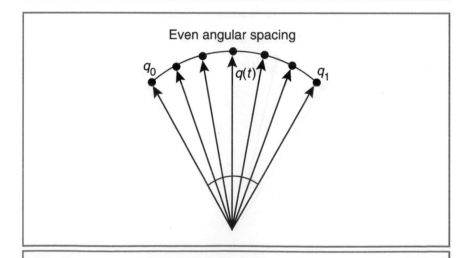

FIGURE 10.5 *Spherical, length-preserving interpolant of a path on the circle* $\mathbf{S}^1$.

The SLERP has the goal of *automatically* adjusting the value of the interpolated vector so that it is *guaranteed* to lie on the circle and to maintain constant angular velocity. We have already seen something similar to this, in our circle-preserving algebra visualization (Figure 7.1). The method for meeting the requirements of the SLERP, which extends trivially to any dimension once we have worked it out for the circle $\mathbf{S}^1$ in 2D, is based on the classic *Gram–Schmidt* procedure. We begin as before by taking two points $\mathbf{q}_0$ and $\mathbf{q}_1$ on the circle, or any sphere, obeying $\|\mathbf{q}_0\| = \|\mathbf{q}_1\| = 1$. We impose one restriction, namely, that the angle ϕ describing the angle between the two vectors, where

$$\cos\phi = \mathbf{q}_0 \cdot \mathbf{q}_1,$$

satisfies $0 \leqslant \phi < \pi$. These conditions can be relaxed if care is taken, but this range turns out to be all that is needed for most quaternion applications.

10.1.2   GRAM–SCHMIDT DERIVATION OF THE SLERP

To find an interpolated unit vector that is guaranteed to remain on the sphere, and thus preserve its length of unity, we first assume that one vector, say $\mathbf{q}_0$, is the starting point of the interpolation. To apply a standard rotation formula using a rigid length-preserving orthogonal transformation in the plane of $\mathbf{q}_0$ and $\mathbf{q}_1$, we need a unit vector *orthogonal* to $\mathbf{q}_0$ that contains some portion of the direction $\mathbf{q}_1$. Defining

$$\mathbf{q}'_1 = \frac{\mathbf{q}_1 - \mathbf{q}_0(\mathbf{q}_0 \cdot \mathbf{q}_1)}{\|\mathbf{q}_1 - \mathbf{q}_0(\mathbf{q}_0 \cdot \mathbf{q}_1)\|},$$

we see that by construction

$$\mathbf{q}'_1 \cdot \mathbf{q}_0 = 0,$$
$$\mathbf{q}'_1 \cdot \mathbf{q}'_1 = 1,$$

where we of course continue to require $\mathbf{q}_0 \cdot \mathbf{q}_0 = 1$. The denominator of this expression has the curious property that

$$\|\mathbf{q}_1 - \mathbf{q}_0(\mathbf{q}_0 \cdot \mathbf{q}_1)\|^2 = 1 - 2\cos^2\phi + \cos^2\phi$$
$$= \sin^2\phi,$$

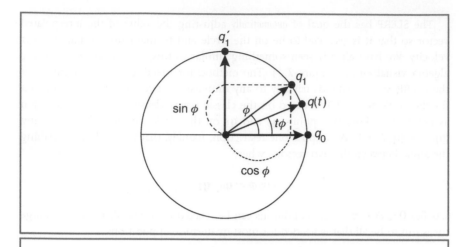

FIGURE 10.6 *The length-preserving spherical interpolation framework, showing the construction of a new orthonormal basis* $(\mathbf{q}_0, \mathbf{q}'_1)$ *that allows us simply to rotate the unit vector rigidly using sines and cosines.*

where we recall that $\cos\phi = \mathbf{q}_0 \cdot \mathbf{q}_1$. Note that because we have imposed $0 \leqslant \phi < \pi$, the sine is always nonnegative, and we can replace $|\sin\phi|$ with $\sin\phi$, which we will find convenient in the following. When $\phi = 0$, there is no interpolation to be done in any event.

Referring to the graphical construction shown in Figure 10.6, we next rephrase the unit-length-preserving rotation using the angle $t\phi$, where $0 \leqslant t \leqslant 1$ takes us from a unit vector aligned with $\mathbf{q}_0$ at $t = 0$ to one aligned with $\mathbf{q}_1$ at $t = 1$. Using our new orthonormal basis, we thus have

$$\mathbf{q}(t) = \mathbf{q}_0 \cos t\phi + \mathbf{q}'_1 \sin t\phi$$

$$= \mathbf{q}_0 \cos t\phi + (\mathbf{q}_1 - \mathbf{q}_0 \cos\phi)\frac{\sin t\phi}{\sin\phi}$$

$$= \mathbf{q}_0 \frac{\cos t\phi \sin\phi - \sin t\phi \cos\phi}{\sin\phi} + \mathbf{q}_1 \frac{\sin t\phi}{\sin\phi}$$

$$= \mathbf{q}_0 \frac{\sin(1-t)\phi}{\sin\phi} + \mathbf{q}_1 \frac{\sin t\phi}{\sin\phi}. \qquad (10.1)$$

This is the SLERP *formula*, which guarantees that

$$\mathbf{q}(t) \cdot \mathbf{q}(t) \equiv 1$$

by construction. The fact that $\mathbf{q}(0) = \mathbf{q}_0$ is obvious, whereas the fact that $\mathbf{q}(1) = \mathbf{q}_1$ is slightly more subtle, recognition of which depends on our observing in Figure 10.6 that $\cos\phi$ is the component of $\mathbf{q}_1$ projected onto the $\mathbf{q}_0$ axis (remember again the definition of $\cos\phi$), and $\sin\phi$ can only be the remaining component in the orthogonal direction. In summary:

SLERP Properties

> The SLERP interpolator rotates one unit vector into another, keeping the intermediate vector in the mutual plane of the two limiting vectors while guaranteeing that the interpolated vector preserves its unit length throughout and therefore always remains on the sphere. The formula is true *in any dimension whatsoever* because it depends only on the local 2D plane determined by the two limiting vectors.

10.1.3 † ALTERNATIVE DERIVATION

Another way of understanding the SLERP is directly in terms of a linear algebra problem (e.g., see Eberly [41]). Let q be a unit vector on a sphere (we will be thinking of quaternions, but it does not matter what dimension the sphere is). We assume that q is located partway between two other unit vectors q_0 and q_1, with the location defined by some constants c_0 and c_1 as

$$q = c_0 q_0 + c_1 q_1. \tag{10.2}$$

As shown in Figure 10.7, q must partition the angle ϕ between q_0 and q_1, where $\cos\phi = q_0 \cdot q_1$, into two subangles, ϕ_0 and ϕ_1, where $\cos\phi_0 = q \cdot q_1$ and $\cos\phi_1 = q \cdot q_0$ and $\phi = \phi_0 + \phi_1$. The apparently backward labeling is in fact intentional: it is chosen so that $\phi_0 = \phi$ makes $q = q_0$ and $\phi_1 = \phi$ makes $q = q_1$. No matter what the dimension of the unit-length qs, taking two dot products reduces this to a solvable linear system, as follows:

$$q \cdot q_0 = \cos\phi_1 = c_0 + c_1 \cos\phi,$$
$$q \cdot q_1 = \cos\phi_0 = c_0 \cos\phi + c_1.$$

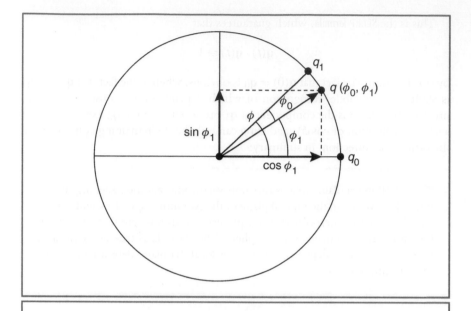

FIGURE 10.7 *Alternate length-preserving spherical interpolation framework, applying linear algebra to a partition of the angles with* $\phi = \phi_0 + \phi_1$.

Using Cramer's rule, we immediately find c_0 and c_1:

$$c_0 = \frac{\det \begin{bmatrix} \cos\phi_1 & \cos\phi \\ \cos\phi_0 & 1 \end{bmatrix}}{\det \begin{bmatrix} 1 & \cos\phi \\ \cos\phi & 1 \end{bmatrix}}$$

$$= \frac{\cos\phi_1 - \cos\phi_0 \cos\phi}{1 - \cos^2\phi},$$

$$c_1 = \frac{\det \begin{bmatrix} 1 & \cos\phi_1 \\ \cos\phi & \cos\phi_0 \end{bmatrix}}{\det \begin{bmatrix} 1 & \cos\phi \\ \cos\phi & 1 \end{bmatrix}}$$

$$= \frac{\cos\phi_0 - \cos\phi_1 \cos\phi}{1 - \cos^2\phi}.$$

When we replace ϕ by substituting $\phi = \phi_0 + \phi_1$, after a little trigonometry we see that

$$c_0 = \frac{(\cos\phi_1 \sin\phi_0 + \cos\phi_0 \sin\phi_1)\sin\phi_0}{\sin^2\phi}$$

$$= \frac{\sin\phi_0}{\sin\phi}$$

$$= \frac{\sin t_0\phi}{\sin\phi},$$

$$c_1 = \frac{(\cos\phi_1 \sin\phi_0 + \cos\phi_0 \sin\phi_1)\sin\phi_1}{\sin^2\phi}$$

$$= \frac{\sin\phi_1}{\sin\phi}$$

$$= \frac{\sin t_1\phi}{\sin\phi},$$

where we have defined $t_0 = \phi_0/\phi$ and $t_1 = \phi_1/\phi$ to obtain a partition of unity, $t_0 + t_1 = 1$. Choosing, for example, $t_0 = 1 - t$ and $t_1 = t$, we recover the standard SLERP formula (Equation 10.1).

10.2   QUATERNION INTERPOLATION

Because the SLERP formula (Equation 10.1), applies to any sphere in any dimension, we can use the same formula for simple quaternion interpolation along a geodesic or great circle arc between two quaternion points q_0 and q_1 with $\cos\phi = q_0 \cdot q_1$.

$$q(t) = q_0\frac{\sin(1-t)\phi}{\sin\phi} + q_1\frac{\sin t\phi}{\sin\phi}. \qquad [10.3]$$

However, there are several special tricks that we will mention here, and apply in detail later on. In particular, comparing quaternion interpolation operations to ordinary linear interpolations reveals that even the simple ideas of the identity, addition, subtraction, and multiplication need to be handled in a new way, as specified by the following.

- Identity: In linear interpolation, $x = 0$ is the identity. In quaternion interpolation, it is important to remember that quaternion multiplication forms a group and thus has an identity. The quaternion identity is *not zero* but is analogous to the complex identity value $(1, 0)$:

$$q_{\text{Identity}} = (1, 0, 0, 0).$$

- *Addition:* Addition is replaced by quaternion multiplication. In this way, we preserve such properties as the need to get back the same value if we add something to the identity. We thus transform addition in quaternion space as follows.

$$x_1 + x_2 \quad \Longrightarrow \quad q_1 \star q_2.$$

All of the basic properties are then preserved, and membership in the unit three-sphere is intact as well.

- *Subtraction:* In linear interpolation, we often subtract two objects to get an incremental form of the interpolation—as in, for example, $t(x_1 - x_0)$. Subtraction has no meaning for quaternion rotation representations, but in most cases where a difference would be used in a linear interpolation we can replace subtraction by a *quaternion inverse displacement*. The idea is that if $x_0 = 0$ the difference must reduce to the object itself, but there must be a way of getting smoothly from one quaternion point to another. The technique is to premultiply by the inverse of the object you would subtract in ordinary arithmetic. Thus,

$$(x_1 - x_0) \quad \Longrightarrow \quad (q_0)^{-1} \star q_1.$$

- *Multiplication:* Just as subtraction is mapped into multiplication, multiplication is mapped into exponentiation. The quaternion analog of an expression such as $t(x_1 - x_0)$ uses the factor t as a power, so that $t = 0$ is the null operation and $t = 1$ reaches the end via

$$t(x_1 - x_0) \quad \Longrightarrow \quad \left(q_0^{-1} \star q_1\right)^t$$

$$\Longrightarrow \quad e^{t * \log(q_0^{-1} \star q_1)},$$

where quaternion logarithms are treated in the preceding section.

- *Iteration:* Iterative procedures such as the de Casteljau construction can be converted into Bezier splines and B-splines just as is done for linear splines by substituting different control points. For almost any case of this sort, following the quaternion substitution rules for the linear interpolation arithmetic analogies just given works straightforwardly.

We may, in fact, write the SLERP equation now with a completely alternative derivation based on the arithmetic analogies just presented. Following these rules, we would convert the Euclidean expression

$$x(t) = x_0 + t(x_1 - x_0)$$

to the analogous quaternion expression

$$q(t) = q_0 \star \left(q_0^{-1} \star q_1\right)^t$$
$$= q_0 \star \exp t \log\left(q_0^{-1} \star q_1\right).$$

To see the equivalence of this notation to the others presented, we first note that the scalar component of the quaternion product $Q = q_0^{-1} \star q_1$ is in fact

$$Q_0 = q_0 \cdot q_1 = \cos \phi. \tag{10.4}$$

Second, we see that we can isolate all essential features of the interpolation by looking at the special case $q_0 = $ identity $= (1, 0, 0, 0)$. Then we can simply replace $(q_0^{-1} \star q_1)$ with an equivalent q_1 whose scalar component agrees with Equation 10.4 and whose vector component is in the direction $\hat{\mathbf{n}}$. Thus, $q_1 = (\cos \phi, \hat{\mathbf{n}} \sin \phi)$. Then, we find

$$q(t) = q_0 \frac{\sin(1-t)\phi}{\sin \phi} + q_1 \frac{\sin t\phi}{\sin \phi}$$

$$= (1, \mathbf{0}) \frac{\sin(1-t)\phi}{\sin \phi} + (\cos \phi, \hat{\mathbf{n}} \sin \phi) \frac{\sin t\phi}{\sin \phi}$$

$$= \left(\frac{\sin \phi \cos t\phi - \sin t\phi \cos \phi + \sin t\phi \cos \phi}{\sin \phi}, \hat{\mathbf{n}} \frac{\sin t\phi \sin \phi}{\sin \phi} \right)$$

$$= (\cos t\phi, \hat{\mathbf{n}} \sin t\phi)$$

$$= \exp(0, t\phi \hat{\mathbf{n}})$$

$$= (q_1)^t,$$

where we used $\log(q_1) = (0, t\phi \hat{\mathbf{n}})$. Displacing this to start the interpolation at any arbitrary q_0 and using the value of $\hat{\mathbf{n}}$ computed from $(q_0^{-1} \star q_1)$ immediately gives the standard SLERP formula (Equation 10.3).

10.3   EQUIVALENT 3×3 MATRIX METHOD

Although it is often claimed that the smooth orientation interpolations desired for computer graphics applications can only be obtained using quaternions, this is not strictly correct. Although the quaternion hypersphere $\mathbf{S}^3$ provides us with many capabilities that are most naturally described in terms of quaternions, and in particular quaternion distances for measuring the closeness of families of orientation frames, many individual tasks support alternative approaches that are equivalent to a quaternion approach.

The SLERP and the nested SLERP used to describe higher-order quaternion interpolations do in fact have a completely equivalent formulation that follows from the treatment in the previous section. All we need to do is replace the quaternions in the previous section by ordinary 3×3 rotation matrices.

† The logarithm of a 3×3 matrix is defined in a group-theoretic context as the associated 3×3 antisymmetric matrix (an element of the $\mathbf{so}(3)$ Lie algebra) that when exponentiated generates the desired $\mathbf{SO}(3)$ matrix corresponding to $\mathbf{R}(\theta, \hat{\mathbf{n}})$. This turns out to be easy to compute using essentially the same technology we use to extract θ and $\hat{\mathbf{n}}$ from any 3×3 rotation matrix $\mathbf{R}$. If we let $\hat{\mathbf{r}}$ be a basis for the set of antisymmetric matrices, one can show that

$$\mathbf{R}(\theta, \hat{\mathbf{n}}) = \exp(\theta \hat{\mathbf{r}} \cdot \hat{\mathbf{n}}).$$

Assuming that we can compute the necessary logarithms and axis-angle components of the chosen 3×3 rotation matrices $\mathbf{R}_0$, $\mathbf{R}_1$, and their products, the entire set of arguments follows through in exactly the same formal way, so that we may write

$$\mathbf{R}(t) = \mathbf{R}_0 \cdot \left(\mathbf{R}_0^{-1} \cdot \mathbf{R}_1\right)^t$$
$$= \mathbf{R}_0 \cdot \exp t \log\left(\mathbf{R}_0^{-1} \cdot \mathbf{R}_1\right),$$

where we have explicitly denoted the 3×3 matrix multiplication by a dot ($\cdot$) to indicate how matrix multiplication replaces quaternion multiplication in this context. With this formalism, we may replace any iterated SLERP operations on quaternions by iterated 3D matrix interpolations. However, in general it will be easier technically to implement the algebra using quaternion notation and use Equation 6.8 to compute $\mathbf{R}(t)$.

Looking at Elementary
Quaternion Frames

The quaternion framework allows us to analyze the relationships among coordinate frames in a variety of ways. In preparation for the more complex situations that will soon arise in subsequent chapters, we pause for a moment to review a few very simple cases that exploit quaternion visualization. In the following we will look at single frames, the relationships between two or more discrete frames, and smoothly changing sequences of frames.

11.1   A SINGLE QUATERNION FRAME

The simplest possible frame is the identity frame. If we take coordinate labels for points on $\mathbf{S}^3$ to be (w, x, y, z)—standing for the Euler-eigenvector rotation parameterization $q = (\cos(\theta/2), \hat{\mathbf{n}}\sin(\theta/2))$—the 3D identity frame can be represented by either of the two possible quaternions, $q = (1, 0, 0, 0)$ and $q = (-1, 0, 0, 0)$.

Figure 11.1a represents the standard positive hemisphere of the vector part of the quaternions, and the black dot at the origin is the quaternion identity $q = (1, 0, 0, 0)$. We cannot see $q = (-1, 0, 0, 0)$ in this visualization because we can only draw one hemisphere of $\mathbf{S}^3$ at a time projected to this coordinate system. On the other hand, if we switch from the standard (x, y, z) projection to the (x, y, w) projection (shown in Figure 11.1b), we can see both alternative signs of the identity quaternion, with the black dot showing $q = (1, 0, 0, 0)$ and the white dot showing $q = (-1, 0, 0, 0)$.

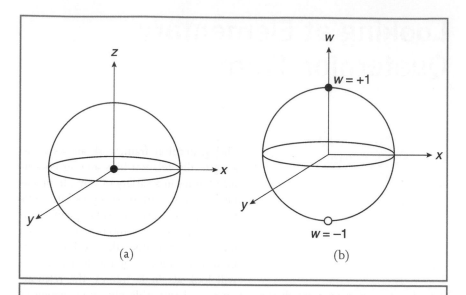

FIGURE 11.1 (a) The identity frame, $q = (1, 0, 0, 0)$, is a point at the origin in our standard "vector-only" projection. (b) Looked at from the side, we see that this quaternion is of unit length, with the only nonzero component $w = 1$, as we can infer from the fact that all quaternion frames must have unit length.

11.2   SEVERAL ISOLATED FRAMES

If we wish to examine two isolated frames q_1 and q_2 and their relationship, the first thing we need to do is to remember that we should always simplify our lives in this situation by making a good "anchor" for our visualization. We can accomplish this by simply translating q_1 back to the identity frame using

$$\bar{q}_1 \star q_1 = (1, \mathbf{0}),$$

(11.1)

and applying the same transformation to q_2 to produce a *new* quaternion p that has the *same relation* to the identity as q_2 does to q_1:

$$\bar{q}_1 \star q_2 = p.$$

(11.2)

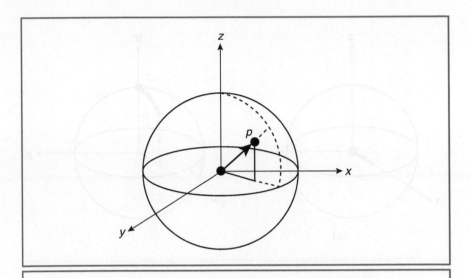

FIGURE 11.2 *The relationship between two arbitrary quaternions, simplified by transforming q_1 back to the identity, leaving the relationship defined completely by $p = \bar{q}_1 \star q_2$.*

Then, as shown in Figure 11.2, we can simply plot (p_x, p_y, p_z) in the standard eigenvector coordinate system. The relationship to the identity transform (the origin of the coordinate system) is immediate: If we denote the vector part of the quaternion $p = \bar{q}_1 \star q_2$ by $\mathbf{p} = (p_x, p_y, p_z)$, then q_1 and q_2 are related by a rotation about the fixed axis in the direction of

$$\hat{\mathbf{p}} = \frac{\mathbf{p}}{\|\mathbf{p}\|},$$

and with a total angle of rotation about $\hat{\mathbf{p}}$ given by

$$\theta = 2\arcsin\|\mathbf{p}\|.$$

11.3   A ROTATING FRAME SEQUENCE

The simplest example of a rotating frame is a parameterized sequence of rotation matrices. For example, this might be a rotation about the $\hat{\mathbf{x}}$ axis by an angle ranging

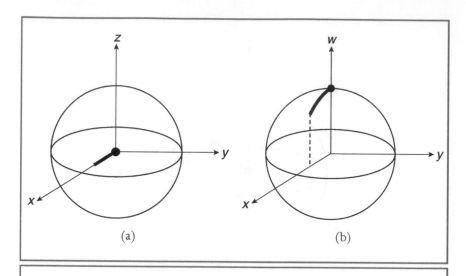

FIGURE 11.3 *A sequence of frames formed by rotating about the* $\hat{\mathbf{x}}$ *axis. (a) Usual projection. (b) w projection.*

from 0 (the identity frame) to θ using intermediate angles $t\theta$ with $0 \leqslant t \leqslant 1$:

$$\begin{bmatrix} 1 & 0 & 0 \\ 0 & \cos t\theta & -\sin t\theta \\ 0 & +\sin t\theta & \cos t\theta \end{bmatrix}.$$

The corresponding family of quaternions is simply the line segment given by

$$q(t) = (\cos t\theta/2,\ \sin t\theta/2, 0, 0).$$

In the standard coordinate system, the visualization of this family of frames is the straight line on the $\hat{\mathbf{x}}$ axis, shown in Figure 11.3a. Viewing this curve from the side to show the (redundant) w coordinate explicitly yields Figure 11.3b.

To replace the rotation about the $\hat{\mathbf{x}}$ axis by a rotation about an arbitrary axis $\hat{\mathbf{n}}$, we simply reorient the straight line on the $\hat{\mathbf{x}}$ axis (Figure 11.3a) to point in the direction $\hat{\mathbf{n}}$. Figures 11.4a and 11.4b show the corresponding general cases for a sequence of frames generated by a rotation about the $\hat{\mathbf{n}}$ axis.

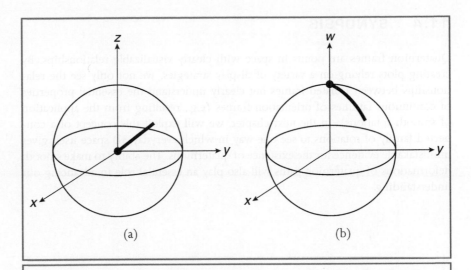

FIGURE 11.4 *A sequence of frames formed by rotating about the $\hat{\mathbf{n}}$ axis. (a) Usual projection. (b) w projection.*

Using the $\hat{\mathbf{x}}$ axis as a reference frame: In conclusion, it is important to note that for the case of the simple rotation about the $\hat{\mathbf{n}}$ axis we really do not need to consider anything but the $\hat{\mathbf{x}}$-axis rotation! Just as we understood the relationship between q_1 and q_2 more clearly by translating q_1 back to the identity frame, there is a direct construction that transforms a rotation about the $\hat{\mathbf{n}}$ axis back to a "standard" equivalent rotation about the $\hat{\mathbf{x}}$ axis, which one can adopt as the "reference origin" for studying these simple rotations. The explicit transformation is achieved by rotating the $\hat{\mathbf{n}}$ direction back to the $\hat{\mathbf{x}}$ direction using the fixed rotation axis $\hat{\mathbf{m}}$ and angle ϕ derived from

$$\hat{\mathbf{m}} = \frac{\hat{\mathbf{n}} \times \hat{\mathbf{x}}}{|\hat{\mathbf{n}} \times \hat{\mathbf{x}}|},$$

$$\cos\phi = \hat{\mathbf{n}} \cdot \hat{\mathbf{x}}.$$

11.4   SYNOPSIS

Quaternion frames are points in space with clearly visualizable relationships. By creating plots relying on a variety of display strategies, we not only see the relationships between isolated frames but clearly understand the essential properties of continuous families of orientation frames (e.g., resulting from the application of smooth rotations). In the next chapter, we will exploit this concept of a connected family of rotations to see one way in which everyday 3D space itself gives unmistakable evidence for the existence of quaternions. The ability to make smooth deformations of quaternion paths will also play an essential role in advancing our understanding.

Quaternions and the Belt Trick: Connecting to the Identity

12

There is an ancient parlor trick, whose popularization in the physics community has long been attributed to Dirac (e.g., see Hart et al. [87] and Misner et al. [126]), that catapults the reality of the quaternion into an everyday context seemingly innocent of such esoteric constructs. We have already met the belt trick in Chapter 2, and now we are ready to follow the details of the quaternion visualization.

The belt trick begins with two people holding opposite ends of an ordinary leather belt. (In one of many variants, sometimes known as the Dirac string trick, two or more strings replace the edges of the belt.) As illustrated in the initial frames of Figures 12.1 and 12.2, one person then twists the belt about its long axis by either 360 degrees (2π radians) or 720 degrees (4π radians). The main rule of the game is that the *absolute orientation* of the last inch of the two ends of the belt cannot change. The two people holding the ends of the belt can move the location of the belt end anywhere they like, but they cannot twist or rotate their end of the belt relative to, say, the walls of the room. As long as the belt is not torn or cut, the players can pass the end of the belt from hand to hand if desired, provided the orientation of the coordinate frame of the last inch does not change.

The belt-end holders are then given the goal of trying to untwist the belt subject to the rules. The "trick" then consists of the following pair of observations.

- *With a 360-degree twist:* With a 360-degree clockwise twist, no matter what the two belt holders do when they return to the original position the belt

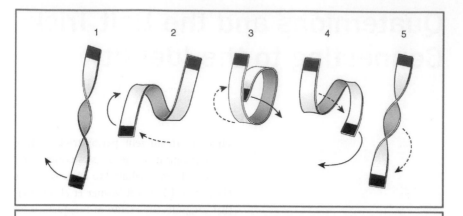

FIGURE 12.1 The result of the belt-trick move for a 360-degree twist changes the orientation sequence from clockwise to counterclockwise, but cannot untwist the belt.

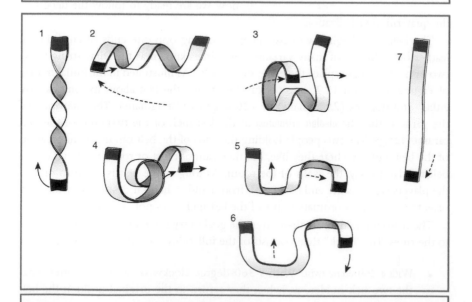

FIGURE 12.2 A 720-degree twist can be continuously deformed to an untwisted belt.

is either unchanged or possibly changed to a 360-degree counterclockwise twist. However, it is never untwisted. (See Figure 12.1.)

- *With a 720-degree twist:* With a 720-degree twist, it is possible (even easy) to find a motion that leaves the belt entirely flat and untwisted in its initial position, as illustrated in Figure 12.2.

12.1 VERY INTERESTING, BUT WHY?

We are now led to pose two questions: how can we explain the belt trick and, interesting as it is, how could it have anything to do with quaternions?

12.1.1 THE INTUITIVE ANSWER

What is astonishing, and not at all obvious, is that *continuous sequences* of ordinary orientation frames can be arranged in such a way that we can actually *detect the distinction* in our everyday lives between the pairs of quaternion frames that *appear the same* if we look at them as isolated instances of an object positioned in 3D space. If we cut the belt, it is impossible to tell whether one end has been twisted relative to the other by 0, 360, or 720 degrees, but when we keep a record of the orientation frame sequence (the belt itself remembers), there is something in the real world that is different and that allows us to tell the difference. Only quaternions can help us clearly explain *why* we can tell the difference. The belt trick proves, in some sense, that quaternions are real and are not an irrelevant abstraction. This is because they tell us *how* the belt is "keeping track" of continuous sequences of orientation frames, each differing only infinitesimally from its neighbors.

Considering another example, we are accustomed to looking at a pair of isolated objects, such as two teapots, and concluding that if they look exactly the same they have indistinguishable orientations. In fact, the belt trick is showing that in a continuous sequence of orientation frames the *relationship* between the frames differing by 360 degrees is fundamentally different from the *relationship* between the frames differing by 720 degrees, even though in isolation they cannot be distinguished.

12.1.2 † THE TECHNICAL ANSWER

The precise mathematical answer is that the group corresponding to quaternions is the group $\mathbf{SU}(2)$, with a topological space $\mathbf{S}^3$ that is simply connected, and that 720-

degree rotations correspond to a closed path in the group $\mathbf{SU}(2)$ that is smoothly deformable to the identity. Ordinary rotations correspond to the group $\mathbf{SO}(3)$, with a topological space $\mathbf{RP}^3$ that is not simply connected. Although a 360-degree rotation has a path in $\mathbf{SO}(3)$ that connects an identity orientation to an identity orientation, this path gets "hung up" in such a way that a smooth deformation of all points on the path to the identity is impossible. We already know that $\mathbf{SU}(2)$ is the double cover of $\mathbf{SO}(3)$; that is, there are two distinct quaternions for each distinct orientation frame in 3D space. The belt trick reflects this double-valued relationship, distinguishing a one-circuit 360-degree rotation from the equivalent two-circuit 720-degree rotation. In the following, we will work out exactly how this happens, and we will clearly visualize the unmistakable difference as it shows up in the quaternion coordinates for the frame sequences.

12.2   THE DETAILS: HOW QUATERNION VISUALIZATION EXPLAINS THE BELT TRICK

Using quaternions, we can construct a visualization of the belt and the belt trick that is both mathematically exact and intuitively appealing. In addition, we can actually create a computer simulation of the motions made by the belt holders doing the belt trick (although for computational simplicity the simulation turns out to be much easier using an elastic belt or elastic strings).

The basic idea is that because each small piece of the belt (a line drawn on the belt in its shortest direction) and the vector perpendicular to the belt form a 3D *frame*, and because each such frame is a *point* in quaternion space, the entire belt is representable as a *connected path of points* in the quaternion space $\mathbf{S}^3$. The initial twisting of the belt and the motions of the belt holders attempting to untwist the belt are then nothing more than curves in quaternion space, and are immediately visualizable using appropriate projections of the curves.

When the belt is untwisted, all of its frames are the identity frame, and these are therefore all stacked up as a huge pile of identical quaternion points at $w = +1$ corresponding to the connected frames of the untwisted belt. If we let t parameterize the belt, with $t = 0$ being the start of the belt and $t = 1$ being the end of the belt, then

$$q(t) = (1, 0, 0, 0)$$

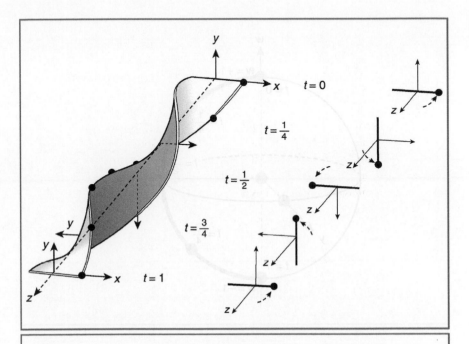

FIGURE 12.3 *Each section of the belt defines a 3D frame based on the line cutting across the belt and the normal direction to the belt surface at that point.*

and there is no change in the frame as t varies. When we twist the belt about its long axis (say, the z axis), for a total twist θ in 3D space we have

$$q(t) = \left(\cos \frac{t\theta}{2}, \, 0, \, 0, \, \sin \frac{t\theta}{2} \right).$$

In Figure 12.3, we show the belt itself as a sequence of frames, along with a sampling of isolated frames that correspond explicitly to

$$\begin{bmatrix} \cos t\theta & -\sin t\theta & 0 \\ \sin t\theta & \cos t\theta & 0 \\ 0 & 0 & 1 \end{bmatrix}.$$

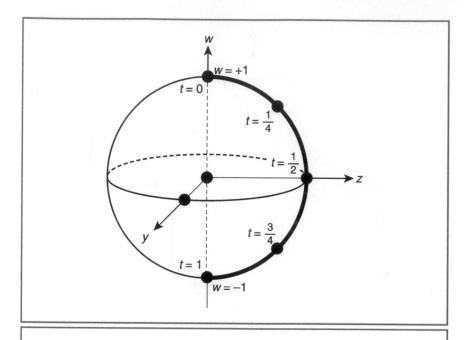

FIGURE 12.4 *The set of frames of a twisted belt forms a curve in quaternion space. The 360-degree twist along the z axis forms a half circle in $\mathbf{S}^3$, connecting the North and South Poles at $w = +1$ and $w = -1$, respectively.*

Then we see the following, both analytically and visually.

- $\theta = 2\pi$: *360 degrees:* If we twist by a single full rotation, 2π or 360 degrees, then

$$q(0) = (+1, 0, 0, 0),$$

$$q(t) = (\cos t\pi, 0, 0, \sin t\pi),$$

$$q(1) = (-1, 0, 0, 0),$$

and the belt's frames correspond to the quaternion curve shown in Figure 12.4, which runs between the North and South Poles.

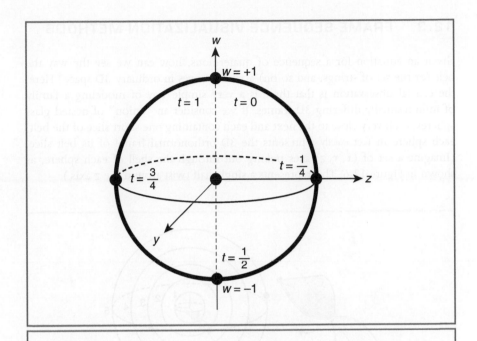

FIGURE 12.5 In quaternion *space*, the 720-degree twist along the *z* axis forms a full circle in **S**³, making a closed curve that connects the North Pole at $w = +1$ back to itself in a continuous path.

- $\theta = 4\pi$: *720 degrees*: If we twist by a double full rotation, 4π or 720 degrees, then

$$q(0) = (+1, 0, 0, 0),$$

$$q(t) = (\cos 2t\pi, 0, 0, \sin 2t\pi),$$

$$q(1) = (+1, 0, 0, 0).$$

The belt's frames correspond to a closed curve, as shown in Figure 12.5.

12.3   FRAME-SEQUENCE VISUALIZATION METHODS

Given an equation for a sequence of quaternions, how can we see the way the belt (or the set of strings, and so on) actually moves in ordinary 3D space? Here, the crucial observation is that there is a very simple way of modeling a family of infinitesimally differing 3D frames. If we consider an "onion" of nested glass spheres, each very close to the next and each containing one short slice of the belt, each sphere in fact *exactly* represents the 3D orthonormal frame of its belt slice. (Imagine a set of (x, y, z) axes poking out through the shell of each sphere, as shown in Figure 12.6. This represents a single half twist around the z axis.)

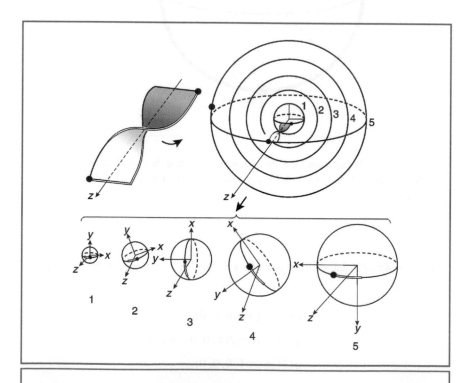

FIGURE 12.6 *A sequence of frames shown both as a twisted belt, and as a series of nested glass spheres, each representing the frame of the belt at a selected orientation.*

The sequence of frames corresponds also, as we know, to a sequence of connected points forming a smooth curve $q(t)$ in quaternion space. Thus, if we want to represent a deformation of this curve, say $q(t, \alpha)$, corresponding to a sequence of belt deformations we can simply take each quaternion point in the deformation equations to correspond, via the quadratic map, to the frame of a glass sphere. Thus, we simply start at an initial belt state (e.g., $\alpha = 0$) for which the family of nested frames represented by the glass spheres are twisted to some extent about the z axis as t varies from 0 to 1. As we slowly vary α, the family of spheres varies as well, and the belt goes along for the ride (with a little stretching). Because each point on each sphere differs only infinitesimally from the corresponding point on its neighbor, one can fill the volume of the original nested spheres with belts, strings, or tubes and still maintain a continuous and tearing-free motion for any smooth quaternion deformation path. This is the basis of the animation "Air on the Dirac Strings" [144], described in detail by Hart, Francis, and Kaufmann [87]. Next, we will write down in complete detail explicit examples of exactly how to accomplish examples of the deformations of interest.

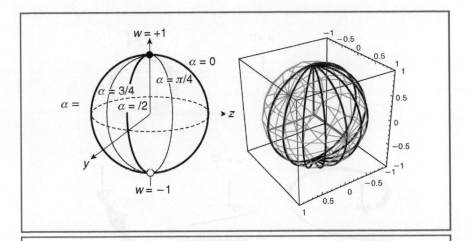

FIGURE 12.7 The function $q(t, \alpha)$ showing the deformation of the single-twist quaternion path to smoothly change from a clockwise twist about the z axis to a counterclockwise twist. The plot is a 3D projection of the quaternion w, y, z components, omitting the vanishing x component.

12.3.1 ONE ROTATION

According to the rules of the belt trick, the frames of the ends of the belt *cannot change*, and thus for $\theta = 2\pi = 360°$ the belt holders can deform the curve itself any way they like, but they can never detach the ends from their locked positions at $w = +1$ and $w = -1$. As shown in Figure 12.1, the largest possible change is to change a clockwise to a counterclockwise rotation, and this is simply a deformation of the path to $q(t) = (\cos t\pi, 0, 0, -\sin t\pi)$, which can take any route that preserves $q \cdot q = 1$. An example of this is

$$q(t, \alpha) = (\cos t\pi, 0, \sin\alpha \sin t\pi, \cos\alpha \sin t\pi),$$

with $0 \leqslant \alpha \leqslant \pi$. Figure 12.7 shows the 3D (w, y, z) projection of this deformation in quaternion space, and Figure 12.8 shows the corresponding nested-sphere frame-sequence visualization for selected values of α.

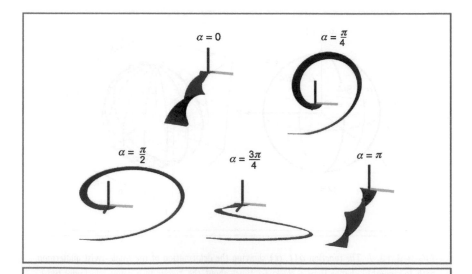

FIGURE 12.8 *Selected frames from the deformation sequence of the single-twist belt, using the nested glass-sphere visualization mechanism for each single belt configuration.*

12.3.2 TWO ROTATIONS

For the $\theta = 4\pi = 720$-degree twist, the rules of the belt trick do not restrict us in any way as long as the single point $w = +1$ remains fixed on the curve. The belt holders can deform the curve so that every single point shrinks (maintaining the quaternion condition $q \cdot q = 1$) slowly toward $w = +1$ and every frame of the belt returns to the identity frame $q = (1, 0, 0, 0)$, as illustrated in Figure 12.2. An explicit example of such a deformation (following a method suggested by Hart et al. [87]) is

$$q(t, \beta) = \left(\sin^2 \beta + \cos^2 \beta \cos 2t\pi, 0, \cos \beta \sin \beta (1 - \cos 2t\pi), \cos \beta \sin 2t\pi\right)$$

which obeys $q \cdot q = 1$, reduces to the full double twist for $\beta = 0$, and is the single point $q = (1, 0, 0, 0)$ (that is, $w = +1$) for $\beta = \pi/2$. One can of course make any desired deformation shrinking the original circle smoothly to a point to obtain an arbitrary sequence of belt cross-section frame changes that straighten out the belt

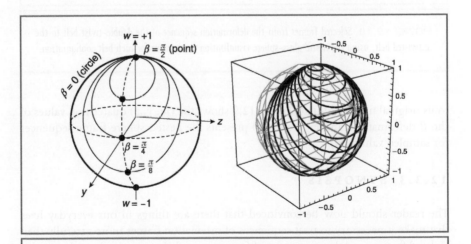

FIGURE 12.9 *The function* $q(t, \beta)$ *showing the deformation of the quaternion path for a double-twist about the* z *axis to collapse all points smoothly to the quaternion identity at* $w = +1$. *The plot is a 3D projection of the quaternion* w, y, z *components, omitting the vanishing* x *component.*

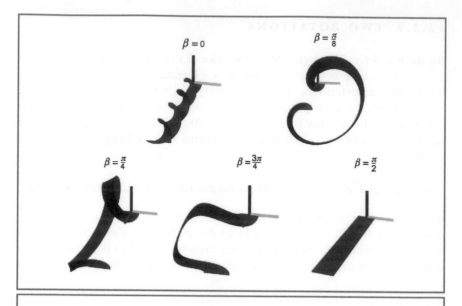

FIGURE 12.10 *Selected frames from the deformation sequence of the double-twist belt to the untwisted belt, using the nested glass-sphere visualization mechanism for each belt configuration.*

to its original untwisted state. Figure 12.9 shows the (w, y, z) quaternion values of the β deformation, and Figure 12.10 presents the nested-sphere frame sequences for sampled values of β.

12.3.3 SYNOPSIS

The reader should now be convinced that there are things in our everyday lives that make it necessary to treat sequences of orientation frames more carefully than isolated frames, and that sequences of frames can only be completely understood if we visualize them in the back of our minds as *curves in quaternion space*. It is intriguing to ask what this might imply about isolated objects rotating in *spacetime*.

Quaternions and the Rolling Ball: Exploiting Order Dependence

13

We return now to the mystery of the rolling ball from Chapter 2. What we saw there was that we can place a baseball on a table under the palm of our hand and make it rotate around an axis *perpendicular* to the table, pointing right through the palm of our hand, even though we never twist our wrist in that direction. Somehow, just moving the hand parallel to the table in small "rubbing" circles makes the ball rotate *as though* we were twisting our wrist! We know this happens, but we had no explanation why. Now that we have learned a bit about quaternions and how to use them to understand what happens when things rotate in space, we can try to "see" what is happening with the rolling ball [66].

13.1 ORDER DEPENDENCE

The basic fact is that although the order of two rotations that take place in the same plane (rotating about the same axis) does not matter, two rotations in different planes (about different axes) will give different results if their order is reversed.

† Noncommutativity: The mathematical term for operations that are order dependent is *noncommutativity*, which is useful when we need a precise terminology for this phenomenon.

123

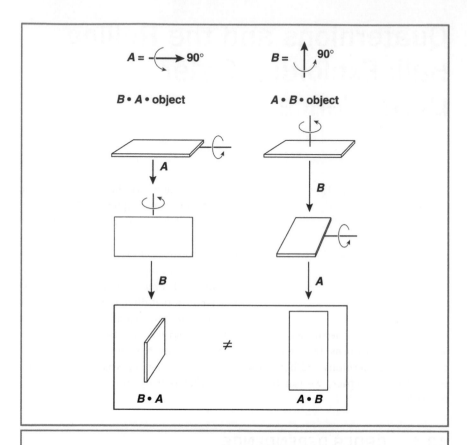

FIGURE 13.1 Rotation order matters. *If we implement two simple rotations A and B about different axes and look at the results of one order—say A · B versus the opposite order, B · A—we can see plainly that the results differ!*

If we take a simple object and rotate it as shown in Figure 13.1, we find that when we rotate first in one direction and then the other we get a particular final orientation. However, if we perform the rotation operations in the *opposite order* the final orientation is quite different!

In fact, we can make the order dependence work for us once we understand what it is doing. We will accomplish this by examining a deceptively simple way of controlling the orientation of an object in space, and watch what it does both in ordinary space and using the quaternion viewpoints we have now learned.

13.2   THE ROLLING BALL CONTROLLER

Figure 13.2 shows a particular example of a set of order-dependent rotations that consist only of operations that can be done with a ball on a table and the flat palm of your hand. Try it yourself: you will see that a mark on the equator of the ball will move exactly as shown, as though your entire hand had been spun about the vertical axis, and yet no such motion ever occurred (the hand stays rigidly oriented with respect to the vertical axis).

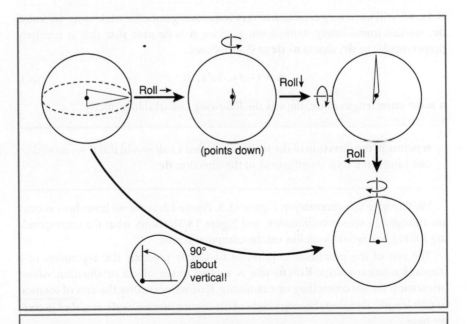

Roll →   Roll ↓   (points down)   Roll   90° about vertical!

FIGURE 13.2 *Sketch of how the rolling ball generates rotation about the orthogonal axis.*

We can write a computer program that implements this very easily based on what we now know regarding rotations about a particular axis of "spin." Our "table" is now going to correspond to two equivalent flat spaces: (1) the flat mouse pad on which the mouse moves in response to the user and (2) the flat computer screen containing a cursor controlled by the mouse. We will consider these spaces to be essentially the same. You need to be comfortable with this before the rest of the argument will make sense.

Each time the mouse moves on the mouse pad, the computer interface will provide us with two numbers: the amount dx telling us how far the mouse cursor moved in the screen-horizontal direction since the last motion, and the amount dy that the mouse cursor moved in the screen-vertical direction. We will make a 3D vector out of this pair of numbers that describes the incremental motion of the mouse cursor on the screen, and which we write as follows:

$$\mathbf{dr} = (dx, dy, 0).$$

Next, we will use a very ancient and very important geometric trick: once we know $\mathbf{dr}$, we can immediately write down a vector $\mathbf{n}$ in *the screen plane* that is precisely perpendicular to $\mathbf{dr}$, that is $\mathbf{n} \cdot \mathbf{dr} \equiv 0$, as follows.

$$\mathbf{n} = (-dy, +dx, 0). \tag{13.1}$$

$\mathbf{n}$ is the same length as $\mathbf{dr}$, but has the following remarkable property.

> $\mathbf{n}$ points in the direction of the rotation axis that a ball would roll if we moved our palm (in screen coordinates) in the direction $\mathbf{dr}$.

Rolling Ball Axis

We illustrate this geometry in Figure 13.3. Figure 13.3a shows how these vectors are arranged in screen coordinates, and Figure 13.3b shows what the corresponding rolling ball would look like on the computer screen.

The rest of the procedure requires us to specify precisely the arguments to a standard rotation matrix $\mathbf{R}(\theta, \hat{\mathbf{n}})$ that is applied to the object or situation whose orientation we are controlling or examining. First we normalize the axis of rotation to unit length. Defining the magnitude of the mouse motion as $dr = \sqrt{dx^2 + dy^2}$, we have

$$\hat{\mathbf{n}} = \frac{\mathbf{n}}{\|\mathbf{dr}\|} = \left(\frac{-dy}{dr}, \frac{+dx}{dr}, 0 \right),$$

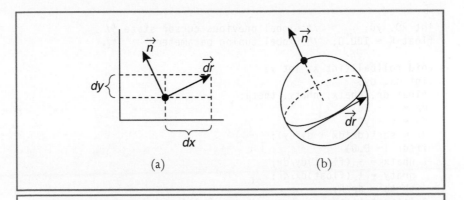

FIGURE 13.3 (a) *The geometric components of the rolling ball controller.* (b) *The general rotation of the controlled object about the screen-constrained axis* $\mathbf{n} = (-dy, +dx)$ *in response to the controller motion* (dx, dy).

which satisfies $\hat{\mathbf{n}} \cdot \hat{\mathbf{n}} \equiv 1$, as required. Note: You do not want to divide by zero. If the hardware has for some reason called your program with vanishing mouse motion, be sure to handle this situation as a null operation.

Finally, to determine the angle of rotation we need only decide what type of virtual ball we are rolling. A marble will rotate very quickly for a given hand motion, and a beach ball will hardly rotate at all. The *effective radius* R of the ball relative to the scale of the mouse motion parameters is what determines the sensitivity of the control. If we choose an appropriate value of R, the amount by which the ball rotates is determined by

$$\tan\theta = \frac{dr}{R},$$

$$\cos\theta = \frac{R}{\sqrt{R^2 + dr^2}},$$

$$\sin\theta = \frac{dr}{\sqrt{R^2 + dr^2}}.$$

In practice, we might choose $R \approx 50 \times \langle \text{average } dr \rangle$, and thus there is no reason to compute expensive transcendental functions because for small θ we know that

```
int x0, y0;        /* Global previous cursor state */
float R = 100.0;   /* Global tuning parameter      */

void rollball(int x, int y)
{int dx, dy;
 float dr, nhatx, nhaty, theta;
 dx = x - x0;
 dy = y - y0;
 dr = sqrt(dx*dx + dy*dy);
 if(dr != 0.0)
 { nhatx = - (float)dy/dr;
   nhaty = + (float)dx/dr;
   theta = dr/R;
   leftrotate(theta, nhatx, nhaty, 0.0);
 }
}
```

TABLE 13.1 *Rolling ball implementation code fragment.*

$\sin\theta \approx \theta$ and we can often simply choose

$$\theta = \frac{dr}{R}$$

and achieve perfectly acceptable interface behavior. Different values of R can be selected to adjust the sensitivity. The rotation to be applied is then simply

$$\mathbf{R}(\theta, \hat{\mathbf{n}}).$$

An example code fragment implementing this behavior is shown in Table 13.1.

13.3   ROLLING BALL QUATERNIONS

The following describes how we can put all this into context with a quaternion visualization. To translate the rolling ball algorithm into quaternion notation, we

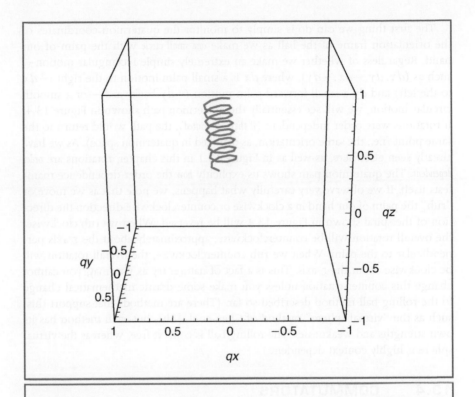

FIGURE 13.4 *The quaternion path of a series of small circular motions of the rolling ball controller. This generates a curve in space that does not close but ends a small distance away from the origin. This signifies that we have made a small rotation about the orthogonal axis.*

simply compute the values of θ and $\hat{\mathbf{n}}$ exactly as before and choose the corresponding quaternion to be

$$q(\theta, \hat{\mathbf{n}}) = \left(\cos(\theta/2), \sin(\theta/2)\hat{\mathbf{n}}\right).$$

Successive values of this incremental quaternion are left-multiplied onto the cumulative quaternion orientation using quaternion multiplication to find each point on the quaternion path.

The first thing we can do is simply to monitor the quaternion coordinates of the orientation frame of the ball as we make *one small circle* with the palm of our hand. Regardless of whether we make an extremely simple rectangular motion—such as $(dx, dy, -dx, -dy)$, where dx is a small palm motion to the right ($-dx$ to the left) and dy a small forward palm motion ($-dy$ backward)—or a smooth circular motion, we will see essentially the quaternion path shown in Figure 13.4. If rotations were order independent (if *they commuted*), the path would return to the same point (i.e., the same orientation, as depicted in quaternion space). As we have already seen elsewhere, as well as in Figure 13.1 in this chapter, rotations are *order dependent*. The quaternion path shows us explicitly *how* the order dependence manifests itself. If we observe very carefully what happens, we note that as we move or "rub" the palm of our hand in a clockwise or counterclockwise direction the direction of the spiral shown in Figure 13.4 will be reversed. When we rub clockwise, the overall rotation will be counterclockwise, approximately about the z axis perpendicular to the palm. When we rub counterclockwise, the overall rotation will be clockwise about the z axis. This is a fact of nature: try as you may, you cannot change this counter-rotation unless you make some drastic mathematical change to the rolling ball method described so far. (There are methods that support this, such as the "virtual sphere" method of Chen et al. [30], but each method has its own strengths and weaknesses. The rolling ball is context free, whereas the virtual sphere is highly context dependent.)

13.4   † COMMUTATORS

The Lie algebra of rotations is generated by three abstract operators (see, for example, Gilmore [56]), which we write as

$$X = (X_x, X_y, X_z).$$

The properties of these operators make mathematically precise statements about the noncommutativity of the rotation group because we can actually *write down the way in which the order matters* by defining the *commutator* of two operators as $[A, B] = AB - BA$ and writing the properties of the generators of the Lie algebra of rotations as

$$[X_y, X_z] = -X_x,$$

$$[X_z, X_x] = -X_y,$$

$$[X_x, X_y] = -X_z,$$

with all other combinations vanishing. The first line means that when we move the palm of our hand in a small clockwise circle in the (x, y) plane the entire ball will spin by a small amount in the *counterclockwise* direction about the z axis, perpendicular to our palm. The *sign* of this rotation is fixed by the immutable laws of mathematics. There is nothing we can do to make the rolling ball controller rotate the ball *clockwise* when we rub our palm in a clockwise direction. The ball will *always* counter-rotate.

13.5 THREE DEGREES OF FREEDOM FROM TWO

The final visualization that makes the entire context of the rolling ball perfectly clear (with or without any knowledge of Lie group theory) is shown in Figure 13.5. Performing a continuous small, circular rubbing motion in the (x, y) plane produces a quaternion path that is a spiral that *does indeed return* to its original approximate position in orientation space, but only after *two complete revolutions* of the ball around the z axis! The average path in quaternion space, the center of the spiral in Figure 13.5, is exactly the same as a simple rotation about the z axis implemented by $\mathbf{R}(t, \hat{\mathbf{z}})$ as $t : 0 \to 4\pi$. If the circular motions in the xy plane are "left-handed" (clockwise), the quaternion path starts in the "right-handed" (counterclockwise, or *positive z*) direction and proceeds around a great circle. If the circular motions in the xy plane are "right-handed" (counterclockwise), the quaternion path starts in the "left-handed" (clockwise, or *negative z*) direction and proceeds around a great circle in the opposing direction. Because exactly the same thing happens in any plane chosen for the rolling ball controller, the particular case we have examined tells us what happens in all cases. There is nothing more you can know about order dependence of rotations and how they are utilized in the rolling ball controller that one cannot deduce from Figure 13.5. The most important fact from a practical standpoint is simply:

Three Degrees of Freedom from Two Variables

> With only two controller variables, dx and dy, we can explore the entire three-degree-of-freedom space of rotations due to the availability of the z-axis rotations (shown in Figure 13.5).

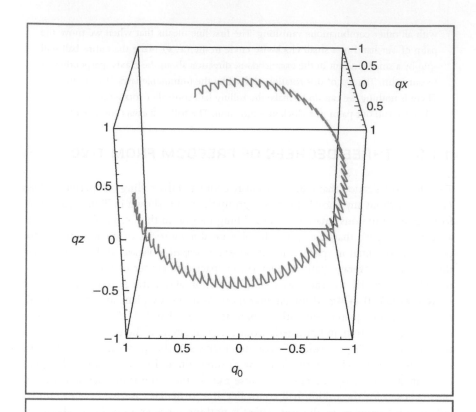

FIGURE 13.5 The quaternion path of a large number of small circular motions of the rolling ball controller. This generates an "orbit" of the simple quaternion rotation path describing a rotation about the orthogonal axis. A few more motions returns the state to its starting point at $q = (1, 0, 0, 0)$.

Quaternions and Gimbal Lock: Limiting the Available Space

One of the classic problems with using ordinary 3×3 matrices to represent rotations is the possibility of gimbal lock, a phenomenon we explored qualitatively in our introduction. Now we have enough tools to look more carefully at the phenomenon, and to work toward exploring how to understand it, when to expect it, and how to avoid it.

There are basically two related, but quite distinct, phenomena that are referred to as gimbal lock. These are discussed in the sections that follow.

14.1 GUIDANCE SYSTEM SUSPENSION

Figure 14.1 shows the basic configuration of a set of rotatable rings that typically house an inertial guidance system. There is a very specific mechanical problem that can occur when two of the axes line up. The problem is critical when one has three rings, the minimal number for free rotation of the guidance system, but no matter how many concentric rotating rings there are, you can still line them all up. As soon as all the rings are coplanar, as shown in Figure 14.2, any rotation about the perpendicular axis generates a torque on the guidance system. This is a serious problem, in that the rotating rings (or gimbals) are supposed to move freely and prevent any torques from acting on the guidance system. At this point, either the guidance system or the vehicle experiences a destructive force, because the guidance system cannot accommodate the attempted change in direction without applying a huge backward torque on the framework surrounding it. (See also Figure 14.3.)

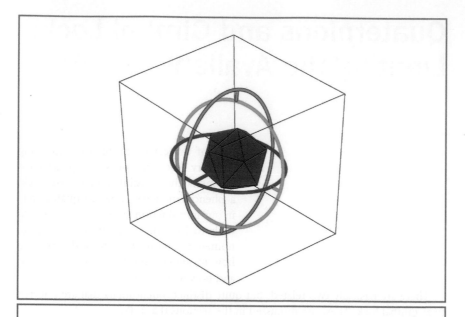

FIGURE 14.1 *The basic gimbal configuration, with three orthogonal rotation axes correspond-ing initially to elementary rotations about the orthogonal Cartesian axes* ($\hat{\mathbf{x}}$, $\hat{\mathbf{y}}$, $\hat{\mathbf{z}}$).

14.2 MATHEMATICAL INTERPOLATION SINGULARITIES

The physical phenomenon of gimbal lock results from having coplanar axes of rota-tion. When trying to control continuous changes in the orientation of an object in 3D space, we can encounter essentially the same results. When we have a sequence of orientations that are smoothly changing, and the sequence reaches a point at which the fixed axes of the rotation line up in a single plane, there is no way to perform a rotation about the axis perpendicular to the plane.

14.3 QUATERNION VIEWPOINT

We can see how all the gimbal lock phenomena are related by expressing the prob-lem in terms of quaternions. We simply take the innocent-looking three-degree-

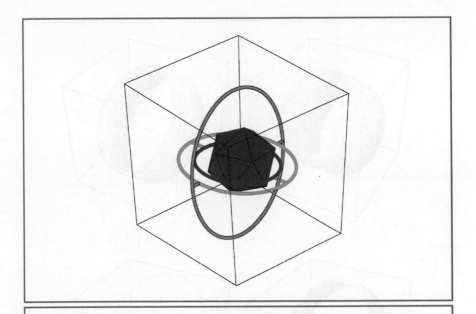

FIGURE 14.2 *A gimbal lock situation occurs when the concentric rings rotate into configuration where a degree of freedom is lost.*

of-freedom rotation sequence

$$M(\theta, \phi, \psi) = \mathbf{R}(\theta, \hat{x}) \cdot \mathbf{R}(\phi, \hat{y}) \cdot \mathbf{R}(\psi, \hat{z})$$

and write down its quaternion counterpart

$$Q(\theta, \phi, \psi) = q(\theta, \hat{x}) \star q(\phi, \hat{y}) \star q(\psi, \hat{z}).$$

In Figure 14.3, we show the quaternion surfaces that result when we fix various values of ϕ and plot the surface $Q(\theta, \phi, \psi)$ as a function of the parameters (θ, ψ). As $\phi \to \pi/2$, the surface loses its 2D character and becomes a 1D ring, even though there are two parameters still attempting to describe it. As illustrated in Figure 14.3d, this is gimbal lock.

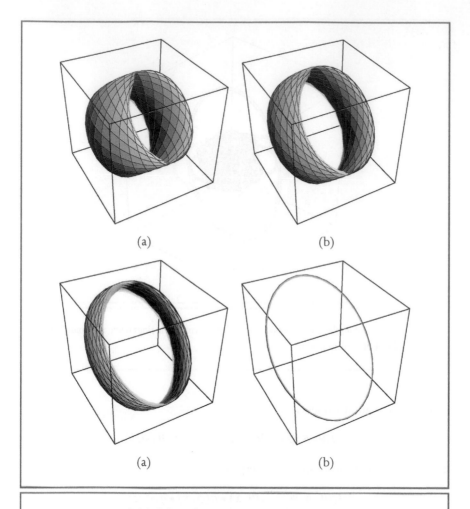

$$(a) \qquad\qquad (b)$$

$$(a) \qquad\qquad (b)$$

FIGURE 14.3 Gimbal lock from the quaternion viewpoint. For a given ϕ and a rotation sequence $\mathbf{R}(\theta, \hat{\mathbf{x}}) \cdot \mathbf{R}(\phi, \hat{\mathbf{y}}) \cdot \mathbf{R}(\psi, \hat{\mathbf{z}})$, the remaining orientation degrees of freedom are plotted here in the $\mathbf{q}$ projection as a function of (θ, ψ). (a) $\phi = 0$, the largest available space. (b and c) $\phi = \pi/6$ and $\phi = \pi/3$, approaching gimbal lock. (d) At $\phi = \pi/2$, all signs of two degrees of freedom disappear, leaving only a 1D circle (this is the quaternion picture of what happens in the gimbal-lock state).

Advanced Quaternion Topics

The remaining chapters in this book focus on more advanced topics that should be informative to a variety of readers with specialized interests. Most of the material in these chapters would qualify to be marked with the dagger (†) designation used in Part I.

Alternative Ways of Writing Quaternions

15

We have taken advantage so far of the clarity of a notation in which a quaternion is represented simply as a four-vector—a list of four variables that obeys certain rules. There are several other completely equivalent notations, including Hamilton's original formulas, that incorporate the rules of quaternion multiplication as an intrinsic part of the notation itself. As far as we are concerned, *these add no new information* to the notation we have been using and thus they are not essential to the overall problem of trying to visualize quaternions. Nevertheless, they do relate to notations used in *other* fields of mathematics and physics and thus have value for their ability to expose those connections.

Basic four-vector: To provide a comparative basis, we repeat our own basic notation for a quaternion, which is

$$q = (q_0, q_1, q_2, q_3) = (q_0, \mathbf{q}). \tag{15.1}$$

In this way, a quaternion q can be regarded simply as a list of four numbers—possibly a row vector, possibly a column vector, or as an object that has one scalar part q_0 and one three-vector part $\mathbf{q}$.

15.1   HAMILTON'S GENERALIZATION OF COMPLEX NUMBERS

When he invented quaternions, Hamilton had as his goal to create a generalization of complex numbers that permitted a division operation in which any nonzero number could divide any other number and produce a sensible answer. Hamilton's original notation is then closely tied to complex numbers, and provides three separate copies of a quantity that strongly resembles the $i = \sqrt{-1}$ of ordinary complex arithmetic. We denote these as (i, j, k) and assign them the multiplication rules

$$i^2 = -1,$$
$$j^2 = -1,$$
$$k^2 = -1,$$
$$ijk = -1, \qquad\qquad (15.2)$$

which lead to the more detailed rules of the form

$$ij = k,$$
$$jk = i,$$
$$ki = j,$$

$$ij + ji = 0,$$
$$jk + kj = 0,$$
$$ki + ik = 0,$$

$$ij - ji = 2k,$$
$$jk - kj = 2i,$$
$$ki - ik = 2j.$$

Any quaternion can then be written, using the vector $\mathbf{i} = (i, j, k)$, in a way that implies its multiplication rule in exactly the same way that $z = x + iy$ implies the complex algebra. Hamilton's form is then

$$q = q_0 + iq_1 + jq_2 + kq_3$$
$$= q_0 + \mathbf{i} \cdot \mathbf{q}, \qquad\qquad (15.3)$$

and the reader can easily verify that Equation 15.2 implies all of the properties we have been referring to as the quaternion algebra.

Note in addition that if we examine quaternions with arbitrary q_0 and only $q_1 \neq 0$, or only $q_2 \neq 0$, or only $q_3 \neq 0$ in turn they are *indistinguishable* from complex numbers. As long as we keep the other components equal to zero, a quaternion can be equivalent to a complex number in three different ways. When we consider this in light of the relationship of ordinary complex numbers to 2D rotations (detailed in Section 6.1), we see that these three cases correspond to the quaternion representation of 3D rotations restricted to the following three independent 2D subspaces.

- $q_0 + iq_1$: leaving $\hat{\mathbf{x}}$ fixed (yz-plane rotations)
- $q_0 + jq_2$: leaving $\hat{\mathbf{y}}$ fixed (zx-plane rotations)
- $q_0 + kq_3$: leaving $\hat{\mathbf{z}}$ fixed (xy-plane rotations)

15.2   PAULI MATRICES

Because of the relationship between quaternions and 3D rotations, we might expect that there would be a way of writing down quaternions in terms of matrix elements and reproducing the quaternion algebra in terms of matrix multiplication properties. There are an infinite number of ways of doing this, related to the theory of group representations. Despite the mathematical terminology, the basic ideas can be simply understood in practical terms. We could start by writing down the simplest matrices, which are 2×2, and play with them to see what happens. What we would discover is that a set of three matrices (known as the *Pauli matrices*, used in physics to describe the angular momentum of elementary particles such as electrons) do the job perfectly. The Pauli matrices are written by convention as follows, where i is the usual imaginary number with $i^2 = -1$.

$$\sigma_1 = \begin{bmatrix} 0 & 1 \\ 1 & 0 \end{bmatrix},$$

$$\sigma_2 = \begin{bmatrix} 0 & -i \\ i & 0 \end{bmatrix},$$

$$\sigma_3 = \begin{bmatrix} 1 & 0 \\ 0 & -1 \end{bmatrix}.$$

We can immediately verify that the Pauli matrices obey the following relationships.

$$+I_2 = (\sigma_1)^2 = (\sigma_2)^2 = (\sigma_3)^2,$$

$$\sigma_1\sigma_2 = i\sigma_3,$$

$$\sigma_2\sigma_3 = i\sigma_1,$$

$$\sigma_3\sigma_1 = i\sigma_2. \tag{15.4}$$

Now any quaternion can be written

$$q = q_0 I_2 - i\sigma_1 q_1 - i\sigma_2 q_2 - i\sigma_3 q_3$$

$$= \begin{bmatrix} q_0 - iq_3 & -iq_1 - q_2 \\ -iq_1 + q_2 & q_0 + iq_3 \end{bmatrix}$$

$$= q_0 - i\boldsymbol{\sigma} \cdot \mathbf{q},$$

where in the last expression q_0 abbreviates the implicit multiplication by I_2, the 2×2 identity matrix. We can immediately see that $\mathbf{i} = (i, j, k)$ and $-i\boldsymbol{\sigma} = -i\,(\sigma_1, \sigma_2, \sigma_3)$ have exactly the same properties.

Physicists and mathematicians are fond of two pieces of shorthand, known as the Kronecker delta δ_{jk} and the Levi-Civita symbol ϵ_{jkl} (the totally antisymmetric pseudotensor), which we describe in more detail in Appendix F for those who are interested. With this notation, the set of equations in Equation 15.4 collapses into

$$\sigma_j\sigma_k = \delta_{jk} + i\sum_{l=1}^{3} \epsilon_{jkl}\sigma_l,$$

where $\delta_{jk} = 0$ if $j \neq k$, $\delta_{jk} = 1$ if $j = k$, $\epsilon_{jkl} = 0$ if any two indices are equal, $\epsilon_{jkl} = +1$ if the indices are in cyclic order, and $\epsilon_{jkl} = -1$ if the indices are in anticyclic order.

15.3   OTHER MATRIX FORMS

One can easily find sets of three matrices with larger dimensions that can be used in place of $\boldsymbol{\sigma}$ to produce equivalent ways of writing down a quaternion as a matrix.

One example we have already encountered is the orthogonal 4×4 matrix form, which also exactly reproduces the algebra. To see this, we define

$$
\Sigma_1 = \begin{bmatrix} 0 & -1 & 0 & 0 \\ +1 & 0 & 0 & 0 \\ 0 & 0 & 0 & -1 \\ 0 & 0 & +1 & 0 \end{bmatrix},
$$

$$
\Sigma_2 = \begin{bmatrix} 0 & 0 & -1 & 0 \\ 0 & 0 & 0 & +1 \\ +1 & 0 & 0 & 0 \\ 0 & -1 & 0 & 0 \end{bmatrix},
$$

$$
\Sigma_3 = \begin{bmatrix} 0 & 0 & 0 & -1 \\ 0 & 0 & -1 & 0 \\ 0 & +1 & 0 & 0 \\ +1 & 0 & 0 & 0 \end{bmatrix},
$$

where

$$
-I_4 = \Sigma_1{}^2 = \Sigma_2{}^2 = \Sigma_3{}^2,
$$
$$
\Sigma_1 \Sigma_2 = \Sigma_3,
$$
$$
\Sigma_2 \Sigma_3 = \Sigma_1,
$$
$$
\Sigma_3 \Sigma_1 = \Sigma_2.
$$

The quaternion algebra then follows directly if we write the quaternion (using these matrices) as

$$
q = q_0 I_4 + \boldsymbol{\Sigma} \cdot \mathbf{q}
$$

$$
= \begin{bmatrix} q_0 & -q_1 & -q_2 & -q_3 \\ q_1 & q_0 & -q_3 & q_2 \\ q_2 & q_3 & q_0 & -q_1 \\ q_3 & -q_2 & q_1 & q_0 \end{bmatrix}. \tag{15.5}
$$

Efficiency and Complexity Issues

One might hope that because the quaternion representation of a 3D orientation frame or rotation matrix requires only *four* floating-point numbers for its computer representation, and the standard 3×3 matrix requires *nine* floating-point numbers, that the computational complexity of standard computer graphics operations involving rotations would favor quaternions. In fact, one can find many places in which it is claimed that all rotations should be converted to quaternions at the outset and that thereafter one should use only the quaternion libraries for implementing standard operations. The procedures that need to be examined to confirm or deny such efficiency claims include the following.

- *Matrix to quaternion:* Produce a quaternion four-vector from a 3×3 orthogonal rotation matrix.
- *Quaternion to matrix:* Produce a 3×3 orthogonal rotation matrix from a quaternion four-vector.
- Transform a three-vector to a new orientation.
- Transform many three-vectors to the same new orientation.
- Compose two rotations to find a new composite rotation.

Unfortunately, in most situations involving these operations it is actually more efficient to keep mapping the quaternion representation of an orientation matrix back to the 3×3 orthogonal rotation matrix rather than performing pure quaternion operations. The notable exception is the case of optimal interpolation among coordinate frames, where, regardless of efficiency, the quaternion representation is required at the lowest level.

The complexity analysis of the standard quaternion operations listed above has been considered by a number of authors, but we feel that the issues have been misrepresented often enough that it is important for reference purposes to repeat the results in this short section. We will closely follow the analysis presented by Eberly [42] and by Schneider and Eberly [146].

16.1   EXTRACTING A QUATERNION

The basic method of obtaining the appropriate quaternion representation if one is given a particular 3×3 orthogonal rotation matrix $\mathbf{R}$ was outlined in Chapter 6, where we showed that the basic quantities (including the rotation axis and the angle giving the amount of spin about that axis) could be computed from the trace

$$t = \text{Trace } \mathbf{R} = 1 + 2\cos\theta, \tag{16.1}$$

and by subtracting the transpose

$$\mathbf{R} - \mathbf{R}^T = \begin{bmatrix} 0 & -2n_3\sin\theta & +2n_2\sin\theta \\ +2n_3\sin\theta & 0 & -2n_1\sin\theta \\ -2n_2\sin\theta & +2n_1\sin\theta & 0 \end{bmatrix}. \tag{16.2}$$

However, as pointed out by various authors (e.g., see Nielson [132], Shuster [153], and Shuster and Natanson [154]), one can get into trouble making naive assumptions about these equations because, for example, $\theta = \pi$ is a perfectly legitimate rotation parameter but the matrix $\mathbf{R} - \mathbf{R}^T$ is no longer useful.

The completely general basic procedure for extracting a quaternion from a matrix depends on the fact, proved in detail by Shuster and Natanson [154], that there will always be at least *one* component of the diagonal of $\mathbf{R}$ that will be "large." We can find from Equation 16.1 and the half-angle formula that the value of q_0 is

$$q_0 = \cos\frac{\theta}{2} = \frac{1}{\sqrt{2}}\sqrt{\cos\theta + 1} = \frac{1}{2}\sqrt{\text{Trace } \mathbf{R} + 1}. \tag{16.3}$$

Because a quaternion and its negation produce the same rotation matrix, we can choose the positive root in Equation 16.3, as long as we keep the three-vector part of the quaternion consistent with this sign.

16.1.1 POSITIVE TRACE R

Suppose first that Trace $\mathbf{R} > 0$. Then, because $q_0 > 1/2$, we have no trouble dividing by it, and we can examine the quaternion form

$$\mathbf{R} = \begin{bmatrix} q_0^2 + q_1^2 - q_2^2 - q_3^2 & 2q_1q_2 - 2q_0q_3 & 2q_1q_3 + 2q_0q_2 \\ 2q_1q_2 + 2q_0q_3 & q_0^2 - q_1^2 + q_2^2 - q_3^2 & 2q_2q_3 - 2q_0q_1 \\ 2q_1q_3 - 2q_0q_2 & 2q_2q_3 + 2q_0q_1 & q_0^2 - q_1^2 - q_2^2 + q_3^2 \end{bmatrix} \qquad (16.4)$$

and observe that

$$\mathbf{R}_{32} - \mathbf{R}_{23} = 4q_0q_1,$$

$$\mathbf{R}_{13} - \mathbf{R}_{31} = 4q_0q_2,$$

$$\mathbf{R}_{21} - \mathbf{R}_{12} = 4q_0q_3,$$

and thus

$$q_1 = \frac{\mathbf{R}_{32} - \mathbf{R}_{23}}{4q_0},$$

$$q_2 = \frac{\mathbf{R}_{13} - \mathbf{R}_{31}}{4q_0},$$

$$q_3 = \frac{\mathbf{R}_{21} - \mathbf{R}_{12}}{4q_0}.$$

This yields the quaternion $q = (q_0, q_1, q_2, q_3)$.

16.1.2 NONPOSITIVE TRACE R

Now suppose that Trace $\mathbf{R} \leqslant 0$. We now do not actually know that q_0 is a suitable divisor. It could approach zero. We thus now examine the diagonal entries in the quaternion quadratic form (Equation 16.4), observing that if $\mathbf{R}_{11}$ is the largest of the diagonal terms $\mathbf{R}_{ii}$ then q_1 will be the largest of the vector quaternion components q_i. Repeating this for each case, we can extract the values of q_i from the *diagonal* components, avoiding the potentially problematic trace, with the results

$$q_1 = \frac{1}{2}\sqrt{1 + \mathbf{R}_{11} - \mathbf{R}_{22} - \mathbf{R}_{33}},$$

$$q_2 = \frac{1}{2}\sqrt{1 + \mathbf{R}_{22} - \mathbf{R}_{33} - \mathbf{R}_{11}},$$

$$q_3 = \frac{1}{2}\sqrt{1 + \mathbf{R}_{33} - \mathbf{R}_{11} - \mathbf{R}_{22}}.$$

From this basic information on the magnitudes of the diagonals, we can finally determine the remaining values safely using the following family of conditions:

- q_1 has the largest magnitude: Then we choose

$$q_0 = \frac{R_{32} - R_{23}}{4q_1}, \quad q_2 = \frac{R_{21} + R_{12}}{4q_1}, \quad q_3 = \frac{R_{13} + R_{31}}{4q_1}.$$

- q_2 has the largest magnitude: Then we choose

$$q_0 = \frac{R_{13} - R_{31}}{4q_2}, \quad q_1 = \frac{R_{12} + R_{21}}{4q_2}, \quad q_3 = \frac{R_{23} + R_{32}}{4q_2}.$$

- q_3 has the largest magnitude: Then we choose

$$q_0 = \frac{R_{21} - R_{12}}{4q_3}, \quad q_1 = \frac{R_{31} + R_{13}}{4q_3}, \quad q_2 = \frac{R_{23} + R_{32}}{4q_3}.$$

16.2   EFFICIENCY OF VECTOR OPERATIONS

Following the approach in Eberly [42] and in Schneider and Eberly [146], we first define a dictionary of low-level operations in terms of which the complexity of the basic composite operations will be phrased. Here, we will tabulate only the relative complexity of the quaternion operations. A more complete comparison, including axis-angle methods, may be found in [146].

- M: Multiplication
- A: Addition or subtraction
- D: Division
- F: Library function evaluations such as trigonometric functions, which are composite operations often involving a large number of elementary multiplications, divisions, and additions

- C: Comparisons, which can also add significantly to computational cost by interrupting pipelined operations, etc.

Operation	A	M	D	F	C
Quaternion to matrix	12	12			
Matrix to quaternion (Trace > 0)	6	5	1	1	1
Matrix to quaternion (Trace $\leqslant 0$)	6	5	1	1	3

Rotate Vector	A	M	Notes
With rotation matrix	6	9	
With quaternion	24	32	Using generic quaternion multiplies
With quaternion	17	24	Using specialized quaternion multiplies
With quaternion	18	21	Convert to 3×3 matrix first

Rotate n Vectors	A	M	Notes
With rotation matrix	$6n$	$9n$	
With quaternion	$24n$	$32n$	Using generic quaternion multiplies
With quaternion	$17n$	$24n$	Using specialized quaternion multiplies
Quaternion	$12 + 6n$	$12 + 9n$	Convert to 3×3 matrix first

Compose Two Rotations	A	M
Rotation matrix	18	27
Quaternion	12	16

We see that the only place in this set of standard operations where quaternions have reduced complexity is in the composition of large numbers of rotation operations to produce a new composition of the repeated operations. Once this is accomplished, however, it is still typically faster to convert the resulting quaternion to a standard 3×3 matrix before performing additional vector operations.

Advanced Sphere Visualization

17

This chapter presents additional methods that can be used to study spheres. We will start from the basic observations of Chapter 8 and explore a selection of other mathematical methods that can be used in specific applications for representing, manipulating, and visualizing spheres. As usual, we introduce the methods beginning with the circle, $\mathbf{S}^1$, and work our way up to the ordinary sphere $\mathbf{S}^2$ and the unit hypersphere $\mathbf{S}^3$ needed to describe quaternions.

17.1   PROJECTIVE METHOD

An alternative to the Northern–Southern hemisphere method introduced in Chapter 8 is to construct a projection that removes one point of the sphere to infinity and maps the rest to a flat Euclidean space. The advantage of this approach is that it places the entire sphere, with the exception of a single point, into a flat Euclidean framework. The disadvantage is that the metric properties (the ability to accurately estimate distances) are completely distorted. Although there is indeed distortion in the more conventional image of a sphere as well, it is a more familiar and intuitive distortion.

17.1.1 THE CIRCLE $\mathbf{S}^1$

For the circle, the projective method is an interesting alternative to the hemicircle method we used previously to disambiguate the square root in the solution of the

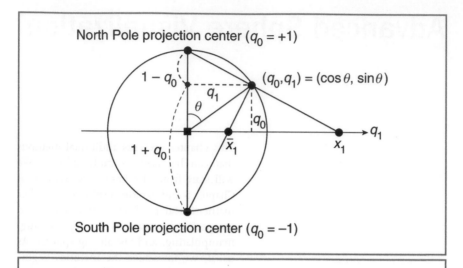

FIGURE 17.1 *Polar projection maps the North Pole* $(+1, 0)$ *or the South Pole* $(-1, 0)$ *of the circle* $\mathbf{S}^1$ *to infinity on the real line.*

algebraic equation of the circle. To project the unit circle from either pole onto the horizontal real line, we choose a point $(q_0, q_1) = (\cos\theta, \sin\theta)$ on the circle (see Figure 17.1). Because the circle is assumed to have unit radius, q_0 provides a partition of the vertical axis into two pieces of length $1 - q_0$ and $1 + q_0$. The right triangle with base of length x_1 is similar to the upper triangle with base q_1, and thus the law of similar triangles shows that the projection from the North Pole is

$$
x_1 = \frac{q_1}{1 - q_0}
$$

$$
= \frac{\sin\theta}{1 - \cos\theta}. \tag{17.1}
$$

To project from the South Pole to the horizontal real line, we employ instead the right triangle whose bottom apex is at the South Pole and whose horizontal edge has length $\bar{x}_1$. This triangle is similar to the larger triangle containing it, with horizontal edge of length q_1, showing that

$$\bar{x}_1 = \frac{q_1}{1 + q_0}$$

$$= \frac{\sin\theta}{1 + \cos\theta}. \qquad (17.2)$$

The resulting polar projections create a point on a straight line for every angle θ on the circle except the poles, but destroy any symmetry between the North Pole and South Pole, as can be seen in Figure 17.1. We also note the following useful inverse formulas, which map a point on the infinite real line to a point on a circle:

$$[\text{North projection center}] \quad q_0 = \frac{r^2 - 1}{r^2 + 1}, \quad q_1 = \frac{2x_1}{r^2 + 1}, \quad r = |x_1|, \quad (17.3)$$

$$[\text{South projection center}] \quad q_0 = \frac{1 - \bar{r}^2}{1 + \bar{r}^2}, \quad q_1 = \frac{2\bar{x}_1}{1 + \bar{r}^2}, \quad \bar{r} = |\bar{x}_1|. \quad (17.4)$$

Here, r and $\bar{r}$ generalize to higher dimensions, taking the value of the distance from the origin to the point being mapped. In the 2D case, $r = |x_1|$ for the North Pole projection and $\bar{r} = |\bar{x}_1|$ for the South Pole projection. Observe that

$$[\text{North}] \quad (q_0)^2 + (q_1)^2 = \frac{x_1^4 + 2x_1^2 + 1}{(x_1^2 + 1)^2} = 1,$$

$$[\text{South}] \quad (q_0)^2 + (q_1)^2 = \frac{\bar{x}_1^4 + 2\bar{x}_1^2 + 1}{(\bar{x}_1^2 + 1)^2} = 1$$

as required. Because the projection of the circle to the infinite real line introduces massive distortions and asymmetries into the geometry, we would use this method only in special circumstances, preferring the more uniform two-part square-root projection.

17.1.2   GENERAL S^N POLAR PROJECTION

The extension of the 2D polar projection, which sends either the North Pole or the South Pole to infinity, basically takes the same form in all higher dimensions, and thus it will be straightforward to handle the S^2 and S^3 cases together.

We now expect that for the two-sphere $\mathbf{S}^2$ the projection will send the North Pole or the South Pole to infinity and map the rest into the flat 2D plane $\mathbb{R}^2$. For the three-sphere $\mathbf{S}^3$, all points except for one pole are flattened out into ordinary 3D Euclidean space $\mathbb{R}^3$. This is the origin of the statement that $\mathbf{S}^3$ can be considered to be the same as Euclidean 3D space $\mathbb{R}^3$, with the addition of a single point. The projective method implements precisely this transformation.

From the previously cited Equations 17.1 and 17.2 for the $\mathbf{S}^1$ polar projection, we can deduce the general polar projection equations for $\mathbf{S}^2$ and $\mathbf{S}^3$ by simply replacing q_1 with $\mathbf{q} = (q_1, q_2)$ or $\mathbf{q} = (q_1, q_2, q_3)$, respectively, and x_1 with $\mathbf{x} = (x_1, x_2)$ or $\mathbf{x} = (x_1, x_2, x_3)$, respectively. Thus, the projection of the N-sphere $\mathbf{S}^N$ to a Euclidean space of dimension N is simply

$$\mathbf{x} = \frac{\mathbf{q}}{1 - q_0}, \tag{17.5}$$

$$\bar{\mathbf{x}} = \frac{\mathbf{q}}{1 + q_0}. \tag{17.6}$$

The inverse transformation takes a point in Euclidean space back to the sphere $\mathbf{S}^N$ embedded in $(N+1)$-dimensional Euclidean space with constant unit radius. We basically need to generalize the distances r and $\bar{r}$ in Equations 17.3 and 17.4 to higher dimensions. For the $\mathbf{S}^2$ embedded in a 3D space with coordinates (q_0, q_1, q_2), we take $r^2 = \mathbf{x}^2 = x^2 + y^2$, and for $\mathbf{S}^3$ we take $r^2 = \mathbf{x}^2 = x^2 + y^2 + z^2$. The result is

$$q_0 = \frac{r^2 - 1}{r^2 + 1} = \frac{1 - \bar{r}^2}{1 + \bar{r}^2}, \tag{17.7}$$

$$\mathbf{q} = \frac{2\mathbf{x}}{r^2 + 1} = \frac{2\bar{\mathbf{x}}}{1 + \bar{r}^2}. \tag{17.8}$$

The basic picture of what happens in the transformation is unchanged from that represented by Figure 17.1.

17.2   DISTANCE-PRESERVING FLATTENING METHODS

As noted, the polar projection method for the sphere reduces the dimension of the space needed to study a sphere to the intrinsic dimension of the sphere itself, but at the price of massive interference with our ability to accurately estimate distances in terms of the arc length on the given sphere. Is it possible to have our cake and eat it too, that is, to create a flat projection that also permits access to accurate distance or size information? The answer is "sort of." In this section, we show how this goal is *exactly* achievable with an ordinary circle and demonstrate that some features of the goal can be achieved for $\mathbf{S}^2$ and $\mathbf{S}^3$ as well.

17.2.1   UNROLL-AND-FLATTEN $\mathbf{S}^1$

Our alternative way of understanding a circle is to forget altogether about the circle $\mathbf{S}^1$ and its embedding in 2D space as a curve of constant distance from the origin. The main feature that is often required (the preservation of distances among points) is obtained if we simply cut the circle at the North Pole and flatten it down to a straight line without stretching or distorting the point-to-point distances. This is shown schematically in Figure 17.2. Just as all the visualization methods have some drawback, there is one here as well: because we cut the circle at some point, we can no longer take a continuous path that goes around and around the circle, returning automatically to the original point after each complete circuit. In the unroll-and-flatten method, this feature has to be imposed externally in some way, either by simply declaring the path to jump to the other end when the traveler reaches one end or by making a periodic tiling of the real line and identifying all points that are 2π units apart.

17.2.2 $\mathbf{S}^2$ FLATTENED EQUAL-AREA METHOD

For $\mathbf{S}^2$, the simple unrolling approach that worked for the circle has no analog. $\mathbf{S}^2$ is intrinsically curved and cannot be flattened. However, there is a trick here as well: if we do not care about avoiding distortion, and we do not care about having to impose periodicity by hand, there exist families of maps from $\mathbf{S}^2$ to the plane that *preserve local area*. The map often used for this purpose in astronomy is called the Aitoff–Hammer map. (There are many classic equal-area maps, and others are equally popular in other disciplines.) This map starts from the latitude ϕ and the longitude λ and describes the 3D points on the sphere by the expression

$$(q_1, q_2, q_3) = (\cos\lambda\cos\phi, \sin\lambda\cos\phi, \sin\phi),$$

where ϕ is the latitude, with $-\pi/2 \leqslant \phi \leqslant +\pi/2$, and λ is the longitude, with $-\pi \leqslant \lambda \leqslant +\pi$. The Aitoff–Hammer projection maps the spherical point parameterized by (ϕ, λ) to a Cartesian point (x, y) in such a way that area is locally

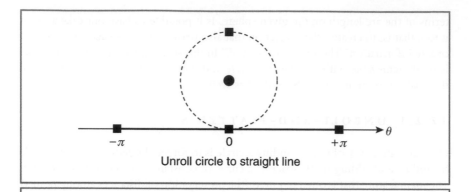

Unroll circle to straight line

FIGURE 17.2 Framework for unwrapping the circle so that it becomes a straight line. Here, unlike the projective method, the metric distances are preserved, but in contrast the periodicity must be artificially imposed.

preserved. Thus, for example, light intensities of spherical sky maps are uniform if they appear uniform in the sky. Using the auxiliary variable

$$t = \sqrt{1 + \cos\phi\cos(\lambda/2)}, \qquad (17.9)$$

we find (x, y), which lie within an oval-shaped region defined by the equations

$$x = \frac{2\sqrt{2}}{t}\cos\phi\sin(\lambda/2), \qquad (17.10)$$

$$y = \frac{\sqrt{2}}{t}\sin\phi. \qquad (17.11)$$

Alternatively, we can map the Cartesian source coordinates (q_1, q_2, q_3) obeying $\mathbf{q} \cdot \mathbf{q} = 1$ into the xy plane directly instead of using the polar coordinates. We let

$$r^2 = (q_1)^2 + (q_2)^2,$$

$$t = \sqrt{1 + \sqrt{(r^2 + rq_1)/2}}$$

so that

$$x = \frac{2\,\mathrm{sign}(q_2)}{t}\sqrt{(r^2 - rq_1)},$$

$$y = \frac{\sqrt{2}}{t}q_3.$$ (17.12)

The inverse map is given by taking the auxiliary variable

$$w = \sqrt{1 - \left(\frac{x}{4}\right)^2 - \left(\frac{y}{2}\right)^2}$$ (17.13)

and performing the transformation

$$\lambda = 2\arctan\left(\frac{xw}{2(2w^2 - 1)}\right),$$ (17.14)

$$\phi = \arcsin(yw).$$ (17.15)

In Figure 17.3, we show the result of the Aitoff–Hammer transformation on a standard latitude–longitude ruled sphere, on the remaining whole triangles of an icosahedral split down a line of constant longitude, and on a random collection of points chosen to be evenly distributed on the surface of a sphere. This is the closest we can come to an advantageous unwrapping of the sphere $\mathbf{S}^2$ onto a flat plane. The relative areas of individual local sections are preserved, providing at least one type of metric property, but Figure 17.3 clearly shows that geodesic paths (the great circles of the longitude lines) are irregularly distorted, and periodicity is lost. However, in certain types of applications this can be an appropriate choice.

17.2.3 $\mathbf{S}^3$ FLATTENED EQUAL–VOLUME METHOD

For $\mathbf{S}^3$, we can again use a variant of the Aitoff–Hammer trick for creating a distorted but volume-preserving representation of $\mathbf{S}^3$ in ordinary flat Euclidean 3D space. The transformation takes the three-sphere coordinates

$$q_0 = \cos\lambda\cos\phi\cos\psi,$$

$$q_1 = \sin\lambda\cos\phi\cos\psi,$$

$$q_2 = \sin\phi\cos\psi,$$

$$q_3 = \sin\psi$$ (17.16)

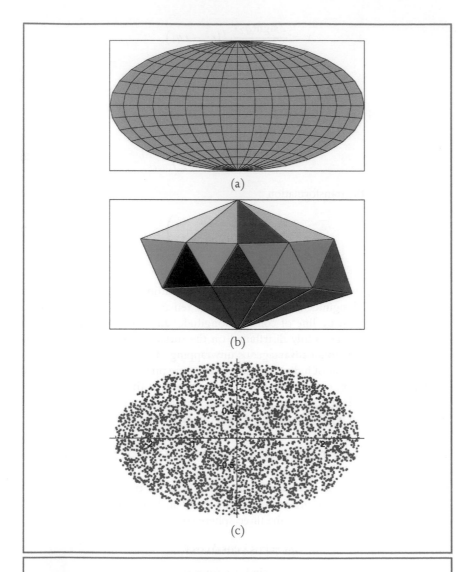

(a)

(b)

(c)

FIGURE 17.3 *Aitoff–Hammer equal-area projections of* $\mathbf{S}^2$. *(a) A standard latitude–longitude ruling. (b) Partition of* $\mathbf{S}^2$ *using the remaining triangles of an icosahedron split down a longitude line. (c) Map of a spherically uniform random distribution of dots.*

to the Cartesian coordinates (x, y, z) while preserving local volume elements. Using the auxiliary variable

$$t = \sqrt{1 + \cos\psi \cos\phi \cos(\lambda/2)}, \qquad (17.17)$$

we find the Cartesian point (x, y, z)—which lies within an ellipsoidal region—from the equations

$$x = \frac{2\sqrt{2}}{t} \cos\psi \cos\phi \sin(\lambda/2), \qquad (17.18)$$

$$y = \frac{\sqrt{2}}{t} \cos\psi \sin\phi, \qquad (17.19)$$

$$z = \frac{\sqrt{2}}{t} \sin\psi. \qquad (17.20)$$

To map the Cartesian coordinates (q_0, q_1, q_2, q_3) into the volume-preserving ellipsoid directly, we take the same auxiliary variables as in the 2D case,

$$r^2 = (q_0)^2 + (q_1)^2,$$

$$t = \sqrt{1 + \sqrt{(r^2 + rq_0)/2}},$$

and choose

$$x = \frac{2\,\text{sign}(q_1)}{t} \sqrt{(r^2 - rq_0)},$$

$$y = \frac{\sqrt{2}}{t} q_2,$$

$$z = \frac{\sqrt{2}}{t} q_3. \qquad (17.21)$$

The inverse map is given by taking the auxiliary variable

$$w = \sqrt{1 - \left(\frac{x}{4}\right)^2 - \left(\frac{y}{2}\right)^2 - \left(\frac{z}{2}\right)^2} \qquad (17.22)$$

and performing the transformation

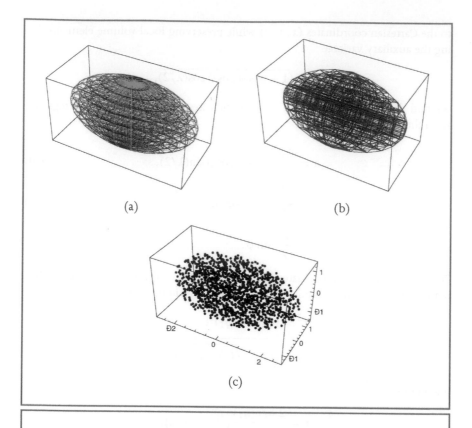

(a)   (b)

(c)

FIGURE 17.4 (a) *The Aitoff–Hammer equal-volume projection of* $\mathbf{S}^3$ *for a standard polar ruling.* (b) *The Aitoff–Hammer equal-volume projection for* $\mathbf{S}^3$ *subdivided using the parameters of the Clifford torus tessellation.* (c) *Map of a uniform random distribution of dots on the hypersphere* $\mathbf{S}^3$.

$$\lambda = 2\arctan\left(\frac{xw}{2(2w^2-1)}\right),$$

$$\phi = \arcsin(yw),$$

$$\psi = \arcsin(zw). \tag{17.23}$$

Whereas there is significant distortion at the outer surface of this transformation, the very center (like the 2D case) is essentially correct and can be used to study the properties of the 3D quaternion space of interest by applying a quaternion transformation to move each point of interest to the origin. In Figure 17.4, we show the overall appearance of the 3D Aitoff–Hammer map for a uniform random sample on $\mathbf{S}^3$, for the standard polar coordinates of Equation 17.16, and for the Clifford torus parameterization, which is given by

$$q_0 = \cos\theta\cos\alpha,$$

$$q_1 = \cos\theta\sin\alpha,$$

$$q_2 = \sin\theta\cos\beta,$$

$$q_3 = \sin\theta\sin\beta. \qquad\qquad (17.24)$$

More on Logarithms and Exponentials

18

The visualization of the calculus of rotations requires that we perceive not only the global properties of individual isolated frames but also their infinitesimal properties, which tell us how nearby frames are related and provide a basis for the calculus of frames, angular velocities, and higher derivatives. In this chapter, we will carry out a more advanced treatment of the properties of infinitesimal rotations and their relationship to the concept of logarithms and exponentials, both for matrices, and for abstract quaternions.

18.1 2D ROTATIONS

Rotations in two dimensions correspond precisely to a unit-length complex vector, which in turn can always be written as the exponential

$$R_2(\theta) = e^{i\theta}. \tag{18.1}$$

By representing a general 2D vector as the complex number

$$V = (r\cos\alpha, r\sin\alpha) = re^{i\alpha},$$

we can then implement rotations as complex multiplication in the following way:

$$V' = R_2(\theta)V = re^{i(\alpha+\theta)}. \tag{18.2}$$

Thus, we may consider the *logarithm* of a 2D rotation to be defined as

$$\log R_2(\theta) \equiv i\theta.$$ (18.3)

This has a particularly interesting implication if we attempt to perform a rotation from one particular 2D frame, say, the frame given by rotating the identity frame by θ_0 so that

$$F_0 = e^{i\theta_0},$$

to a second frame related to the identity frame by the rotation θ_1 so that

$$F_1 = e^{i\theta_1}.$$

To make a smooth parametric rotation from one frame to the other we would require the interpolating rotation $F(t)$ to obey

$$F(0) = F_0,$$

$$F(1) = F_1.$$

Starting from a standard multiplicative interpolation $x(t)$ from $x(0) = a$ to $x(1) = b$ given by

$$x(t) = a\left(\frac{b}{a}\right)^t,$$

we are led to examine

$$F(t) = F_0 \cdot \left(F_0^{-1}F_1\right)^t.$$

Written in terms of the exponentials, this becomes

$$F(t) = e^{i(\theta_0 + t(\theta_1 - \theta_0))}.$$

Linearity in the logarithm: We see that we can use the logarithm to turn a multiplicative interpolation into a linear one:

$$\log\big(F(t)\big) = i\big(\theta_0 + t(\theta_1 - \theta_0)\big).$$

The calculus of the transformation is also easy to investigate, because (assuming that t is a correctly scaled time variable) we have

$$\frac{dF(t)}{dt} = i(\theta_1 - \theta_0)F(t).$$

Because the standard differential equation for an object rotating at a constant angular velocity ω is $\dot{F}(t) = i\omega F(t)$, we see that the angular velocity of the interpolation is

$$\omega = \theta_1 - \theta_0,$$

where we have implicitly assumed that a time unit for t has been carried along as a divisor, so that ω has units of angular measure divided by time.

18.2   3D ROTATIONS

In three dimensions, a rotation in any plane can be rewritten locally (using the unit complex number form of the previous section) precisely as a 2D rotation. What the complex notation cannot account for, of course, is the noncommatativity of 3D rotations, and that is why the quaternion algebra has three separate "imaginary" components with a mutually noncommutative algebra.

To try to understand the concept of a logarithm of a 3D rotation, we will of course wind up using quaternions. However, all fundamental properties of the formulas can be understood first via some very powerful observations regarding matrices and powers of matrices. First, consider the triplet of antisymmetric traceless 3×3 matrices

$$L_x = \begin{bmatrix} 0 & 0 & 0 \\ 0 & 0 & -1 \\ 0 & 1 & 0 \end{bmatrix}, \tag{18.4}$$

$$L_y = \begin{bmatrix} 0 & 0 & 1 \\ 0 & 0 & 0 \\ -1 & 0 & 0 \end{bmatrix}, \tag{18.5}$$

$$L_z = \begin{bmatrix} 0 & -1 & 0 \\ 1 & 0 & 0 \\ 0 & 0 & 0 \end{bmatrix}. \tag{18.6}$$

In the group theory of 3D rotations, these are the infinitesimal generators of the rotations about each Cartesian axis. We can see what this means by looking at powers of the matrices:

$$L_z \cdot L_z = \begin{bmatrix} -1 & 0 & 0 \\ 0 & -1 & 0 \\ 0 & 0 & 0 \end{bmatrix}, \tag{18.7}$$

$$L_z \cdot L_z \cdot L_z = -L_z$$

$$= \begin{bmatrix} 0 & 1 & 0 \\ -1 & 0 & 0 \\ 0 & 0 & 0 \end{bmatrix}. \tag{18.8}$$

This holds similarly for L_x and L_y. If we take an exponential power series of θL_z, we find exactly $\mathbf{R}_3(\theta, \hat{\mathbf{z}})$.

$$I_3 + \sum_{k=1}^{\infty} \frac{(\theta L_z)^k}{k!} \tag{18.9}$$

$$= I_3 + \sum_{k=\text{odd}}^{\infty} (-1)^{(k-1)/2} \frac{\theta^k}{k!} L_z \tag{18.10}$$

$$+ \sum_{k=\text{even}}^{\infty} (-1)^{(k-2)/2} \frac{\theta^k}{k!} L_z \cdot L_z \tag{18.11}$$

$$= \begin{bmatrix} \cos\theta & -\sin\theta & 0 \\ \sin\theta & \cos\theta & 0 \\ 0 & 0 & 1 \end{bmatrix}. \tag{18.12}$$

Because the logarithm is the object that, when inserted into the exponential series produces the function in question, we see that

$$\log R_3(\theta, \hat{\mathbf{z}}) = \theta L_z,$$

and conversely

$$\exp(\theta L_z) = R_3(\theta, \hat{\mathbf{z}}).$$

The standard formula for the 3D rotation matrix $R_3(\theta, \hat{\mathbf{n}})$ (whose Euler-theorem eigenvector is $\hat{\mathbf{n}}$) can thus be shown to have the logarithm

$$\log R_3(\theta, \hat{\mathbf{n}}) = \theta \hat{\mathbf{n}} \cdot \mathbf{L},$$

Log of a Matrix

where $\mathbf{L} = (L_x, L_y, L_z)$. In fact, there is a generally useful way of deducing the form of the matrices $\mathbf{L}$ (or, more technically, $\hat{\mathbf{n}} \cdot \mathbf{L}$) from this set of formulas using calculus to find the result of an infinitesimal transformation. We replace $\theta \to t\theta$, take the derivative with respect to t, and evaluate at $t = 0$, producing a vector in the tangent space of the transformation, which is the mathematical meaning of the matrices $\mathbf{L} = (L_x, L_y, L_z)$. For example,

$$
\left. \frac{d R_3(t\theta, \hat{\mathbf{z}})}{dt} \right|_{t=0} = \left. \begin{bmatrix} -\sin t\theta & -\cos t\theta & 0 \\ \cos t\theta & -\sin t\theta & 0 \\ 0 & 0 & 0 \end{bmatrix} \right|_{t=0}
$$

$$
= \begin{bmatrix} 0 & -1 & 0 \\ +1 & 0 & 0 \\ 0 & 0 & 0 \end{bmatrix}
$$

$$
\equiv +L_z.
$$

We note in conclusion that the components of $\mathbf{L}$ themselves obey the following algebra.

$$
L_y \cdot L_z - L_z \cdot L_y = +L_x, \tag{18.13}
$$

$$
L_z \cdot L_x - L_x \cdot L_z = +L_y, \tag{18.14}
$$

$$
L_x \cdot L_y - L_y \cdot L_x = +L_z. \tag{18.15}
$$

Note: Thus we can identify $\mathbf{L}$ with the conventional Lie algebra with $\mathbf{X} = \mathbf{L}$ (see Chapter 13).

Quaternion exponentiation: Essentially the same argument can be made using the Pauli matrices $\boldsymbol{\sigma}$ instead of the matrices $\mathbf{L}$, except that the Pauli matrices become proportional to the identity matrix when squared:

$$
(\sigma_1)^2 = (\sigma_2)^2 = (\sigma_3)^2 = I_2.
$$

Thus, as noted in Chapter 15, the Pauli matrices become an exact realization of the quaternion algebra with the identification

$$
(-i\sigma_x, -i\sigma_y, -i\sigma_z) = (\mathbf{i}, \mathbf{j}, \mathbf{k}).
$$

Using the quaternion algebra in the abstract—which is the same as the Pauli matrix algebra (or $(\mathbf{i}, \mathbf{j}, \mathbf{k})$ algebra, if one prefers that notation)—we can make essentially the same observation relating exponentials of infinitesimal quaternions to a full quaternion as we did for **L**. Starting with the quaternion $(0, \theta\hat{\mathbf{n}})$, and using quaternion multiplication, we form an exponential power series as follows:

$$\exp(0, \theta\hat{\mathbf{n}}) = (1, \mathbf{0}) + \sum_{k=1}^{\infty} \frac{\star(0, \theta\hat{\mathbf{n}})^k}{k!}. \qquad [18.16]$$

Because

$$(0, \theta\hat{\mathbf{n}}) \star (0, \theta\hat{\mathbf{n}}) = \left(-\theta^2, \mathbf{0}\right)$$

$$(0, \theta\hat{\mathbf{n}}) \star (0, \theta\hat{\mathbf{n}}) \star (0, \theta\hat{\mathbf{n}}) = \left(0, -\theta^3\hat{\mathbf{n}}\right)$$

$$(0, \theta\hat{\mathbf{n}}) \star (0, \theta\hat{\mathbf{n}}) \star (0, \theta\hat{\mathbf{n}}) \star (0, \theta\hat{\mathbf{n}}) = \left(+\theta^4, \mathbf{0}\right),$$

the series repeats every fourth term in a way that is familiar from $e^{i\theta}$, except with components in the quaternion algebra:

$$\exp(0, \theta\hat{\mathbf{n}}) = (\cos\theta, \sin\theta\hat{\mathbf{n}}).$$

Quaternion Exponential

Thus, we conclude that the logarithm of a unit quaternion (the quaternion that when exponentiated produces a generic quaternion) must be

$$\log(\cos\theta, \sin\theta\hat{\mathbf{n}}) = (0, \theta\hat{\mathbf{n}}).$$

Quaternion Logarithm

Note: Recall that to obtain correspondence with ordinary 3D rotations we use the half-angle form $q(\theta, \hat{\mathbf{n}}) \equiv (\cos(\theta/2), \hat{\mathbf{n}}\sin(\theta/2)) = \exp(0, \hat{\mathbf{n}}\theta/2)$.

18.3   USING LOGARITHMS FOR QUATERNION CALCULUS

We can now easily proceed to study powers and derivatives of quaternions using the logarithmic machinery we have just developed. Because

$$\exp(t\theta\hat{\mathbf{n}}) = \big(\cos(t\theta), \sin(t\theta)\hat{\mathbf{n}}\big),$$

and because

$$q^t = \exp(t \log q),$$

we conclude that

$$q^t = \big(\cos(t\theta), \sin(t\theta)\hat{\mathbf{n}}\big).$$

Taking the derivative with respect to t at $t = 0$, we find

$$\frac{dq^t}{dt}\bigg|_{t=0} = (\log q)q^t \bigg|_{t=0} \tag{18.17}$$

$$= \log q \tag{18.18}$$

$$= (0, \theta\hat{\mathbf{n}}). \tag{18.19}$$

When we substitute $(0, \hat{\mathbf{n}})$ back into the exponential, we see that this is the precise analog in quaternion space of the matrix $\hat{\mathbf{n}} \cdot \mathbf{L}$ generating infinitesimal rotations in the space of 3×3 matrices.

18.4   QUATERNION INTERPOLATIONS VERSUS LOG

We have now seen that rotations in three dimensions correspond precisely to the exponentiation of a pure quaternion $(0, \theta\hat{\mathbf{n}})$ to produce a unit-length quaternion four-vector:

$$q(2\theta, \hat{\mathbf{n}}) = e^{(0, \theta\hat{\mathbf{n}})} = \big(\cos(\theta), \sin(\theta)\hat{\mathbf{n}}\big). \tag{18.20}$$

We can now examine a rotation from one particular 3D frame, say, the quaternion frame F_0 given by rotating the identity frame by $2\theta_0\hat{\mathbf{n}}_0$ so that

$$F_0 = e^{(0, \theta_0\hat{\mathbf{n}}_0)},$$

to a second frame F_1, related to the identity frame by the rotation $2\theta_1\hat{\mathbf{n}}_1$, where

$$F_1 = e^{(0, \theta_1 \hat{\mathbf{n}}_1)}.$$

To make a smooth parametric rotation from one frame to the other, the 3D interpolating rotation $F(t)$ must obey

$$F(0) = F_0,$$
$$F(1) = F_1,$$

just as in the 2D case. We again examine

$$F(t) = F_0 \star \left(F_0^{-1} \star F_1\right)^t,$$

which written in terms of the exponentials becomes

$$F(t) = e^{(0, \theta_0 \hat{\mathbf{n}}_0)} \star e^{t\left(\log(F_0^{-1} \star F_1)\right)}.$$

Following essentially the same argument as that of Section 10.2, we can identify this as a SLERP parameterized by the angle ϕ between the quaternions F_0 and F_1, with

$$F_0 \cdot F_1 = \cos\phi = \cos\theta_0 \cos\theta_1 + (\hat{\mathbf{n}}_0 \cdot \hat{\mathbf{n}}_1) \sin\theta_0 \sin\theta_1.$$

If the two rotations are coplanar, we verify that this reduces to $\cos(\theta_1 - \theta_0)$ as expected.

Two-Dimensional Curves

19

Our purpose in this chapter is to introduce the idea of orientation frames in two dimensions, and in particular to study the technology of *moving* orientation frames. We choose two dimensions as our logical starting point because it is the simplest framework available, and there are many basic concepts that generalize to three dimensions and provide insight into quaternion methods.

19.1 ORIENTATION FRAMES FOR 2D SPACE CURVES

Suppose we have a 2D object (a simple box, for example), as shown in Figure 19.1. The orientation frame consists of two orthogonal vectors, denoted by the *tangent* **T** (the direction corresponding to the slope of the "hillside" the box is sitting on) and the *normal* **N**, which is the direction perpendicular to the "hillside" curve and perpendicular to the direction of **T**. This simple example will allow us to study the following two profound concepts:

- *Basic fact 1*: The columns of any rotation matrix are interpretable as *coordinate frame axes*.
- *Basic fact 2*: Rotation matrices belong to a *group*, and because groups have *geometric properties* each matrix is actually a *point* on a geometric object. Therefore, any set of coordinate frame axes is also such a point.

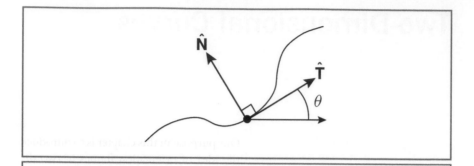

FIGURE 19.1 *Moving frame on a smooth curve defined by the tangent and normal directions. The pair* $(\hat{\mathbf{T}}, \hat{\mathbf{N}})$ *is a rotation matrix defining the relation of the curve's tangent frame to the orientation described by the* 2×2 *identity matrix.*

We will now begin the task of more deeply understanding rotations, their associated coordinate frame axes, and the phenomenon of moving frames in 2D. We will continue to find new ways to exploit the fact that there exists a geometric correspondence between rotations and the geometry of *complex numbers*. Because complex numbers are a special subspace of quaternions, 2D rotations can be used to introduce us to quaternion frames and their geometric meaning.

19.1.1 2D ROTATION MATRICES

The first visualization of a rotation matrix we will use is based on the idea of a smooth curve in the plane, as shown in Figure 19.1. At each point of the curve, we can see that there is a tangent vector. We will let one end of the curve be the "beginning," so that the tangent vector points unambiguously from the beginning toward the end. In precise mathematical terms, if the curve is specified by a pair of parametric functions, which we write as the two-vector

$$\mathbf{f}(t) = \big(x(t), y(t)\big),$$

the tangent vector at each point of the curve is the derivative

$$\mathbf{T}(t) = \frac{d\mathbf{f}(t)}{dt} = \big(\dot{x}(t), \dot{y}(t)\big).$$

The reader should be convinced that this derivative and the slope of the curve are the same thing. (Hint: Try some simple examples, such as $\mathbf{f}(t) = (t, 0)$, $\mathbf{f}(t) = (t, t)$, $\mathbf{f}(t) = (t, t^3)$, and so on.)

We take the convention that if the direction $\mathbf{T}$ is the "virtual x axis" of a 2D coordinate frame there is a unique "virtual y axis" dictated by the right-handed orientation convention. We call this the *normal direction* and denote it by

$$\mathbf{N}(t) = \times\mathbf{T} = \left(-\dot{y}(t), +\dot{x}(t)\right).$$

Clearly, $\mathbf{T}(t) \cdot \mathbf{N}(t) = 0$ for all t, and therefore the tangent and the normal together can be normalized to form an *orthonormal moving frame*:

$$\hat{\mathbf{T}} = \frac{\mathbf{T}}{\|\mathbf{T}\|}, \qquad \hat{\mathbf{N}} = \frac{\mathbf{N}}{\|\mathbf{N}\|}.$$

The tangent and normal change continuously as we move along the 2D curve, as implied by Figure 19.1.

19.1.2   THE FRAME MATRIX IN 2D

The varying orientation of the frame is described at each point (or time) by the 2D rotation matrix

$$R_2(\theta) = [\hat{\mathbf{T}} \quad \hat{\mathbf{N}}]$$
$$= \begin{bmatrix} \cos\theta & -\sin\theta \\ \sin\theta & \cos\theta \end{bmatrix}, \tag{19.1}$$

as well as the two more abstract forms we have already introduced,

$$R_2(A, B) = \begin{bmatrix} A & -B \\ B & A \end{bmatrix}.$$

(with the constraint $A^2 + B^2 = 1$) and

$$R_2(a, b) = \begin{bmatrix} a^2 - b^2 & -2ab \\ 2ab & a^2 - b^2 \end{bmatrix},$$

with the constraint $a^2 + b^2 = 1$.

19.1.3 FRAME EVOLUTION IN 2D

Next let us examine the time evolution of the 2D frame, which will be of great interest in 3D. First, using the coordinates $\theta(t)$ for the frame we have

$$[\hat{\mathbf{T}} \quad \hat{\mathbf{N}}] = \begin{bmatrix} \cos\theta(t) & -\sin\theta(t) \\ \sin\theta(t) & \cos\theta(t) \end{bmatrix}. \tag{19.2}$$

Differentiating the columns separately, we find the frame equations

$$\dot{\hat{\mathbf{T}}}(t) = +\kappa\hat{\mathbf{N}},$$

$$\dot{\hat{\mathbf{N}}}(t) = -\kappa\hat{\mathbf{T}}, \tag{19.3}$$

where $\kappa(t) = d\theta/dt$ is the *curvature*.

This is the 2D analog of the 3D *Parallel Transport Frame*, which handily maps any continuously differentiable curve into its corresponding tangent frame. The set of Equations 19.3 is often arranged to make a "vector matrix equation:"

$$\begin{bmatrix} \dot{\hat{\mathbf{T}}}(t) & \dot{\hat{\mathbf{N}}}(t) \end{bmatrix} = \begin{bmatrix} \hat{\mathbf{T}}(t) & \hat{\mathbf{N}}(t) \end{bmatrix} \begin{bmatrix} 0 & -\kappa(t) \\ +\kappa(t) & 0 \end{bmatrix}. \tag{19.4}$$

19.2   WHAT IS A MAP?

We are going to start studying objects by creating new objects that are closely related to them, usually with a point-by-point correspondence, but which may not correspond exactly to the objects themselves. This approach is known as a *map*. Creating a map amounts to taking a point $\mathbf{p}$ on a geometric object and giving a rule that tells us how to compute a new point $\mathbf{x}$ that is *not* necessarily even of the same dimension as $\mathbf{p}$. Formally, we might write

$$f \quad \text{maps} \quad \mathbf{p} \to \mathbf{x},$$

$$\mathbf{f}(\mathbf{p}) = \mathbf{x},$$

with typical examples being, respectively, the constant map, a scalar field, the identity map, or a tensor map:

$$f(\mathbf{p}) = 1,$$

$$f(\mathbf{p}) = \mathbf{p} \cdot \mathbf{p},$$

$$f(\mathbf{p}) = \mathbf{p},$$

$$f(\mathbf{p}) = \mathbf{p}_i \mathbf{p}_j.$$

The task of the map $f(\mathbf{p})$ is to perform some computation depending on the point $\mathbf{p}$ and to return some object, possibly a vector of the same dimensions, possibly not, that is related to each point $\mathbf{p}$.

19.3 TANGENT AND NORMAL MAPS

In Figure 19.2, we show an example of a 2D curve with its tangent and normal fields. The *normalized* tangent and normal fields have only one degree of freedom, which we denote by the angle $\theta(t)$. The column vectors $\hat{\mathbf{T}}$ and $\hat{\mathbf{N}}$ then represent a moving orthonormal coordinate frame, as in Equation 19.2.

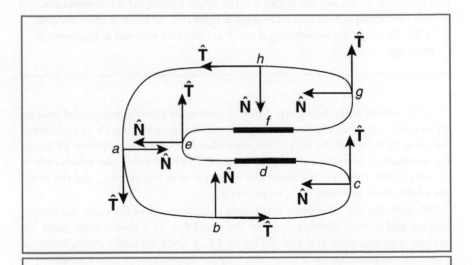

FIGURE 19.2 *A smooth 2D curve with its normal and tangent frame fields. The segments d and f are intended to be straight.*

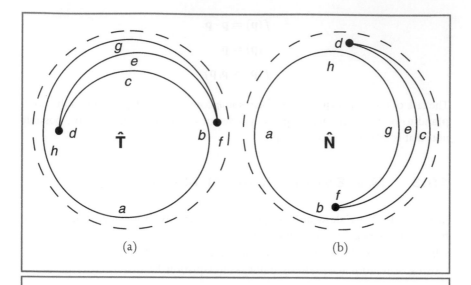

FIGURE 19.3 *2D Gauss map sketches of (a) the tangent directions and (b) the normal directions corresponding to the U-shaped curve shown in Figure 19.2. All vectors lie on the unit circle in 2D. The straight line segments along d and f in Figure 19.2 correspond to single points in both maps.*

A 2D version of the Gauss map [46,61] used in the classical differential geometry of surfaces follows when we discard the original curve (Figure 19.2) and restrict our view to show only the path of the normalized tangents (as in Figure 19.3a) or the normalized normals, as in Figure 19.3b. Both vector fields take values only in the unit circle. These are the *tangent map* of the curve in Figure 19.2, and the *normal map* of the curve in Figure 19.2, respectively.

We note that any sufficiently small open neighborhood of the curve has unique tangent and normal directions, up to the possibility of a shared limit point for straight segments such as d and f (Figure 19.2). Over the entire curve, however, particular neighborhoods of directions may be repeated many times, resulting in an overlapping nonunique 2D map, as indicated schematically in Figure 19.3. We will accept this as a feature, not necessarily a deficiency, of the construction.

19.4   SQUARE ROOT FORM

19.4.1   FRAME EVOLUTION IN (a, b)

With some amount of prescience about quaternions, we can ask what happens if we express the standard angular form of the 2D frame equations we have been using not in the form of the $(A = \cos\theta, B = \sin\theta)$ variables but in terms of $a = \cos(\theta/2)$, $b = \sin(\theta/2)$ (i.e., $A = a^2 - b^2$, $B = 2ab$). When we carry through this alternative parameterization, we find, as usual,

$$R_2(a, b) = \begin{bmatrix} a^2 - b^2 & -2ab \\ 2ab & a^2 - b^2 \end{bmatrix},$$

where orthonormality implies $(a^2 + b^2)^2 = 1$, which reduces back to

$$a^2 + b^2 = 1.$$

We observe that the pair (a, b) provides a *double-valued* parameterization of the frame:

$$[\hat{\mathbf{T}} \quad \hat{\mathbf{N}}] = \begin{bmatrix} a^2 - b^2 & -2ab \\ 2ab & a^2 - b^2 \end{bmatrix}. \tag{19.5}$$

Here, (a, b) describes precisely the *same frame* as $(-a, -b)$.

Remark: Yet another one-parameter representation. The redundant parameter can be eliminated locally by using projective coordinates such as $c = b/a = \tan(\theta/2)$, where $-\infty < c < +\infty$, to get another parameterization:

$$[\hat{\mathbf{T}} \quad \hat{\mathbf{N}}] = \frac{1}{1 + c^2} \begin{bmatrix} 1 - c^2 & -2c \\ 2c & 1 - c^2 \end{bmatrix}.$$

(*Note:* there is a useful equivalent of this formula for quaternions.)

19.4.2   SIMPLIFYING THE FRAME EQUATIONS

Using the basis $(\hat{\mathbf{T}}, \hat{\mathbf{N}})$ for the frame, we have *four equations* with *three constraints* from orthonormality, yielding *one true degree of freedom*. However, a major simplification occurs when we carry out the calculation in (a, b) coordinates! Differentiating, we get

$$\dot{\mathbf{T}} = 2 \begin{bmatrix} a\dot{a} - b\dot{b} \\ a\dot{b} + b\dot{a} \end{bmatrix} = 2 \begin{bmatrix} a & -b \\ b & a \end{bmatrix} \begin{bmatrix} \dot{a} \\ \dot{b} \end{bmatrix}.$$

However, this formula for $\dot{\mathbf{T}}$ is just $\kappa\hat{\mathbf{N}}$, where

$$\kappa\hat{\mathbf{N}} = \kappa \begin{bmatrix} -2ab \\ a^2 - b^2 \end{bmatrix} = \kappa \begin{bmatrix} a & -b \\ b & a \end{bmatrix} \begin{bmatrix} -b \\ a \end{bmatrix}$$

or

$$\kappa\hat{\mathbf{N}} = \kappa \begin{bmatrix} a & -b \\ b & a \end{bmatrix} \begin{bmatrix} 0 & -1 \\ 1 & 0 \end{bmatrix} \begin{bmatrix} a \\ b \end{bmatrix}.$$

Now, because

$$\begin{bmatrix} a & -b \\ b & a \end{bmatrix}$$

is an orthogonal matrix, we can simply multiply the pair of equations for $\dot{\mathbf{T}}$ and $\kappa\hat{\mathbf{N}}$ by the transpose, so that $\dot{\mathbf{T}} = +\kappa\hat{\mathbf{N}}$ becomes precisely equivalent to

$$\begin{bmatrix} \dot{a} \\ \dot{b} \end{bmatrix} = \frac{1}{2} \begin{bmatrix} 0 & -\kappa(t) \\ \kappa(t) & 0 \end{bmatrix} \begin{bmatrix} a \\ b \end{bmatrix}. \tag{19.6}$$

Thus, $(a(t), b(t))$ with the constraint $a^2 + b^2 = 1$ acts as the *square root* of $(\hat{\mathbf{T}}, \hat{\mathbf{N}})$, and the differential equation given by Equation 19.6 contains all properties needed to compute and reconstruct the behavior of the frame described in a less elegant way by Equations 19.3 and 19.4.

Three-Dimensional Curves

20

The methods of 3D frames are taught as a basic area of study in differential geometry. However, classical methods do not take advantage of the quaternion techniques that can be used to clarify the nature of moving frames. In this chapter we review the classical differential geometry of moving frames on 3D space curves, and then introduce the equivalent quaternion formulation along with the *quaternion tangent map*, which allows us to see global properties of a framing choice at a single glance.

The fundamental difference between 2D space curves and 3D space curves is that although the tangent direction is still determinable directly from the space curve, there is an additional degree of rotational freedom in the normal plane, the portion of the frame perpendicular to the tangent vector. This is indicated schematically in Figure 20.1.

20.1 INTRODUCTION TO 3D SPACE CURVES

Dense families of space curves can be generated by many applications, including time-dependent particle flow fields, static streamlines generated by integrating a volume vector field, and deformations of a solid coordinate grid. Our fundamental approach singles out space curves, although variations could be used to treat individual point frames (see Alpern et al. [3]), stream surfaces (see Hultquist [99]), and orientation differences (which are themselves orientation fields). Thus, we begin with the properties of a curve $\mathbf{x}(t)$ in 3D space parameterized by the unnor-

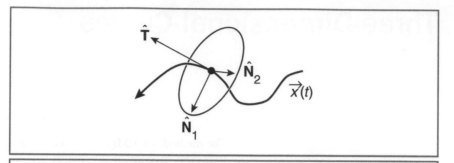

FIGURE 20.1 *General form of a moving frame for a 3D curve $\mathbf{x}(t)$, with the tangent direction $\hat{\mathbf{T}}$ determined directly from the curve derivative and the exact orientation of the basis $(\hat{\mathbf{N}}_1, \hat{\mathbf{N}}_2)$ for the normal plane determined only up to an axial rotation about $\hat{\mathbf{T}}$.*

malized arc length t. If $\mathbf{x}(t)$ is once-differentiable, the normalized tangent vector at any point is

$$\hat{\mathbf{T}}(t) = \frac{\mathbf{x}'(t)}{\|\mathbf{x}'(t)\|}.$$

The standard arc-length differential is typically expressed as

$$v(t)^2 = \left(\frac{ds}{dt}\right)^2 = \left(\frac{d\mathbf{x}(t)}{dt} \cdot \frac{d\mathbf{x}(t)}{dt}\right) = \|\mathbf{x}'(t)\|^2.$$

In practice, we never have smooth curves in numerical applications, but only piecewise linear curves that are presumed to be approximations to differentiable curves. Thus, for a curve given by the set of points $\{\mathbf{x}_i\}$ we might typically take

$$\hat{\mathbf{T}}_i = \frac{\mathbf{x}_{i+1} - \mathbf{x}_i}{\|\mathbf{x}_{i+1} - \mathbf{x}_i\|},$$

or any corresponding formula with additional sampling points and desirable symmetries. We would typically use a five-point formula to get a smoother result. One could also produce finer intermediate states by spline interpolation.

If the curve is locally straight—that is, $\mathbf{x}''(t) = 0$ or $\hat{\mathbf{T}}_{i+1} = \hat{\mathbf{T}}_i$—there is no locally determinable coordinate frame component in the plane normal to $\hat{\mathbf{T}}$. A nonlocal definition must be used to decide on the remainder of the frame once $\hat{\mathbf{T}}$ is determined. In the following, we formulate our two alternate coordinate frames, one of which (the Frenet frame) is completely local but is indeterminable where the curve is locally straight. The other coordinate frame, the parallel transport frame, is defined everywhere but depends on a numerical integration over the entire curve.

Tangent map examples: The tangent direction of a 3D curve at each point is given simply by taking the algebraic or numerical derivative of the curve at each sample point and normalizing the result. Each tangent direction thus has two degrees of freedom and lies on the surface of the two-sphere $\mathbf{S}^2$. The curve resulting from joining the ends of neighboring tangents is the *tangent map* of the curve. As in the 2D case treated earlier, the tangent map of a 3D curve is not necessarily single valued except in local neighborhoods, and may have limit points (e.g., if there are straight segments). In Figures 20.2a and 20.2b we show examples of two classic 3D curves—one a closed knot, the $(2, 3)$ trefoil knot lying on the surface of a torus, and the other the open helix:

$$\mathbf{x}_{\text{torus}}(p, q)(r, a, b)(t) = (r + a\cos(qt))\cos(pt)\hat{\mathbf{x}}$$

$$+ (r + a\cos(qt))\sin(pt)\hat{\mathbf{y}} + b\sin(qt)\hat{\mathbf{z}}, \quad (20.1)$$

$$\mathbf{x}_{\text{helix}}(a, b, c)(t) = a\cos(t)\hat{\mathbf{x}} + b\sin(t)\hat{\mathbf{y}} + ct\hat{\mathbf{z}}. \quad (20.2)$$

For the torus knot, p and q are normally relatively prime integers, r is the radius of the torus, and a and b describe the elliptical cross section of the toroidal tube. Differentiating these curves yields the tangent maps shown in Figure 20.2c.

20.2   GENERAL CURVE FRAMINGS IN 3D

The evolution properties of all possible frames for a 3D curve $\mathbf{x}(t)$ can be written in a unified framework. The basic idea is to consider an arbitrary frame to be represented in the form of columns of a 3×3 orthonormal rotation matrix:

$$\text{Curve Frame} = \begin{bmatrix} \hat{\mathbf{T}} & \hat{\mathbf{N}}_1 & \hat{\mathbf{N}}_2 \end{bmatrix}. \quad (20.3)$$

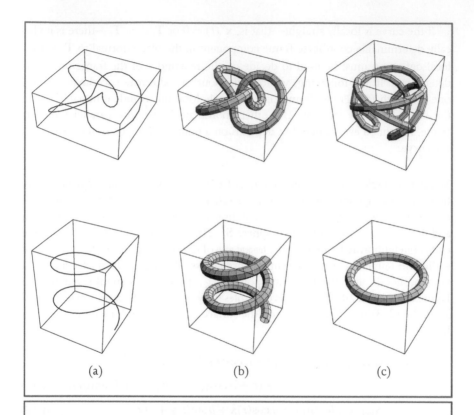

(a)   (b)   (c)

FIGURE 20.2 *Tangent maps. (a) The $(2, 3)$ torus knot and the helix as 3D line drawings. (b) Illustrating an application of tubing to make the 3D curves more interpretable. (c) The corresponding normalized tangent maps determined directly from the curve geometry. These are curves on the two-sphere, which have been tubed to improve visibility.*

Here, $\hat{\mathbf{T}}(t) = \mathbf{x}'(t)/\|\mathbf{x}'(t)\|$ is the normalized tangent vector determined directly by the curve geometry and is thus unalterable. $(\hat{\mathbf{N}}_1(t), \hat{\mathbf{N}}_2(t))$ is a pair of orthonormal vectors spanning the plane perpendicular to the tangent vector at each point of the curve. Because $\|\hat{\mathbf{T}}\|^2 = \|\hat{\mathbf{N}}_1\|^2 = \|\hat{\mathbf{N}}_2\|^2 = 1$ and all other inner products vanish by definition, *any change in a basis vector must be orthogonal to itself and thereby expressible in terms of the other two basis vectors.* Thus, the most general possible form for the frame evolution equations is

$$\begin{bmatrix} \hat{\mathbf{T}}'(t) & \hat{\mathbf{N}}_1'(t) & \hat{\mathbf{N}}_2'(t) \end{bmatrix}$$

$$= \begin{bmatrix} \hat{\mathbf{T}}(t) & \hat{\mathbf{N}}_1(t) & \hat{\mathbf{N}}_2(t) \end{bmatrix} v(t) \begin{bmatrix} 0 & -k_y(t) & +k_x(t) \\ +k_y(t) & 0 & -k_z(t) \\ -k_x(t) & +k_z(t) & 0 \end{bmatrix}, \quad [20.4]$$

where $v(t) = \|\mathbf{x}'(t)\|$ is the velocity of the curve if we are not using a unit-speed parameterization. (See Equation 19.4 for comparison.)

The particular choice of notation and signs for the curvatures $\mathbf{k}$ in Equation 20.4 is compellingly motivated by the quaternion algebra treatment to be presented later. The natural properties of the curve-frame evolution are also exposed using the Darboux form of the equations

$$\hat{\mathbf{T}}' = v(t)\mathbf{F} \times \hat{\mathbf{T}},$$
$$\hat{\mathbf{N}}_1' = v(t)\mathbf{F} \times \hat{\mathbf{N}}_1, \quad [20.5]$$
$$\hat{\mathbf{N}}_2' = v(t)\mathbf{F} \times \hat{\mathbf{N}}_2,$$

where $\mathbf{F}$ generalizes the Darboux vector field (e.g., see Gray [61, p. 205]):

$$\mathbf{F} = k_x\hat{\mathbf{N}}_1 + k_y\hat{\mathbf{N}}_2 + k_z\hat{\mathbf{T}}. \quad [20.6]$$

The square magnitude of the total "force" acting on the frame is $\|\mathbf{F}\|^2 = k_x^2 + k_y^2 + k_z^2$, and we will see in the following that this is a minimum for the parallel transport frame. The arbitrariness of the basis $(\hat{\mathbf{N}}_1(t), \hat{\mathbf{N}}_2(t))$ for the plane perpendicular to $\hat{\mathbf{T}}(t)$ can be exploited as desired to eliminate any one of the (k_x, k_y, k_z) (e.g., see Bishop [19]). For example, if

$$\hat{\mathbf{M}}_1 = \hat{\mathbf{N}}_1 \cos\theta - \hat{\mathbf{N}}_2 \sin\theta,$$
$$\hat{\mathbf{M}}_2 = \hat{\mathbf{N}}_1 \sin\theta + \hat{\mathbf{N}}_2 \cos\theta, \quad [20.7]$$

differentiating and substituting from Equation 20.4 yields

$$\hat{\mathbf{M}}_1' = \hat{\mathbf{M}}_2(k_z - \theta') - \hat{\mathbf{T}}(k_x \sin\theta + k_y \cos\theta), \quad [20.8]$$
$$\hat{\mathbf{M}}_2' = -\hat{\mathbf{M}}_1(k_z - \theta') + \hat{\mathbf{T}}(k_x \cos\theta - k_y \sin\theta). \quad [20.9]$$

Thus, the angle $\theta(t)$ may be chosen to cancel the angular velocity k_z in the plane spanned by $\hat{\mathbf{N}}_1(t)$ and $\hat{\mathbf{N}}_2(t)$. The same argument holds for any other pair. Attempting to eliminate additional components produces new mixing, leaving at least two independent components in the evolution matrix.

20.3   TUBING

We remark that to generate a ribbon or tube such as those we commonly use to display curves one simply sweeps the chosen set of frames through each curve point $\mathbf{p}(t)$ to produce a connected tube:

$$\mathbf{x}(t, \theta) = \mathbf{p}(t) + \cos\theta\,\hat{\mathbf{N}}_1(t) + \sin\theta\,\hat{\mathbf{N}}_2(t).$$

The resulting structure is sampled in t and over one full 2π period in θ to produce a tessellated tube. Arbitrary functions of (t, θ) can be introduced instead of the cosine and sine to produce ribbons and general linear structures. A typical example of a tube-based modeling operation, sometimes known as the generalized cone, takes a particular 2D cross-section shape and moves it along a framed path, possibly varying in size or shape as it travels. The result is the class of tube-like geometric objects shown in Figure 20.3.

20.4   CLASSICAL FRAMES

We now note a variety of approaches to assigning frames to an entire 3D space curve, each with its own peculiar advantages. Figure 20.4 compares the tubings of the $(2, 3)$ trefoil knot and the helix for each of the three frames described in the following. For further details, see (for example) Eisenhart [46] and Gray [61].

20.4.1   FRENET–SERRET FRAME

The classical Frenet–Serret frame, also commonly referred to as the "Frenet frame," is determined by local conditions at each point of the curve but is undefined whenever the curvature vanishes (e.g., when the curve straightens out or has an inflection point). For the Frenet frame, we make the choices

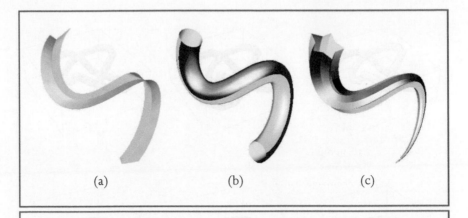

(a)   (b)   (c)

FIGURE 20.3 *Moving orientation control of extruded shapes. (a) A ribbon generated by a pair of vectors. (b) A tube with a circular cross section. (c) A tube with a star-shaped cross section and a varying radius.*

$$k_x = 0,$$
$$k_y = \kappa(t),$$
$$k_z = \tau(t),$$

where $\kappa(t)$ is the curvature (which geometrically can be understood as the inverse radius of curvature at a curve point) and $\tau(t)$ is the torsion, which mixes the two normal vectors in their local plane. This identification produces the Frenet–Serret equations

$$\left[\hat{\mathbf{T}}'(t) \quad \hat{\mathbf{N}}'(t) \quad \hat{\mathbf{B}}'(t)\right]$$

$$= \left[\hat{\mathbf{T}}(t) \quad \hat{\mathbf{N}}(t) \quad \hat{\mathbf{B}}(t)\right] v(t) \begin{bmatrix} 0 & -\kappa(t) & 0 \\ +\kappa(t) & 0 & -\tau(t) \\ 0 & +\tau(t) & 0 \end{bmatrix}. \qquad (20.10)$$

Note that the squared Darboux vector is thus $\|\mathbf{F}\|^2 = \kappa^2 + \tau^2 \geqslant \kappa^2$.

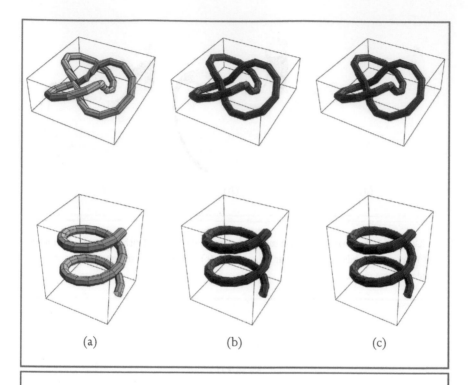

(a)   (b)   (c)

FIGURE 20.4 *Curve framings for the $(2, 3)$ torus knot and the helix based on (a) Frenet frame, (b) geodesic reference frame (minimal tilt from North Pole), and (c) parallel transport frame, which is not periodic like the other frames.*

If $\mathbf{x}(t)$ is any thrice-differentiable space curve, we can identify the triad of normalized Frenet frame vectors directly with the local derivatives of the curve,

$$\hat{\mathbf{T}}(t) = \frac{\mathbf{x}'(t)}{\|\mathbf{x}'(t)\|},$$

$$\hat{\mathbf{N}}_1 = \hat{\mathbf{N}}(t) = \hat{\mathbf{B}}(t) \times \hat{\mathbf{T}}(t),$$

$$\hat{\mathbf{N}}_2 = \hat{\mathbf{B}}(t) = \frac{\mathbf{x}'(t) \times \mathbf{x}''(t)}{\|\mathbf{x}'(t) \times \mathbf{x}''(t)\|}, \qquad \text{(20.11)}$$

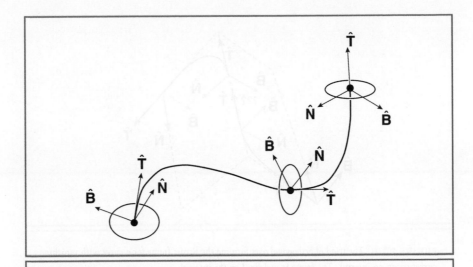

FIGURE 20.5 *The triad of orthogonal axes forming the Frenet frame for a curve with nonvanishing curvature.*

where one can identify the curvature and torsion in Equation 20.10 with

$$\kappa(t) = \frac{\|\mathbf{x}'(t) \times \mathbf{x}''(t)\|}{\|\mathbf{x}'(t)\|^3}$$

and

$$\tau(t) = \frac{\mathbf{x}'(t) \times \mathbf{x}''(t) \cdot \mathbf{x}'''(t)}{\|\mathbf{x}'(t) \times \mathbf{x}''(t)\|^2},$$

respectively.

We illustrate the standard Frenet frame configuration in Figure 20.5. When the second derivative vanishes on some interval, the Frenet frame is temporarily undefined, as illustrated in Figure 20.6. Attempts to work around this problem involve various heuristics [147] but can be resolved rigorously using the Bishop frame [19], also known as the parallel transport frame, which we will describe in a moment.

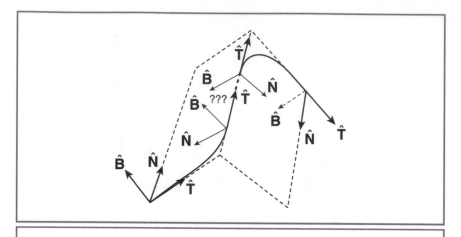

FIGURE 20.6 *The triad of orthogonal axes forming the Frenet frame for a curve with vanishing curvature on an interval. The frame is undefined on the interval.*

20.4.2 PARALLEL TRANSPORT FRAME

The parallel transport frame is equivalent to a heuristic approach that has been frequently used in graphics applications (e.g., see Bloomenthal [21], Klock [114], Max [121], Shani and Ballard [147]). A careful mathematical treatment by Bishop [19] presents its differential properties in a form that can be easily compared with the standard features of the Frenet frame. The parallel transport frame is distinguished by the fact that it uses the smallest possible rotation at each curve sample to align the current tangent vector with the next tangent vector. The current orientation of the plane normal to the tangent vector depends on the history of the curve, starting with an arbitrary initial frame, and thus one is essentially integrating a differential equation for the frame change around the curve. The frame depends on the initial conditions, and unlike the Frenet frame cannot be determined locally on the curve. The algorithm with the best limiting properties [119] for computing this frame involves determining the normal direction $\hat{\mathbf{N}} = \mathbf{T}_i \times \mathbf{T}_{i+1}/\|\mathbf{T}_i \times \mathbf{T}_{i+1}\|$ to the plane of two successive tangents to the curve, finding the angle $\theta = \arccos(\hat{\mathbf{T}}_i \cdot \hat{\mathbf{T}}_{i+1})$ and rotating the current frame to the next frame using the 3×3 matrix $R(\theta, \hat{\mathbf{N}})$ or its corresponding quaternion,

$$q(\theta, \hat{\mathbf{N}}) = q\big(\arccos(\hat{\mathbf{T}}_i \cdot \hat{\mathbf{T}}_{i+1}), \mathbf{T}_i \times \mathbf{T}_{i+1}/\|\mathbf{T}_i \times \mathbf{T}_{i+1}\|\big). \qquad \text{[20.12]}$$

(Recall that we define $q(\theta, \hat{\mathbf{N}}) = (\cos(\theta/2), \hat{\mathbf{N}}\sin(\theta/2))$.) If the successive tangents are collinear, one leaves the frame unchanged. If the tangents are anticollinear, a result can be returned but it is not uniquely determined. To identify the parallel transport frame with Equation 20.4, we set

$$
\begin{aligned}
k_y &\Rightarrow k_1(t), \\
-k_x &\Rightarrow k_2(t), \\
k_z &\Rightarrow 0
\end{aligned}
$$

to avoid unnecessary mixing between the normal components (effectively the definition of parallel transport). This choice produces Bishop's frame equations:

$$
\begin{aligned}
& \begin{bmatrix} \hat{\mathbf{T}}'(t) & \hat{\mathbf{N}}'_1(t) & \hat{\mathbf{N}}'_2(t) \end{bmatrix} \\
&= v(t)\begin{bmatrix} \hat{\mathbf{T}}(t) & \hat{\mathbf{N}}_1(t) & \hat{\mathbf{N}}_2(t) \end{bmatrix} \begin{bmatrix} 0 & -k_1(t) & -k_2(t) \\ k_1(t) & 0 & 0 \\ k_2(t) & 0 & 0 \end{bmatrix}. \qquad \text{[20.13]}
\end{aligned}
$$

This frame choice behaves entirely differently from the Frenet frame because, as illustrated in Figure 20.7, it remains continuous and well defined for a curve with vanishing curvature on a segment.

Because $\|\hat{\mathbf{T}}'\|^2 = (k_1)^2 + (k_2)^2$ is an invariant independent of the choice of the normal frame, Bishop identifies the curvature, orientation, and angular velocity as

$$\kappa(t) = \big((k_1)^2 + (k_2)^2\big)^{1/2},$$

$$\theta(t) = \arctan\left(\frac{k_2}{k_1}\right),$$

$$\omega(t) = \frac{d\theta(t)}{dt}.$$

k_1 and k_2 thus correspond to a Cartesian coordinate system for the curvature polar coordinates (κ, θ), with $\theta = \theta_0 + \int \omega(t)\,dt$ and $\omega(t)$ is effectively the classical

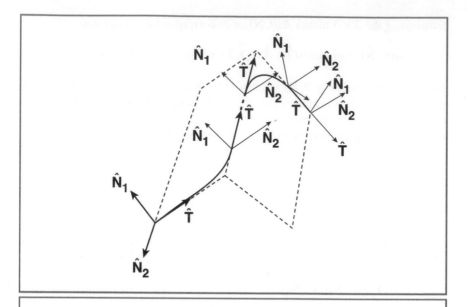

FIGURE 20.7 *The parallel transport curve frame for the curve shown in Figure 20.6 [19].
This frame, unlike the Frenet frame shown in Figure 20.6, is continuous along the "roof peak"
(where the curvature vanishes).*

torsion $\tau(t)$ appearing in the Frenet equations. A fundamental ambiguity in the parallel transport frame compared to the Frenet frame thus arises from the arbitrary
choice of an integration constant for θ, which disappears from τ due to the differ-
entiation. Note that the squared Darboux vector $\|\mathbf{F}\|^2 = \|\hat{\mathbf{T}}'\|^2 = k_1^2 + k_2^2 = \kappa^2$ is
now a frame invariant. It is missing the torsion component present for the Frenet
frame, and thus *assumes its minimal value.*

A numerical method of computing the parallel transport frame with the desired properties works as follows. Given a frame at $\mathbf{x}_{i-1}$, compute two neighbor-
ing tangents $\mathbf{T}_i$ and $\mathbf{T}_{i-1}$ and their unit vectors $\hat{\mathbf{T}}_i = \mathbf{T}_i/\|\mathbf{T}_i\|$. Find the angle
$\theta = \arccos(\hat{\mathbf{T}}_i \cdot \hat{\mathbf{T}}_{i-1})$ between them and find the perpendicular to the plane of the
tangents given by $\mathbf{V} = (\hat{\mathbf{T}}_{i-1} \times \hat{\mathbf{T}}_i)$. Finally, rotate the frame at $\mathbf{x}_{i-1}$ by θ about $\hat{\mathbf{V}}$
to get the frame at point $\mathbf{x}_i$.

Just as for the Frenet frame, one can begin with a curve $\mathbf{x}(t)$ and an initial frame, or a pair of functions $(k_1(t), k_2(t))$ and an initial frame, or a frame over the entire curve, and then integrate where needed to compute the missing variables. It is also worthwhile noting that $(k_1(t), k_2(t))$ form a 2D Cartesian vector field at each point of the curve, and thus allow a natural alternate characterization to Gray's (κ, τ) curve properties [31,60].

20.4.3 GEODESIC REFERENCE FRAME

We will often need a frame that is guaranteed to have a particular axis in one direction, but we will not care about the remaining axes because they will be considered as a space of possibilities. A convenient frame with these properties can always be constructed starting from the assumption that there exists a canonical reference frame in which, say, the $\hat{\mathbf{z}}$ axis corresponds to the preferred direction. Thus, if $\hat{\mathbf{v}}$ is the desired direction of the new axis, we can simply tilt the reference axis $\hat{\mathbf{z}}$ into $\hat{\mathbf{v}}$ along a minimal, geodesic curve using an ordinary rotation $R(\theta, \hat{\mathbf{n}})$ or its corresponding quaternion:

$$q(\theta, \hat{\mathbf{n}}) = q\left(\arccos(\hat{\mathbf{z}} \cdot \hat{\mathbf{v}}), \ \hat{\mathbf{z}} \times \hat{\mathbf{v}}/\|\hat{\mathbf{z}} \times \hat{\mathbf{v}}\|\right). \qquad (20.14)$$

Clearly any reference frame, including frames related to the viewing parameters of a moving observer, could be used instead of $\hat{\mathbf{z}}$. This frame has the drawback that it is ambiguous whenever $\hat{\mathbf{v}} = -\hat{\mathbf{z}}$. Sequences of frames passing through this point will not necessarily be smoothly varying because only a single instance of a one-parameter family of frames can be returned automatically by a context-free algorithm. Luckily, this is of no consequence for most of our applications. As we will discuss later in the quaternion framework, this property is directly related to the absence of a global vector field on the two-sphere.

20.4.4 GENERAL FRAMES

When possible, we will work with the top-level framework for coordinate frames of arbitrary generality, rather than choosing conventional frames or hybrids of the frames described so far (e.g., see Klock [114]). Although the classical frames have many fundamentally appealing mathematical properties, we are not restricted to the use of any one of them. Keeping the tangent vector field intact, we may modify the angle of rotation about the tangent vector at will to produce an application-dependent frame assignment. An example of such an application is a closed curve with inflection points: the Frenet frame is periodic but not globally defined, the parallel transport frame will not be periodic in general, and the geodesic reference frame will be periodic but may have discontinuities for antipodal orientations. Thus, to obtain a satisfactory smooth global frame we need something close to a parallel transport frame but with a periodic boundary condition. An example of an ad hoc solution is to take the parallel transport frame and impose periodicity by adding to each vertex's axial rotation a fraction of the angular deficit of the parallel transport frame after one circuit. However, this is highly heuristic and depends strongly on the chosen parameterization. Chapter 22 introduces a more comprehensive approach.

20.5   MAPPING THE CURVATURE AND TORSION

Any individual space curve implicitly contains additional information that is derivable from its shape by exploiting its derivatives. The classic examples of such measures, which we discuss for completeness on our way to a quaternion treatment, are the torsion and the curvature. The treatment given here is available in almost any differential geometry textbook (e.g., see Eisenhart [46] and Gray [61]), whereas the use of torsion and curvature maps for visualization has been particularly emphasized by Gray [61,31].

The standard treatment is based on the *Frenet frame*, which we recall is defined by first assuming that we are given $\mathbf{x}(t)$, an arbitrary thrice-differentiable parameterization of the (unique) position of the curve in space. If we are given a nonvanishing curvature and a torsion as smooth functions of t, we can theoretically integrate the system of equations to find the unique numerical values of the corresponding space curve $\mathbf{x}(t)$ (up to a rigid motion). (See Appendix H.)

We recall that the classical torsion and curvature for an arbitrary curve are computed as

$$\kappa(t) = \frac{\|\mathbf{x}'(t) \times \mathbf{x}''(t)\|}{\|\mathbf{x}'(t)\|^3}, \tag{20.15}$$

$$\tau(t) = \frac{(\mathbf{x}'(t) \times \mathbf{x}''(t)) \cdot \mathbf{x}'''(t)}{\|\mathbf{x}'(t) \times \mathbf{x}''(t)\|^2} \tag{20.16}$$

and therefore require not only a nonzero first derivative but also that the curve have nonvanishing curvature. To compute the torsion, we must in addition have a well-defined third derivative. Given such information, we can encode their values

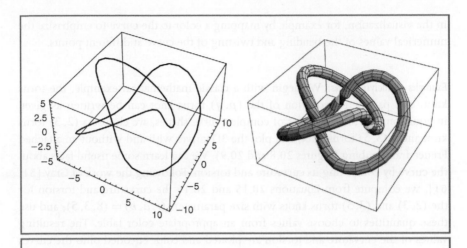

FIGURE 20.8 $(2, 3)$ torus knot.

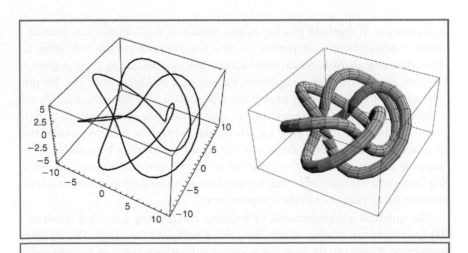

FIGURE 20.9 $(3, 5)$ torus knot.

in the visualization, for example by mapping a color to the curve to emphasize the numerical values of the bending and twisting of the curve at different points.

Examples of analytic curves: We begin with a classic mathematical example, the torus knot. The parametric equation of the (p, q) torus knot can be written as given in Equation 20.1. As examples of complex curve shapes, we choose a $(2, 3)$ torus knot and a $(3, 5)$ torus knot, and plot the 3D curves with and without an attached Frenet-frame tubing (Figures 20.8 and 20.9). We can learn some useful facts about the curve by computing its curvature and torsion. Following the work of Gray [31, 61], we compute from Equations 20.15 and 20.16 the curvature and torsion for the $(2, 3)$ and $(3, 5)$ torus knots with size parameters $(r, a, b) = (8, 3, 5)$, and use these quantities to choose values from an appropriate color table. The resulting values of the curvature and torsion are plotted and color encoded onto the curves themselves in Figures 20.10 and 20.11.

20.6   THEORY OF QUATERNION FRAMES

It is awkward to represent moving frames visually in high-density data because a frame consists of three 3D vectors (or nine components) yet has only three independent degrees of freedom. Some approaches to representing these degrees of freedom in a 3D space were suggested, for example, by Alpern et al. [3]. We propose instead to systematically exploit the representation of 3D orientation frames in four dimensions using equivalent unit quaternions that correspond, in turn, to points on the three-sphere (see, e.g., Shoemake [149]). For example, a collection of oriented frames such as those of a crystal lattice can thus be represented by mapping their orientations to a point set in the 4D quaternion space. The moving frame of a 3D space curve can be transformed into a path in quaternion space corresponding pointwise to the 3D space curve.

The quaternion representation of rotations reexpressing a moving frame of a 3D space curve is an elegant unit four-vector field over the curve. The resulting quaternion frames can be displayed as curves in their own right, or can be used in combination with other methods to enrich the display of each 3D curve (e.g., by assigning a coded display color representing a quaternion component).

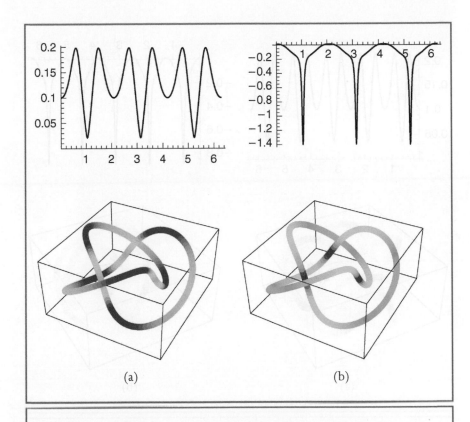

(a)   (b)

FIGURE 20.10 *A* $(2, 3)$ *torus knot with size parameters* $(r, a, b) = (8, 3, 5)$: *(a) the curvature plot and a color encoding and (b) the torsion plot and a color encoding.*

20.6.1 GENERIC QUATERNION FRAME EQUATIONS

Quaternions permit the nine matrix elements with six orthonormality constraints comprising a 3D orientation frame to be succinctly summarized in terms of four quaternion equations with the single constraint of unit length. Here we will derive the general features of these equations and how they are used to transform the standard 3D frame evolution equations into the more elegant quaternion form, which is essentially a square root of the conventional form.

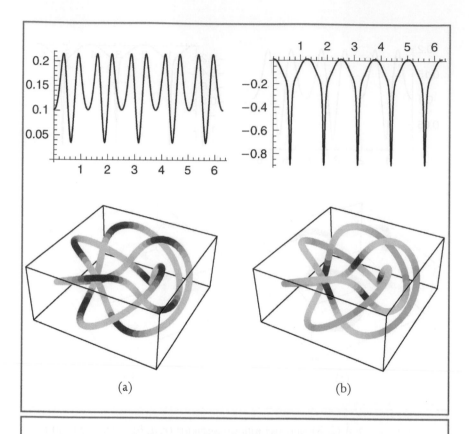

(a)   (b)

FIGURE 20.11 *A $(3, 5)$ torus knot with size parameters $(r, a, b) = (8, 3, 5)$: (a) the curvature plot and a color encoding and (b) the torsion plot and a color encoding.*

We begin with a common formula for the correspondence between 3×3 matrices R_{ij} and quaternions q:

$$\left[\mathbf{R}(q) \cdot \mathbf{V}\right]^i = \sum_j R_{ij} V_j = q \star (0, V_i) \star q^{-1}. \qquad (20.17)$$

[Recall that we use the notation $\star$ to distinguish quaternion multiplication, and will use the dot $(\cdot)$ when necessary to denote ordinary Euclidean inner products.] Next, we express each orthonormal frame component as a column of $\mathbf{R}$ by using an arbitrary quaternion to rotate each of the three Cartesian reference axes to a new, arbitrary orientation:

$$\hat{\mathbf{T}} = q \star (0, \hat{\mathbf{x}}) \star q^{-1},$$
$$\hat{\mathbf{N}}_1 = q \star (0, \hat{\mathbf{y}}) \star q^{-1},$$
$$\hat{\mathbf{N}}_2 = q \star (0, \hat{\mathbf{z}}) \star q^{-1}. \qquad (20.18)$$

(Technically speaking, in Equation 20.18 $\hat{\mathbf{T}}$ really means the quaternion $(0, \hat{\mathbf{T}})$ with only a vector part, and so on.) All of this can be transformed into the following explicit representation of the frame vectors as columns of a matrix of quaternion quadratic forms.

$$\begin{bmatrix} [\hat{\mathbf{T}}] & [\hat{\mathbf{N}}_1] & [\hat{\mathbf{N}}_2] \end{bmatrix}$$

$$= \begin{bmatrix} q_0^2 + q_1^2 - q_2^2 - q_3^2 & 2q_1q_2 - 2q_0q_3 & 2q_1q_3 + 2q_0q_2 \\ 2q_1q_2 + 2q_0q_3 & q_0^2 - q_1^2 + q_2^2 - q_3^2 & 2q_2q_3 - 2q_0q_1 \\ 2q_1q_3 - 2q_0q_2 & 2q_2q_3 + 2q_0q_1 & q_0^2 - q_1^2 - q_2^2 + q_3^2 \end{bmatrix}. \qquad (20.19)$$

Taking differentials of $q \star q^{-1} = (1, \mathbf{0})$, we generate expressions of the form

$$dq = q \star \left(q^{-1} \star dq \right) = q \star \frac{1}{2}(0, \mathbf{k}), \qquad (20.20)$$

$$dq^{-1} = \left(dq^{-1} \star q \right) \star q^{-1}$$
$$= -\left(q^{-1} \star dq \right) \star q^{-1}$$
$$= -\frac{1}{2} (0, \mathbf{k}) \star q^{-1}, \qquad (20.21)$$

where

$$\mathbf{k} = 2(q_0 \, \mathbf{dq} - \mathbf{q} \, dq_0 - \mathbf{q} \times \mathbf{dq}). \qquad (20.22)$$

Substituting these expressions into the calculation for the first column, we find

$$d\hat{\mathbf{T}} = dq \star (0, \hat{\mathbf{x}}) \star q^{-1} + q \star (0, \hat{\mathbf{x}}) \star dq^{-1}$$

$$= \frac{1}{2} q \star \big((0, \mathbf{k}) \star (0, \hat{\mathbf{x}}) - (0, \hat{\mathbf{x}}) \star (0, \mathbf{k}) \big) \star q^{-1}$$

$$= q \star (0, \mathbf{k} \times \hat{\mathbf{x}}) \star q^{-1}.$$

The rest of the columns are computed similarly, and a straightforward expansion of the components of the cross products proves the correspondence between Equations 20.4 and 20.20.

To relate the derivative to a specific curve coordinate system, for example, we would introduce the curve velocity normalization $v(t) = \|\mathbf{x}'(t)\|$ and write

$$q' = v(t)\frac{1}{2} q \star (0, \mathbf{k}). \tag{20.23}$$

One of our favorite ways of rewriting this equation follows directly from the full form for the quaternion multiplication rule. Because this multiplication can be written as an orthogonal matrix multiplication on the 4D quaternion space, we could equally well write

$$\begin{bmatrix} q_0' \\ q_1' \\ q_2' \\ q_3' \end{bmatrix} = v(t)\frac{1}{2} \begin{bmatrix} 0 & -k_x & -k_y & -k_z \\ +k_x & 0 & +k_z & -k_y \\ +k_y & -k_z & 0 & +k_x \\ +k_z & +k_y & -k_x & 0 \end{bmatrix} \cdot \begin{bmatrix} q_0 \\ q_1 \\ q_2 \\ q_3 \end{bmatrix}. \tag{20.24}$$

This is the 3D analog of Equation 19.6.

20.6.2 QUATERNION FRENET FRAMES

Using Equation 20.24, we can express all 3D coordinate frames in the form of quaternions. If we assume that the columns of Equation 20.24 are the vectors $(\mathbf{T}, \mathbf{N}, \mathbf{B})$, respectively, one can show from Equation 20.10 that $[q'(t)]$ takes the form (see Hanson [68])

$$\begin{bmatrix} q_0' \\ q_1' \\ q_2' \\ q_3' \end{bmatrix} = \frac{v}{2} \begin{bmatrix} 0 & -\tau & 0 & -\kappa \\ \tau & 0 & \kappa & 0 \\ 0 & -\kappa & 0 & \tau \\ \kappa & 0 & -\tau & 0 \end{bmatrix} \cdot \begin{bmatrix} q_0 \\ q_1 \\ q_2 \\ q_3 \end{bmatrix}. \qquad (20.25)$$

This equation has the following key properties.

- The matrix on the right-hand side is antisymmetric, so that $q(t) \cdot q'(t) = 0$ by construction. Thus, all unit quaternions remain unit quaternions as they evolve by this equation.
- The number of equations has been reduced from nine coupled equations with six orthonormality constraints to four coupled equations incorporating a single constraint that keeps the solution vector confined to the three-sphere.

We verify that the matrices

$$A = \begin{bmatrix} q_0 & q_1 & -q_2 & -q_3 \\ q_3 & q_2 & q_1 & q_0 \\ -q_2 & q_3 & -q_0 & q_1 \end{bmatrix},$$

$$B = \begin{bmatrix} -q_3 & q_2 & q_1 & -q_0 \\ q_0 & -q_1 & q_2 & -q_3 \\ q_1 & q_0 & q_3 & q_2 \end{bmatrix},$$

$$C = \begin{bmatrix} q_2 & q_3 & q_0 & q_1 \\ -q_1 & -q_0 & q_3 & q_2 \\ q_0 & -q_1 & -q_2 & q_3 \end{bmatrix}$$

explicitly reproduce Equation 20.10,

$$2[A] \cdot [q'] = \mathbf{T}' = v\kappa\mathbf{N},$$

$$2[B] \cdot [q'] = \mathbf{N}' = -v\kappa\mathbf{T} + v\tau\mathbf{B},$$

$$2[C] \cdot [q'] = \mathbf{B}' = -v\tau\mathbf{N},$$

where we have applied Equation 20.25 to obtain the right-hand terms. Just as the Frenet equations may be integrated to generate a unique moving frame with its space curve for nonvanishing $\kappa(t)$, we may integrate the much simpler quaternion Equations (20.25). (See Appendix H.)

20.6.3 QUATERNION PARALLEL TRANSPORT FRAMES

Similarly, a parallel transport frame system given by Equation 20.13 with $(\mathbf{N}_1, \mathbf{T}, \mathbf{N}_2)$ (in that order) corresponding to the columns of Equation 20.24 is completely equivalent to the following parallel transport quaternion frame equation for $[q'(t)]$.

$$
\begin{bmatrix} q_0' \\ q_1' \\ q_2' \\ q_3' \end{bmatrix} = \frac{v}{2} \begin{bmatrix} 0 & -k_2 & 0 & k_1 \\ k_2 & 0 & -k_1 & 0 \\ 0 & k_1 & 0 & k_2 \\ -k_1 & 0 & -k_2 & 0 \end{bmatrix} \cdot \begin{bmatrix} q_0 \\ q_1 \\ q_2 \\ q_3 \end{bmatrix}. \tag{20.26}
$$

Here, antisymmetry again guarantees that the quaternions remain constrained to the unit three-sphere. The correspondence to Equation 20.13 is verified as follows.

$$2[B] \cdot [q'] = \mathbf{T}' = vk_1\mathbf{N}_1 + vk_2\mathbf{N}_2,$$

$$2[A] \cdot [q'] = \mathbf{N}_1' = -vk_1\mathbf{T},$$

$$2[C] \cdot [q'] = \mathbf{N}_2' = -vk_2\mathbf{T}.$$

20.7   ASSIGNING SMOOTH QUATERNION FRAMES

Given a particular curve, we are next faced with the task of assigning quaternion values to whatever moving frame sequence we have chosen.

20.7.1 ASSIGNING QUATERNIONS TO FRENET FRAMES

The Frenet frame equations are pathological, for example, when the curve is perfectly straight for some distance or when the curvature vanishes momentarily. Thus, real numerical data for space curves will frequently exhibit behaviors that make the assignment of a smooth Frenet frame difficult, unstable, or impossible. In addition, because any given 3×3 orthogonal matrix corresponds to two quaternions that differ in sign, methods of deriving a quaternion from a Frenet frame are intrinsically ambiguous. Therefore, we prescribe the following procedure for assigning smooth quaternion Frenet frames to points on a space curve.

1   Select a numerical approach to computing the tangent $\mathbf{T}$ at a given curve point $\mathbf{x}$. This typically depends on the chosen curve model and the number of points one wishes to sample.

2   Compute the remaining numerical derivatives at a given point and use those to compute the Frenet frame according to Equation 20.11. If any critical quantities vanish, tag the frame as undefined (or as needing a heuristic fix).

3   Check the dot product of the previous binormal $\mathbf{B}(t)$ with the current value. If it is near zero, choose a correction procedure to handle this singular point. Among the correction procedures we have considered are (1) simply jump discontinuously to the next frame to indicate the presence of a point with very small curvature, (2) create an interpolating set of points and perform a geodesic interpolation [149], or (3) deform the curve slightly before and after the singular point to "ease in" with a gradual rotation of the frame or apply an interpolation heuristic (e.g., see Shani and Ballard [147]). Creating a jump in the frame assignment is our default choice, since it does not introduce any new information.

4   Apply a suitable algorithm (e.g., see Chapter 16) to compute a candidate for the quaternion corresponding to the Frenet frame.

5   If the 3×3 Frenet frame is smoothly changing, make one last check on the 4D inner product of the quaternion frame with its own previous value. If there is a sign change, choose the opposite sign to keep the quaternion smoothly changing (this will have no effect on the corresponding 3×3 Frenet frame). If this inner product is near zero instead of ±1, you have detected a radical change in the Frenet frame that should have been noticed in the previous tests.

6   If the space curves of the data are too coarsely sampled to give the desired smoothness in the quaternion frames but are still close enough to give consistent qualitative behavior, one may choose to smooth out the intervening frames using the desired level of recursive SLERPing [145,149] to get smoothly splined intermediate quaternion frames. (More details can be found in Chapter 25.)

In Figure 20.12, we plot an example of a torus knot (a smooth space curve with everywhere nonzero curvature) together with its associated Frenet frames, its quaternion frame values, and the path of its quaternion frame field projected from four-space.

Figure 20.13 plots the same information, but this time for a curve with a discontinuous frame that flips instantly at a zero-curvature point. This space curve has two planar parts drawn as though on separate pages of a partly open book and meeting smoothly on the "crack" between pages. We see the obvious jump in the Frenet and quaternion frame graphs at the meeting point. If the two curves are joined by a long straight line, the Frenet frame is ambiguous and is essentially undefined in this segment. Rather than invent an interpolation, we generally prefer to use the parallel transport method described in the following section.

20.7.2 ASSIGNING QUATERNIONS TO PARALLEL TRANSPORT FRAMES

To determine the quaternion frames of an individual curve using the parallel transport method, we follow a similar, but distinct, procedure.

1 Select a numerical approach to assigning a tangent at a given curve point as usual.
2 Assign an initial reference orientation to the initial point on the curve, in the plane perpendicular to the initial tangent direction. The entire set of frames will be displaced from the origin in quaternion space by the corresponding value of this initial orientation matrix, but the shape of the entire curve will be the same regardless of the initial choice. This choice is intrinsically ambiguous and application dependent. However, one appealing strategy is to base the initial frame on the first well-defined Frenet frame, and then proceed from there using the parallel transport frame evolution. This guarantees that identical curves have the same parallel transport frames.
3 Compute the angle between successive tangents, and rotate the frame by this angle in the plane of the two tangents to get the next frame value.
4 If the curve is straight, the algorithm automatically makes no changes.
5 Compute a candidate quaternion representation for the frame, applying consistency conditions (such as forcing positive signs of neighboring quaternion dot products) as needed.

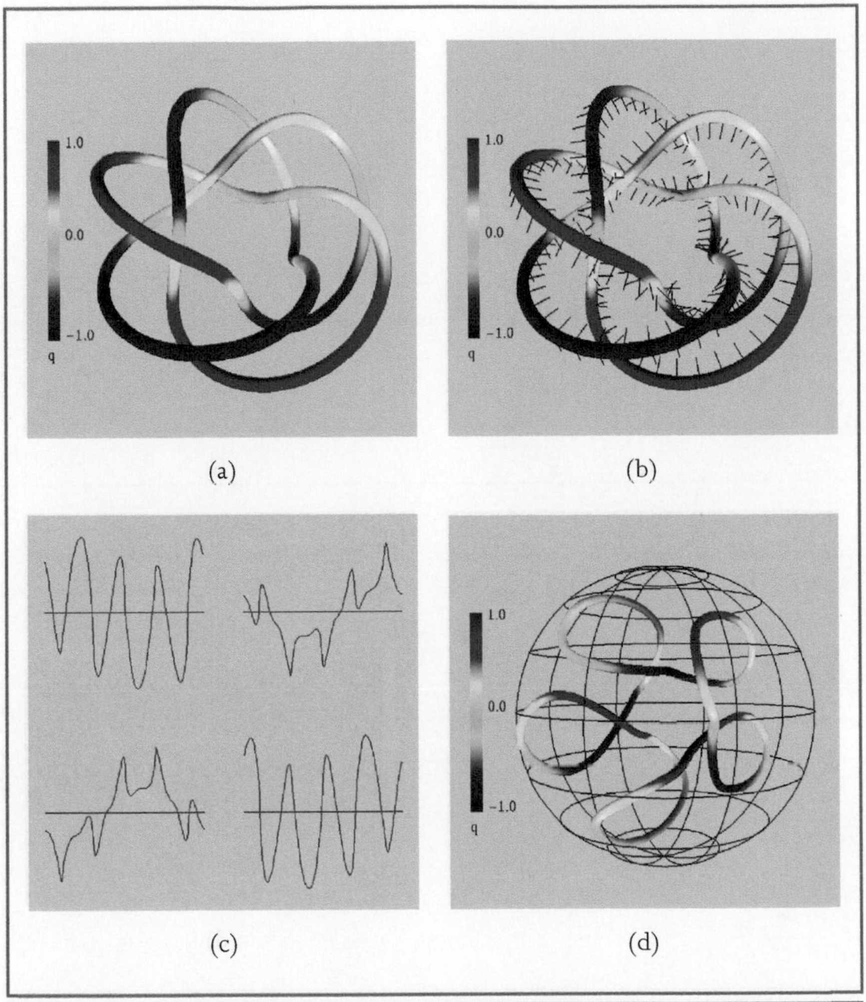

FIGURE 20.12 (a) Projected image of a 3D $(3, 5)$ torus knot. (b) Selected Frenet frame components displayed along the knot. (c) The corresponding smooth quaternion frame components, with q_0, q_1 in the top row and q_2, q_3 in the bottom row. (d) The path of the quaternion frame components in the three-sphere projected from four-space. Color scales indicate the zeroth component of the curve's four-vector frame (upper left-hand graph in c).

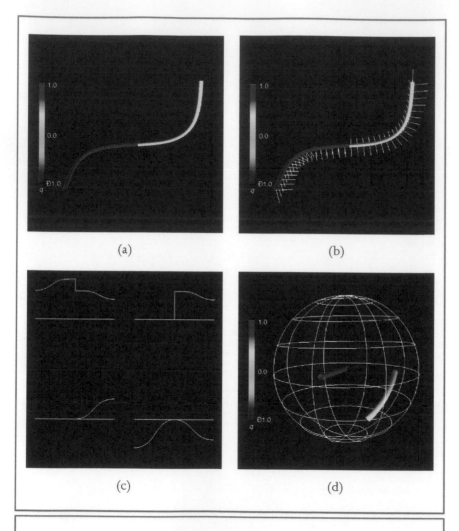

FIGURE 20.13 (a) *Projected image of a pathological curve segment.* (b) *Selected Frenet frame components, showing a sudden change of the normal.* (c) *The quaternion frame components, with* q_0, q_1 *in the top row and* q_2, q_3 *in the bottom row, showing an unacceptable discontinuity in frame values.* (d) *The discontinuous path of the quaternion frame components in the three-sphere. Color scales indicate the zeroth component of the curve's four-vector frame (upper left-hand graph in c).*

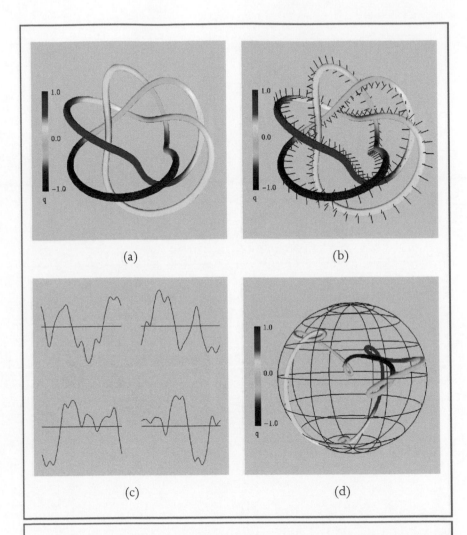

FIGURE 20.14 (a) Projected image of a 3D $(3, 5)$ torus knot. (b) Selected parallel transport frame components displayed along knot. (c) The corresponding smooth quaternion frame components. (d) The path of the quaternion frame components in the three-sphere projected from four-space. Color scales indicate the zeroth component of the curve's four-vector frame (upper left-hand graph in c).

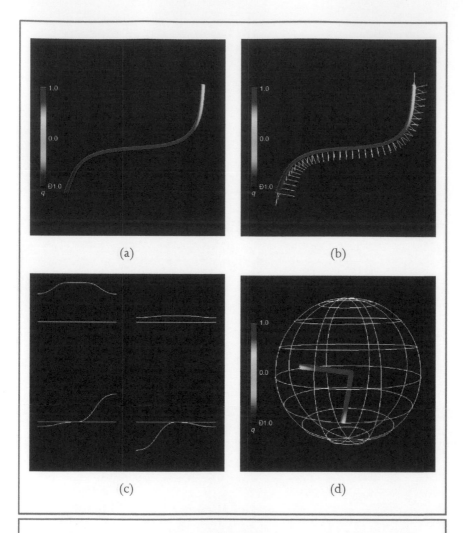

(a)

(b)

(c)

(d)

FIGURE 20.15 (a) *Projected image of a pathological curve segment. (b) Selected parallel transport frame components, showing smooth change of the normal. (c) The quaternion frame components, showing continuity in values. (d) The continuous path of the quaternion frame components in the three-sphere. Color scales indicate the zeroth component of the curve's four-vector frame (upper left-hand graph in c).*

Note that the initial reference orientation and all discrete rotations can be represented *directly in terms of quaternions*, and thus quaternion multiplication can be used directly to apply frame rotations. Local consistency is then automatic.

An example is provided in Figure 20.14, which shows the parallel transport analog of Figure 20.12 for a torus knot. Figure 20.15 is the parallel transport analog of the pathological case shown in Figure 20.13, but this time the frame is continuous when the curvature vanishes.

20.8   EXAMPLES: TORUS KNOT AND HELIX QUATERNION FRAMES

The torus knot and the helix described previously (Equations 20.1 and 20.2) and shown in Figure 20.2 can now be studied in quaternion space using each of the representative frames. The comparison of each of our standard frames (Frenet, Bishop/parallel-transport, and geodesic reference) is depicted in Figure 20.16 for each of the curves.

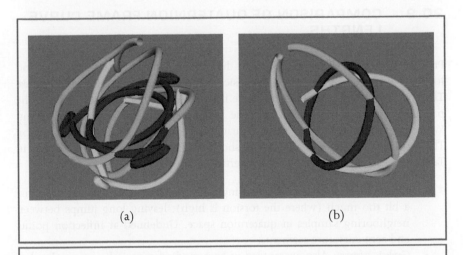

(a)   (b)

FIGURE 20.16 Frenet (green), parallel transport (cyan), and geodesic reference (red) quaternion frames in "standard" 3D vector projection. (a) For the (2, 3) torus knot. (b) For the helix.

Labeled Red in Figure 20.16 is the geodesic reference framing. This is planar by construction, in that all 3D points must lie in the plane perpendicular to the reference axis. The 3D origin is at the centroid of the red curve. Labeled Green in the figure is the Frenet frame. The Frenet frame is actually cyclic, but to see this easily for this $(2, 3)$ torus knot the mirror image of the current frame would need to be added, giving effectively a double traversal of the curve. Finally, we show the quaternion path (labeled Cyan) of the parallel transport frame. The parallel transport frame must be given a starting value, which here is chosen to coincide with the starting Frenet frame (at the top center of the image). The parallel transport frame is not cyclic, but is the shortest path, with three very noticeable tight loops. The same selection of quaternion frames is shown also for the helix. Again, the Red geodesic reference curve is planar (and cycles back on itself twice for this helix). The Green Frenet frame takes a longer path that will return to its original orientation, and the Cyan parallel transport frame, seen starting at the same orientation as the Frenet frame, will not ordinarily return to the same orientation but will have the shortest 4D path length. (The hidden double circuit of the geodesic reference frame for this helix in fact makes it longer.)

20.9   COMPARISON OF QUATERNION FRAME CURVE LENGTHS

Previously (Figure 20.4) we compared the tubings for the $(2, 3)$ torus knot and for the helix based on the Frenet, geodesic reference, and parallel transport frames. The corresponding quaternion paths are illustrated together in Figure 20.16. The parallel transport frame shown uses the initial Frenet frame as a starting point. We could, however, use any starting quaternion with the correct tangent vector. The relative path lengths of the curves shown in Figure 20.16 are summarized in Table 20.1. We note the following properties.

- *Frenet*: Periodic for periodic nonsingular curves and has a tendency to twist a bit too much (where the torsion is high), leaving long jumps between neighboring samples in quaternion space. Undefined at inflection points and zero curvature segments.
- *Geodesic reference*: Also guaranteed to be periodic for periodic curves, but has the odd property that it always lies in a plane perpendicular to the reference axis in our preferred 3D quaternion projection. Ambiguous and therefore potentially not smooth for frames opposing the reference frame direction.

Curve Lengths	(2, 3) Torus Knot	Helix
Frenet frame	14.3168	6.18501
Geodesic reference frame	14.6468	7.82897
Parallel transport frame	10.1865	6.06301

TABLE 20.1 *Relative lengths (in radians) of the quaternion frame maps for various frame choices describing the (2, 3) torus knot and the helix. The parallel transport frame is the shortest possible frame map.*

- *Parallel transport:* This is the quaternion frame with minimal 4D length, though it may be difficult to see this feature immediately in our standard projection. It is not in general a periodic path. Different choices of starting frame produce curves of identical length differing by rigid 4D motions.

3D Surfaces

21

Classical 3D differential geometry deals with the properties of surfaces as well as curves. The various tools used to analyze curves become more complex when we turn to surfaces. The role of the tangent is taken by the surface normal, and the entire treatment becomes more involved. From our point of view, we will focus from the outset not only on the surface normal but on the entire frame at each point, leading ultimately to an alternative quaternion treatment for the entire subject of surface geometry.

21.1   INTRODUCTION TO 3D SURFACES

This section contains a brief outline of the features of classical differential geometry of surfaces that we will require to proceed to our quaternion treatment. For more details, we refer the reader to Eisenhart [46] or Gray [61], or to any other traditional differential geometry treatment.

If we are given a surface patch $\mathbf{x}(u, v)$ with some set of nondegenerate coordinates (u, v), we may determine the normals at each point by computing

$$\mathbf{N}(u, v) = \mathbf{x}_u \times \mathbf{x}_v, \tag{21.1}$$

where $\mathbf{x}_u = \partial \mathbf{x}/\partial u$ and $\mathbf{x}_v = \partial \mathbf{x}/\partial v$. For surfaces defined numerically in terms of vertices and triangles, we would choose a standard procedure (such as averaging the normals of the faces surrounding each vertex) to determine the vertex normal. Alternatively, if we have an implicit surface described by the level-set function

$f(\mathbf{x}) = 0$, the normals may be computed directly from the gradient at any point $\mathbf{x}$ satisfying the level set equation

$$\mathbf{N}(\mathbf{x}) = \nabla f(\mathbf{x}).$$

The normalized normal is defined as usual by $\hat{\mathbf{N}} = \mathbf{N}/\|\mathbf{N}\|$.

For 3D curves, the geometry of the curve determines the tangent vector $\hat{\mathbf{T}}$ and leaves a pair of normal vectors $(\hat{\mathbf{N}}_1, \hat{\mathbf{N}}_2)$, with one extra degree of freedom to be determined in the total frame $[\hat{\mathbf{T}} \quad \hat{\mathbf{N}}_1 \quad \hat{\mathbf{N}}_2]$. The analogous observation for surfaces is that the geometry fixes the *normal* $\hat{\mathbf{N}}$ at each surface point, leaving a pair of *tangent* vectors $(\hat{\mathbf{T}}_1, \hat{\mathbf{T}}_2)$, with one extra degree of freedom to be determined in the total surface frame:

$$\text{Surface Frame} = [\hat{\mathbf{T}}_1 \quad \hat{\mathbf{T}}_2 \quad \hat{\mathbf{N}}]. \tag{21.2}$$

When a (u, v) surface parameterization is available, the surface partial derivatives $\mathbf{x}_u$ and $\mathbf{x}_v$ can in principle be used to assign a frame $[\hat{\mathbf{T}}_1 \quad \hat{\mathbf{T}}_2 \quad \hat{\mathbf{N}}]$ (using Gram–Schmidt orthonormalization methods if $\mathbf{x}_u \cdot \mathbf{x}_v \neq 0$), but there is no reason to believe that this frame has any special properties in general. In practice, it is extremely convenient to define a rectangular mesh on the surface patch, and a grid parameterized by (u, v) typically serves this purpose.

21.1.1 CLASSICAL GAUSS MAP

We have seen in Chapter 20 examples of maps and their applications to space curves. The traditional surface analog of the tangent map of a curve is the Gauss map, which takes a selection of points on the surface, typically connected by a mesh of some sort, and associates to each point its normalized surface normal. The Gauss map is then the plot of each of these normals in the coordinate system of a unit sphere $\mathbf{S}^2$ in $\mathbb{R}^3$. The Gauss map is guaranteed to be unique in some sufficiently small open set of each point of a regular surface, but may be arbitrarily multiple valued for the entire surface. Note also that many nearby surface points can be mapped to a single point in the Gauss map (e.g., for certain types of planar curves in the surface or for a planar area patch).

Figure 21.1 shows a coordinate mesh on an ellipsoidal surface and its single-valued Gauss map, as well as a quarter of a torus and its Gauss map. The Gauss map of the entire torus would cover the sphere twice, and there are two entire circles on

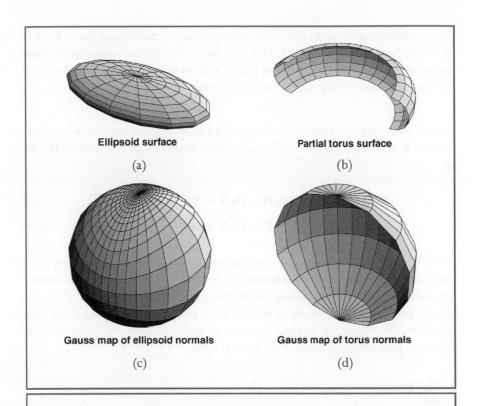

FIGURE 21.1 *Examples of Gauss maps. (a) An ellipsoid and (b) a portion of a torus. (c and d) The corresponding standard Gauss maps projecting the normal vectors of each surface point onto the sphere. Patches with coincident normals (e.g., for the full torus) would overlap in this representation.*

the torus that correspond to single points (the North and South Poles) in the Gauss map.

21.1.2 SURFACE FRAME EVOLUTION

The equations for the evolution of a surface frame follow the same basic structure as those of a space curve, except that the derivatives are now directional, with two linearly independent degrees of freedom corresponding to the tangent basis $(\hat{\mathbf{T}}_1, \hat{\mathbf{T}}_2)$ in the surface. Typically (see [46,61]), one assumes a not-necessarily-orthogonal parameterization (u, v) that permits one to express the tangent space in terms of the partial derivatives $(\mathbf{x}_u, \mathbf{x}_v)$, giving the normals $\hat{\mathbf{N}}(u, v)$ of Equation 21.1.

Standard differential geometry notation: In the standard approach, one writes the local curvatures K and H in terms of any linearly independent pair of vector fields $(\mathbf{U}, \mathbf{V})$ as

$$D_{\mathbf{U}}\hat{\mathbf{N}} \times D_{\mathbf{V}}\hat{\mathbf{N}} = K(\mathbf{U} \times \mathbf{V}), \tag{21.3}$$

$$D_{\mathbf{U}}\hat{\mathbf{N}} \times \mathbf{V} + \mathbf{U} \times D_{\mathbf{V}}\hat{\mathbf{N}} = 2H(\mathbf{U} \times \mathbf{V}). \tag{21.4}$$

With $\mathbf{U} = \mathbf{x}_u \cdot \nabla$ and $\mathbf{V} = \mathbf{x}_v \cdot \nabla$, we get the classical expressions. As Gray [61] succinctly notes, because all derivatives of $\hat{\mathbf{N}}$ are perpendicular to $\hat{\mathbf{N}}$, the entire apparatus amounts to constructing the tangent map of the Gauss map.

If we try to build the geometry of surfaces from a parametric representation, each directional derivative has a vector equation of the form taken by Equation 20.4. Thus, we may write equations of the general form

$$\frac{\partial}{\partial u}\begin{bmatrix}\hat{\mathbf{T}}_1(u, v) & \hat{\mathbf{T}}_2(u, v) & \hat{\mathbf{N}}(u, v)\end{bmatrix}$$

$$= \begin{bmatrix}\hat{\mathbf{T}}_1(u, v) & \hat{\mathbf{T}}_2(u, v) & \hat{\mathbf{N}}(u, v)\end{bmatrix}\begin{bmatrix} 0 & -a_z(u, v) & +a_y(u, v) \\ +a_z(u, v) & 0 & -a_x(u, v) \\ -a_y(u, v) & +a_x(u, v) & 0 \end{bmatrix}$$

$$\tag{21.5}$$

and

$$\frac{\partial}{\partial v}\begin{bmatrix}\hat{\mathbf{T}}_1(u, v) & \hat{\mathbf{T}}_2(u, v) & \hat{\mathbf{N}}(u, v)\end{bmatrix}$$

$$= \begin{bmatrix}\hat{\mathbf{T}}_1(u, v) & \hat{\mathbf{T}}_2(u, v) & \hat{\mathbf{N}}(u, v)\end{bmatrix}\begin{bmatrix} 0 & -b_z(u, v) & +b_y(u, v) \\ +b_z(u, v) & 0 & -b_x(u, v) \\ -b_y(u, v) & +b_x(u, v) & 0 \end{bmatrix}.$$

$$\tag{21.6}$$

The last lines of each of Equations 21.5 and 21.6 are typically combined in textbook treatments to give

$$\left[\frac{\partial \hat{\mathbf{N}}(u, v)}{\partial u} \quad \frac{\partial \hat{\mathbf{N}}(u, v)}{\partial v} \right] = \left[\hat{\mathbf{T}}_1(u, v) \quad \hat{\mathbf{T}}_2(u, v) \right][\mathcal{K}], \qquad (21.7)$$

where the matrix $[\mathcal{K}]$ has eigenvalues that are the principal curvatures k_1 and k_2, and thus

$$K = \det[\mathcal{K}] = k_1 k_2 \qquad (21.8)$$

is the Gaussian curvature and

$$H = \frac{1}{2} \operatorname{tr}[\mathcal{K}] = \frac{1}{2}(k_1 + k_2) \qquad (21.9)$$

is the mean curvature.

21.1.3 EXAMPLES OF SURFACE FRAMINGS

If we are given a description of a surface, we can compute normals and choices of the corresponding frames by various means. Figure 21.2 illustrates three of these for the sphere. The first is derived from the standard orthonormal polar coordinate system, and the second is the extension to surfaces of the geodesic reference frame, which assigns the frame closest to a standard reference axis at the North Pole. The third is a frame based on polar projective coordinates for the sphere. We choose explicitly the parameterization of the South Pole inverse map of Equations 17.7 and 17.8, which yields

$$x(u, v) = \frac{2u}{1 + u^2 + v^2},$$

$$y(u, v) = \frac{2v}{1 + u^2 + v^2},$$

$$z(u, v) = \frac{1 - u^2 - v^2}{1 + u^2 + v^2}. \qquad (21.10)$$

This choice maps the real plane into the unit sphere, with $x^2 + y^2 + z^2 = 1$, except for the point at infinity corresponding to the South Pole. In fact, the polar projective coordinates generate the same assignments as the geodesic reference frame does,

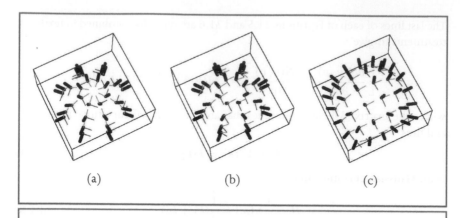

FIGURE 21.2 Examples of frame choices for the upper portion of an ordinary sphere. (a) Frames derived from standard polar coordinates on sphere. (b) Geodesic reference frame for the sphere. Each frame is as close as possible to the canonical coordinate axes at the North Pole. (c) Frames derived from projective coordinates on the sphere, which turn out to be the same frame field as the geodesic reference frame.

and thus except for the difference in locations of the grid sampling these are the same framings.

Note: Do not be confused by alternate *samplings* of the same framings. If a parameterization $\mathbf{x}(u, v)$ gives a frame with $\mathbf{T}_1 = \partial \mathbf{x}(u, v)/\partial u$ and $\mathbf{T}_2 = \partial \mathbf{x}(u, v)/\partial v$, we can change to a polar sampled mesh—$(r = (u^2 + v^2)^{1/2}, \theta = \arctan(v/u))$—yet still retain the same frames at the same points $\mathbf{x}(r, \theta) = \mathbf{x}(u = r \cos \theta, v = r \sin \theta)$. From these basic concepts, we can now proceed to a full quaternion treatment of surface frames that closely follows the quaternion curve framing methods introduced in Chapter 20.

21.2   QUATERNION WEINGARTEN EQUATIONS

21.2.1   QUATERNION FRAME EQUATIONS

Our task is now to rephrase the general properties of curve and surface frames in quaternion language. Ultimately, this will provide us with a well-defined space in which to consider optimizing frame assignments. We begin with the standard definition (Equation 6.13) for the correspondence between 3×3 matrices R^i_j and quaternions q:

$$R_q(\mathbf{V})^i = \sum_j R^i_j V^j = q \star (0, V^i) \star q^{-1}. \qquad \text{(21.11)}$$

Following the pattern of our treatment of curve framings, we express each orthonormal frame component as a column of R^i_j by using an arbitrary quaternion to rotate each of the three Cartesian reference axes to a new, arbitrary orientation:

$$\hat{\mathbf{T}}_1 = q \star (0, \hat{\mathbf{x}}) \star q^{-1},$$
$$\hat{\mathbf{T}}_2 = q \star (0, \hat{\mathbf{y}}) \star q^{-1},$$
$$\hat{\mathbf{N}} = q \star (0, \hat{\mathbf{z}}) \star q^{-1}, \qquad \text{(21.12)}$$

where $\hat{\mathbf{N}}$ really means the quaternion $(0, \hat{\mathbf{N}})$, as usual. All of this can be transformed into the following explicit representation of the frame vectors as columns of a matrix of quaternion quadratic forms.

$$\begin{aligned}
&[[\hat{\mathbf{T}}_1] \quad [\hat{\mathbf{T}}_2] \quad [\hat{\mathbf{N}}]] \\
&= \begin{bmatrix} q_0^2 + q_1^2 - q_2^2 - q_3^2 & 2q_1q_2 - 2q_0q_3 & 2q_1q_3 + 2q_0q_2 \\ 2q_1q_2 + 2q_0q_3 & q_0^2 - q_1^2 + q_2^2 - q_3^2 & 2q_2q_3 - 2q_0q_1 \\ 2q_1q_3 - 2q_0q_2 & 2q_2q_3 + 2q_0q_1 & q_0^2 - q_1^2 - q_2^2 + q_3^2 \end{bmatrix}.
\end{aligned}$$
$$\text{(21.13)}$$

Taking differentials and again using the curve velocity normalization $v(t) = \|\mathbf{x}'(t)\|$ we can write the quaternion differential as

$$q' = v(t) \frac{1}{2} q \star (0, \mathbf{k}), \qquad \text{(21.14)}$$

or equivalently as

$$\begin{bmatrix} q_0' \\ q_1' \\ q_2' \\ q_3' \end{bmatrix} = v(t) \frac{1}{2} \begin{bmatrix} 0 & -k_x & -k_y & -k_z \\ +k_x & 0 & +k_z & -k_y \\ +k_y & -k_z & 0 & +k_x \\ +k_z & +k_y & -k_x & 0 \end{bmatrix} \cdot \begin{bmatrix} q_0 \\ q_1 \\ q_2 \\ q_3 \end{bmatrix}. \qquad \text{(21.15)}$$

This is the 3D analog of Equation 19.6. With

$$\mathbf{k} = 2(q_0 \mathbf{dq} - \mathbf{q} dq_0 - \mathbf{q} \times \mathbf{dq}),$$

we find

$$d\hat{\mathbf{T}}_1 = q \star (0, \mathbf{k} \times \hat{\mathbf{x}}) \star q^{-1},$$

$$d\hat{\mathbf{T}}_2 = q \star (0, \mathbf{k} \times \hat{\mathbf{y}}) \star q^{-1},$$

$$d\hat{\mathbf{N}} = q \star (0, \mathbf{k} \times \hat{\mathbf{z}}) \star q^{-1}.$$

21.2.2 QUATERNION SURFACE EQUATIONS (WEINGARTEN EQUATIONS)

At this point, there are many other directions we could carry this basic structure, but we will not pursue the general theory of quaternion differential geometry further here. We will conclude with a short summary of the quaternion treatment of the classical surface equations. Starting from Equation 21.14, we are led immediately to the quaternion analogs of Equations 21.5 and 21.6:

$$q_u \equiv \partial q / \partial u = \frac{1}{2} q \star (0, \mathbf{a}), \qquad (21.16)$$

$$q_v \equiv \partial q / \partial v = \frac{1}{2} q \star (0, \mathbf{b}). \qquad (21.17)$$

But how shall we express the curvatures in a way similar to the classical formula of Equation 21.7? An elegant form follows by pursuing the quaternion analog of the vector field equations given in Equations 21.3 and 21.4. We write

$$q_u \star q_v^{-1} = -\frac{1}{4} q \star (0, \mathbf{a}) \star (0, \mathbf{b}) \star q^{-1}$$

$$= -\frac{1}{4} q \star (-\mathbf{a} \cdot \mathbf{b}, \mathbf{a} \times \mathbf{b}) \star q^{-1}$$

$$= -\frac{1}{4} \big[-\mathbf{a} \cdot \mathbf{b} \hat{\mathbf{I}} + (\mathbf{a} \times \mathbf{b})_x \hat{\mathbf{T}}_1 + (\mathbf{a} \times \mathbf{b})_y \hat{\mathbf{T}}_2 + (\mathbf{a} \times \mathbf{b})_z \hat{\mathbf{N}} \big],$$

$$(21.18)$$

where we use the quaternion forms in Equation 21.12 with the addition of the quaternion identity element $\hat{\mathbf{I}} = (1, \mathbf{0}) = q \star (1, \mathbf{0}) \star q^{-1}$ for the frame vectors.

We see that the projection to the normal direction gives precisely the determinant $(\mathbf{a} \times \mathbf{b})_z = K$ identified in Equation 21.8 as the scalar curvature. The mean curvature follows from an expression similar to that of Equation 21.4:

$$q \star (0, \hat{\mathbf{x}}) \star q_u^{-1} + q \star (0, \hat{\mathbf{y}}) \star q_v^{-1}$$

$$= -\frac{1}{2}q \star (-\hat{\mathbf{x}} \cdot \mathbf{a} - \hat{\mathbf{y}} \cdot \mathbf{b}, \hat{\mathbf{x}} \times \mathbf{a} + \hat{\mathbf{y}} \times \mathbf{b}) \star q^{-1}$$

$$= -\frac{1}{2}\left[-(a_x + b_y)\hat{\mathbf{I}} + b_z\hat{\mathbf{T}}_1 - a_z\hat{\mathbf{T}}_2 + (a_y - b_x)\hat{\mathbf{N}} \right]. \qquad (21.19)$$

Here, the coefficient of the normal, $(a_y - b_x) = \mathrm{tr}[\mathcal{K}] = 2H$, is again the desired expression. Similar equations can be phrased directly in the 4D quaternion manifold using the forms of Equation 21.15.

21.3   QUATERNION GAUSS MAP

The quaternion Gauss map extends the Gauss map to include a representation of the entire coordinate frame at each surface point, introducing a number of new issues. In particular, there is a useful but mathematically suspect approach (which we might call an "engineering" approach) to the quaternion Gauss map that lets us quickly get informative visualizations for those special cases in which we are given a locally orthogonal parameterization of the surface except perhaps for isolated singularities of the coordinate system.

For these cases, we may construct the precise quaternion analog of the Gauss map by lifting the surface's coordinate mesh into the space of quaternions at each value of the orthonormal coordinatization (u, v) of the surface or surface patch. The correspondence of this map to the Gauss map is *not* directly visible, because (see Equation 21.13) the normal directions of the Gauss map are nontrivial quadratic forms constructed from all quaternion components. However, a projection to a subspace of the quaternion space based on the bilinear action of quaternions on pure vectors may be constructed by imitating the projection of the Hopf fibration of $\mathbf{S}^3$ (e.g., see, Berger [16] and Shoemake [152]). In Figure 21.3, we show two such cases, an ellipsoid with orthonormal polar coordinates singular at the poles and a torus with global, nonsingular coordinates, using our now-standard projections of the quaternion Gauss map to 3D. In each of these cases, a single circuit of the surface generates only half the quaternion surface shown. The symmetric quaternion figure results from traversing the surface twice to adjoin the reflected image of the

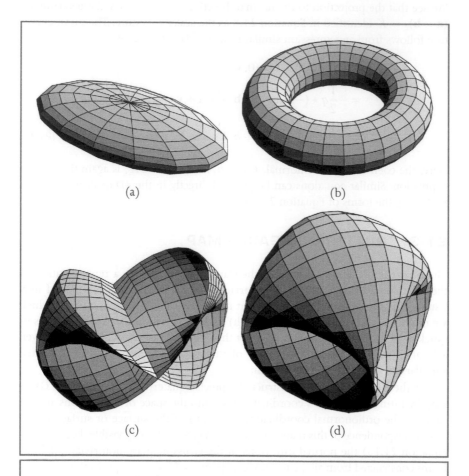

(a)

(b)

(c)

(d)

FIGURE 21.3 Examples of quaternion Gauss maps for surfaces. (a) The ellipsoid and (b) the torus. (c and d) The corresponding quaternion Gauss maps, projected from the three-sphere in 4D. The equatorial direction has been traversed twice in order to get a closed path in the map. The singular poles in the ellipsoid coordinate system correspond to the edges or boundaries of the quaternion-space ribbon.

single-circuit quaternion surface. That is, each point on the 3D surfaces appears twice, once at q, and once at $-q$, in these periodic quaternion Gauss maps.

We see that the singular coordinate system typically used for the ellipsoid is topologically a cylinder. The circles corresponding to the singularities of the coordinate system (circles of normal directions) at the North and South poles correspond to *boundaries* of the quaternion Gauss map. The torus, which has the extremely unusual feature that it possesses a global regular coordinate system, has a (reflection-doubled) quaternion Gauss map that is another 4D torus embedded in the quaternion $\mathbf{S}^3$ space.

21.4   EXAMPLE: THE SPHERE

21.4.1   QUATERNION MAPS OF ALTERNATIVE SPHERE FRAMES

Figure 21.2 showed three alternate sets of frames for the upper half of an ordinary sphere. The assigned coordinate systems may be converted directly into quaternion frames and coerced into consistency in the usual manner. Figure 21.4 shows the results. The geodesic reference frames and the projective coordinates are in fact the same space of frames computed in different ways (both are planes perpendicular to the $\hat{\mathbf{z}}$ axis). The coordinate systems used to compute the quaternion Gauss maps in a and b of the figure are commensurate, and thus we may compare the areas, computed using solid angle on the three-sphere in units of steradians. The results are outlined in Table 21.1.

21.4.2   COVERING THE SPHERE AND THE GEODESIC REFERENCE FRAME SOUTH POLE SINGULARITY

The geodesic reference frame for a surface patch has the peculiarity that it has an ambiguity whenever the vector to be assigned is exactly opposite the reference frame. As we show in Figure 21.5, the tilting from the reference frame in quaternion space (easily seen in ordinary 3D space as well) eventually reaches a quaternion circle representing the ambiguous orientation of the frame with reference direction along the $-\hat{\mathbf{z}}$ axis. This phenomenon is a practical consequence of the fact that the two-sphere does not admit a global vector field. According to classical

Patch Areas	Hemispherical Patch
Polar coordinates	2.1546
Geodesic reference frame	1.9548

TABLE 21.1 *Areas (in steradians) of the quaternion frame maps for the polar coordinate and geodesic reference frame choices on the hemispherical patches of Figure 21.4.*

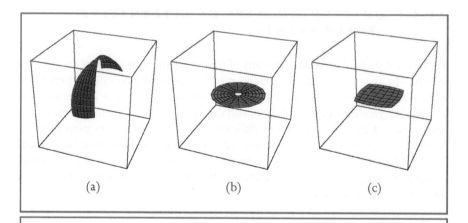

(a)   (b)   (c)

FIGURE 21.4 *Examples of quaternion Gauss maps for the frame choices for the upper portion of an ordinary sphere shown originally in Figure 21.2. (a) Frames derived from standard polar coordinates on sphere. (b) Geodesic reference frame for the sphere. Each frame is as close as possible to the canonical coordinate axes at the North Pole. (c) Frames derived from projective coordinates on the sphere.*

manifold theory (e.g., see, Milnor [123] or Grimm and Hughes [63]), one needs at least two separate patches, one for the North Pole and one for the South Pole,

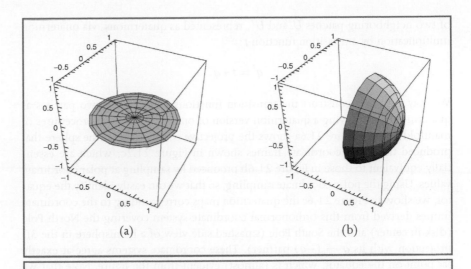

(a)   (b)

FIGURE 21.5 *The geodesic reference frame tilts to an ambiguous result as the tilt angle approaches* π*, the inverted direction of the chosen reference frame. We see two different 3D projections of the quaternion surface: (a) giving the vector coordinates* (q_1, q_2, q_3) *and (b) the coordinates* (q_0, q_1, q_2)*. The center is the North Pole, the middle ring is the equator, and outer circle is in fact the space of possible frames at the South Pole of the sphere. There is no unique way to tilt the North Pole to the orientation of the South Pole, as there is a full circle of arbitrariness in the choice.*

to place a complete set of coordinates (or equivalently, for our problem, a set of frames) on a sphere.

The more mathematical approach requires that interesting surfaces be defined as a collection of patches [45,63], and the spaces of frames for each patch must be matched up and sewn together by assigning a transition function along the boundaries. There are a variety of ways one can approach the problem of taking a manifold and associating fiber bundles with it. The most relevant fiber bundle for the context of the current problem is the *space of moving frames* of the space $\mathbb{R}^3$ in which the surface is embedded [45,156]. We in fact move as usual from the space of frames to the space of associated quaternions. Then at each point **x** of a patch we have frames that are functions from the patch into the topological space $\mathbf{S}^3$ of quaternion frames. We can express the relationship between the frames q and q'

of two neighboring patches U and U', represented as quaternions, via quaternion multiplication by a transition function t:

$$q' = t \star q.$$

We may explicitly construct the transition functions between the two patches as quaternion maps, giving a quaternion version of one of the classical procedures of manifold theory. Figure 21.6a shows the projective coordinates on the sphere that produced the set of coordinate frames shown in Figure 21.2c, which are essentially equivalent to those in Figure 21.6b produced by sampling at polar coordinate values. Using the polar coordinate sampling, so that we can easily identify the equator, we show in Figure 21.6c the quaternion maps corresponding to the coordinate frames derived from this orthonormal coordinate system covering the North Pole (disk in center) and the South Pole (smashed side view of a hemisphere in the 3D projection with its $q \to (-q)$ partner). These coordinate systems agree at exactly one point on the equator, which is (almost) evident from the figure. Note that we have chosen to display the coordinate systems only up to the equator, unlike the patches of Figure 21.5, which cover the entire sphere except for one pole.

To establish a mapping covering the complete sphere, we must write down an explicit correspondence between the quaternion frames for each patch at each shared point on the equator. Figure 21.7a shows the geodesic arcs on $\mathbf{S}^3$ symbolizing the transition rotation

$$t(\theta) = q_{\text{south}}(\theta) \star q_{\text{north}}^{-1}(\theta)$$

at each point on the equatorial circle parameterized by θ. Note carefully the order of quaternion multiplication. With our conventions, a different order will not work. The arcs themselves are actually segments of the space of possible frames, in that the simplest rotation between two frames with the same normal (at the same point on the equator) is a geodesic rotation about that normal. Figures 21.7b and 21.7c show the transition functions $q(\theta)$ sampled at regular intervals in θ and referred to the origin $(1, 0, 0, 0)$ in quaternion space. Each quaternion point at the end of an arc represents a rotation to be applied to a point on the North Pole patch equator to obtain the coordinate frame at the corresponding point on the South Pole patch equator. One point is the identity, and there is some degeneracy due to reflection symmetry across the equator.

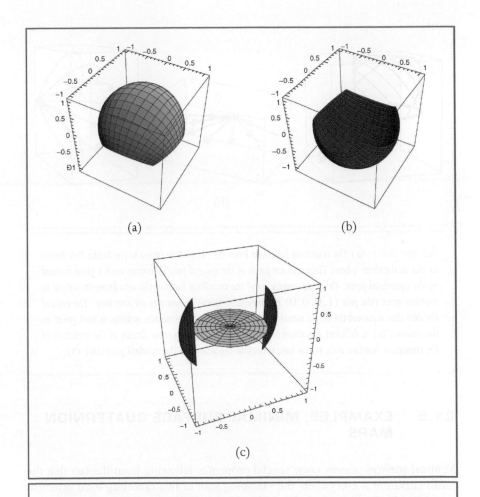

(a)   (b)

(c)

FIGURE 21.6 (a) The North Pole projective coordinatization of the sphere. (b) A similar regular patch for the South Pole. Because of the "no-hair" theorem, no single regular patch can cover the entire sphere. (c) The quaternion mappings of the systems of frames given by the North and South Pole coordinate patches, sampled in polar coordinates. The $q \rightarrow (-q)$ reflected images are included, although the North Pole's images both have the same projection and are thus indistinguishable here. The maps in c extend only to the equator, unlike the patches shown in Figure 21.5.

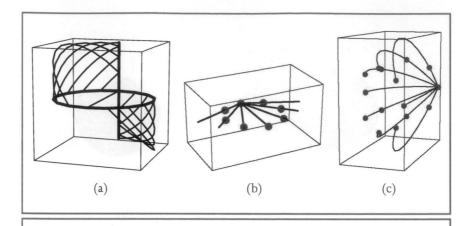

(a)   (b)   (c)

FIGURE 21.7 *(a) The transition functions from the North Pole frame to the South Pole frame as arcs in the three-sphere. These arcs are pieces of the space of possible frames with a given normal on the equatorial point. (b) A representation of the transition functions as arcs from the origin in rotation space (the pole $(1, 0, 0, 0)$ in quaternion space) common to all arcs here. The ends of the arcs thus represent the actual rotation needed to match the coordinate systems at each point on the equator. (c) A different projection from 4D to 3D, showing more details of the structure of the transition function arcs, which have a twofold degeneracy in the standard projection (b).*

21.5   EXAMPLES: MINIMAL SURFACE QUATERNION MAPS

Minimal surfaces possess many special properties following from the fact that the mean curvature is everywhere the vanishing sum of two canceling local principal curvatures [61]. We present a family of classic examples here that is remarkable for the fact that the usual framings are already very close or exactly optimal. Thus, we do not have much work to do except to admire the results, although there may be some interesting theorems implicit that would be beyond our scope to pursue.

Figures 21.8a through 21.8c show the following classical minimal surfaces.

$$\mathbf{x}_{\text{catenoid}}(u, v) = \cos u \cosh v \hat{\mathbf{x}} + \sin u \cosh v \hat{\mathbf{y}} + v \hat{\mathbf{z}}, \qquad (21.20)$$

$$\mathbf{x}_{\text{helicoid}}(u, v) = v \cos u \hat{\mathbf{x}} + v \sin u \hat{\mathbf{y}} + u \hat{\mathbf{z}}, \qquad (21.21)$$

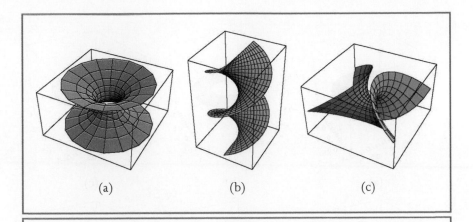

(a)   (b)   (c)

FIGURE 21.8   (a) The catenoid, a classic minimal surface in 3D space with a natural orthonormal parameterization. (b) The helicoid. (c) Enneper surface.

$$\mathbf{x}_{\text{Enneper}}(u, v) = (u - u^3/3 + uv^2)\hat{\mathbf{x}} \tag{21.22}$$

$$+ (v - v^3/3 + vu^2)\hat{\mathbf{y}} + (u^2 - v^2)\hat{\mathbf{z}}. \tag{21.23}$$

The quaternion Gauss map choices determined by these parameterizations and by
the geodesic reference algorithm are shown in Figure 21.9. The coordinate-based
catenoid map and helicoid map are 4π double coverings, whereas the Enneper
surface curiously has a coordinate system map that is exactly identical to the geodesic reference framing. For the periodic framings of the catenoid and helicoid, we
find the noteworthy result that the geodesic reference frame (which has a disjoint
quaternion reflected image) is a minimum (under variations of the surface) that is
distinct from the quaternion frames derived from the coordinate systems (which
are also minima but contain their own $q \to (-q)$ reflected images). The resulting
3D frame triads are shown in Figure 21.10 for comparison. A theoretical analysis
of the general properties of quaternion Gauss maps for minimal surfaces is beyond
our scope, but experimentally we see that there could be very interesting general
properties.

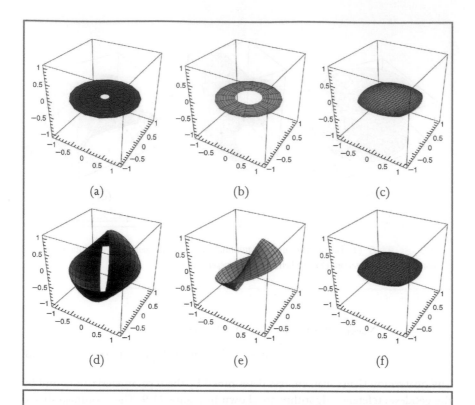

FIGURE 21.9 The geodesic reference quaternion frames of (a) the catenoid, (b) the helicoid, and (c) Enneper surface. (d–f) The corresponding quaternion Gauss maps determined directly from the parameterization. Both the catenoid and the helicoid fail to be cyclic in quaternion space without a 4π turn around the repeating direction, and thus these are doubled maps. The Enneper-surface framing turns out to be identical to its geodesic reference frame.

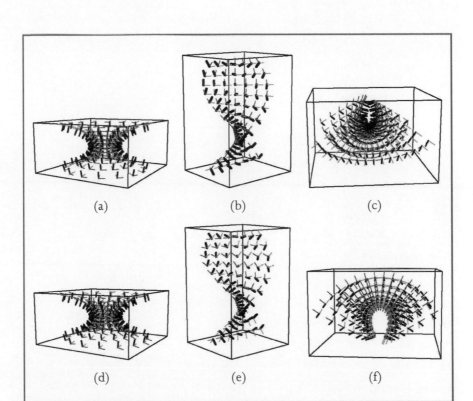

(a) (b) (c)

(d) (e) (f)

FIGURE 21.10 The 3D geodesic reference frames displayed directly on the surfaces of (a) the catenoid, (b) the helicoid, and (c) Enneper surface. (d–f) The 3D frames computed directly from the standard parameterizations. Because the Enneper surface is the same, we show in (f) a different viewpoint of the same frames.

Optimal Quaternion Frames

22

In this chapter we continue to study the nature of orientation frames on curves and surfaces and their corresponding quaternion structures. Our visualizations again exploit the fact that quaternions are points on the three-sphere or hypersphere S^3 embedded in 4D. The methods in this section follow closely techniques introduced in Hanson and Ma [78,80] and Hanson [70,71] for analyzing families of coordinate frames on curves and surfaces using quaternion maps.

22.1 BACKGROUND

General questions involving the specification of curve framings have been investigated in many contexts. For a representative selection of approaches see, for example, Bloomenthal [21], Klock [114], Max [121], and Shani and Ballard [147]. The quaternion Gauss map is a logical extension of the quaternion frame approach to visualizing space curves introduced by Hanson and Ma [78,80]. The formulation of the quaternion form of the differential equations for frame evolution was apparently introduced by Tait [160].

For basic information on orientation spaces and their relationship to quaternions see, for example, Altmann [4], Kuipers [115], and Pletincks [139]. Additional background on the differential geometry of curves and surfaces may be found in sources such as the classical treatise of Eisenhart [46] and in Gray's Mathematica-based text [61].

Our main task is to work out a general framework for selecting optimal systems of coordinate frames that can be applied to the study of curves and surfaces in 3D space. We will see that our preferred optimizations contain minimal-turning parallel transport framings of curves as a special case and extend naturally to situations in which parallel transport is not applicable.

22.2 MOTIVATION

Many graphics problems require techniques for effectively displaying the properties of curves and surfaces. The problem of finding appropriate representations can be quite challenging. Representations of space curves based on single lines are often inadequate for graphics purposes. Significantly better images result from choosing a tubing to display the curve as a graphics object with spatial extent. Vanishing curvature invalidates methods such as the Frenet frame, and alternative approaches such as those based on parallel transport involve arbitrary heuristics to achieve such properties as periodicity. Similar problems occur in the construction of suitable visualizations of complex surfaces and oriented particle systems on surfaces. If a surface patch is represented by a rectangular but nonorthogonal mesh, for example, there is no obvious local orthonormal frame assignment. If the surface has regions of vanishing curvature, methods based on directions of principal curvatures break down as well.

Although we emphasize curves and surfaces to provide intuitive examples, there are several parallel problem domains that can be addressed with identical techniques. Among these are extrusion methods and generalized cones in geometric modeling, the imposition of constraints on a camera-frame axis in keyframe animation, and the selection of a 2D array of camera-frame axis choices as a condition on a constrained-navigation environment (e.g., see Hanson and Wernert [83]).

Figure 22.1 summarizes the basic class of problems involving curves that will concern us here. The line drawing (a) of a $(3, 5)$ torus knot provides no useful information about the 3D structure. Improving the visualization by creating a tubing involves a subtle dilemma that we attempt to expose in the rest of the figure. We cannot use a periodic Frenet frame as a basis for this tubing because inflection points or near-inflection points occur for many nice-looking torus knot parameterizations, and in such cases the Frenet frame is undefined or twists wildly. The parallel transport tubing shown in (b) is well behaved but not periodic. By looking carefully at the magnified portion next to the arrow in Fig-

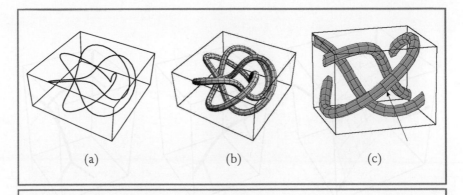

FIGURE 22.1 *The (3, 5) torus knot, a complex periodic 3D curve. (a) The line drawing is nearly useless as a 3D representation. (b) A tubing based on parallel transporting an initial reference frame produces an informative visualization, but is not periodic. (c) The arrow in this close-up exposes the subtle but crucial nonperiodic mismatch between the starting and ending parallel transport frames. This would invalidate any attempt to texture the tube. Quaternion methods provide robust parameterization-invariant principles for resolving such problems.*

ure 22.1c, one can see a gross mismatch in the tessellation (due to the nonperiodicity) that would, for example, preclude the assignment of a consistent texture. Although it would be possible in many applications to ignore this mismatch, it has been the subject of a wide variety of previous papers (e.g., see Bloomenthal [21], Klock [114], and Shani and Ballard [147]), and must obviously be repaired for many other applications, such as those requiring textured periodic tubes.

Figure 22.2 illustrates a corresponding problem for surface patches. Although the normals to the four corners of the patch are always well defined (a), one finds two different frames for the bottom corner, depending on whether one parallel transports the initial frame around the left-hand path (b) or the right-hand path (c). There is no immediately obvious right way to choose a family of frames covering this surface patch. Our goal is to propose a systematic family of optimization methods for resolving problems such as these.

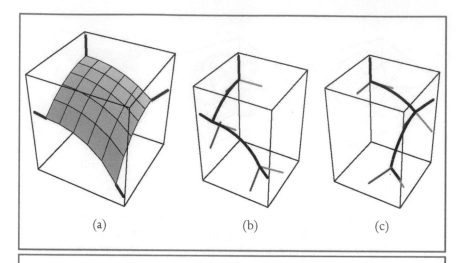

(a)   (b)   (c)

FIGURE 22.2 *(a) A smooth 3D surface patch having a nonorthogonal parameterization, along with its geometrically fixed normals at the four corners. No unique orthonormal frame is derivable from the parameterization. If we imitate parallel transport for curves to evolve the initial frame at the top corner to choose the frame at the bottom corner, we find that the paths shown in b and c result in incompatible final frames at the bottom corner. Our goal is to address the problem of systematically choosing a compatible set of surface frames in situations such as this.*

22.3   METHODOLOGY

We focus on unit quaternion representations of coordinate frames as points on the three-sphere $\mathbf{S}^3$, which admits a natural distance measure for defining optimization problems and supports in addition a variety of regular frame-interpolation methods (e.g., see Kim et al. [111], Nielson [132], Schlag [145], Shoemake [149], and Chapter 25). We do not directly address the related question of optimal freely moving frames treated by the minimal-tangential-acceleration methods (e.g., see Barr et al. [15], Kajiya [106], and Ramamoorthi and Barr [140]). We are instead concerned with closely spaced points on curves and surfaces where one direction of the frame is already fixed and the chosen functional minimization in quaternion space must obey the additional constraint imposed by the fixed family of directions. Additional references of interest, especially regarding the treatment of

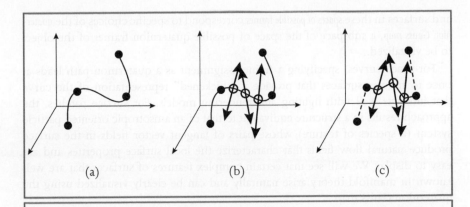

FIGURE 22.3 (a) *The camera frame interpolation problem is analogous to the problem of finding a minimal-bending spline curve through a series of fixed key points.* (b) *The optimal curve frame assignment problem is analogous to fixing the end points of a curve segment and choosing in addition a family of lines along which the intermediate points are constrained to slide during the optimization process. In 3D, the spline path need not pass through the constraint lines.* (c) *In typical practical situations, the sample points are generally close enough together that we can apply the constraints to piecewise linear curves analogous to those shown here.*

surfaces, include [106,138]. Figure 22.3 provides a visualization of the difference between the general interpolation problem and our constrained problem. A typical spline minimizes the bending energy specified by the chosen anchor points. Requiring intermediate points to slide on constrained paths during the minimization modifies the problem. In particular, 3D spline curves need not intersect any of the constraint paths. In addition, we note that we typically have already sampled our curves and surfaces as finely as we need to, and thus piecewise linear curves are generally sufficient for the applications we discuss.

22.3.1 THE SPACE OF POSSIBLE FRAMES

Our solution to the problem is to transform the intrinsic geometric quantities (such as the tangent field of a curve and the normal field of a surface) to quaternion space and to construct the quaternion manifold corresponding to the one remaining degree of rotational freedom in the choice of coordinate frame at each point. Curves and surfaces in these *spaces of possible frames* correspond to specific choices of the *quaternion Gauss map*, a subspace of the space of possible quaternion frames of the object to be visualized.

For space curves, specifying a frame assignment as a quaternion path leads at once to tubular surfaces that provide a "thickened" representation of the curve that interacts well with lighting and rendering models. For surface patches, the approach results in a structure equivalent to that of an anisotropic oriented particle system (a species of texture) whose pairs of tangent vector fields in the surface produce natural flow fields that characterize the local surface properties and are easy to display. We will see that certain complex features of surfaces that are well known in manifold theory arise naturally and can be clearly visualized using the quaternion Gauss map.

We will typically exploit our standard $\mathbf{S}^3$ method of visualizing the geometry of the space of quaternions in which quaternion Gauss maps and the spaces of possible quaternion frames are represented. We show how to compute the required subspaces of allowed frames in practice, and how to express this information in a form that can be used to optimize an energy measure, thereby leading to optimal frame choices.

22.3.2 PARALLEL TRANSPORT AND MINIMAL MEASURE

Our approach is to constrain each quaternion frame with one degree of freedom to its own circular quaternion path (the axial degree of rotational freedom), and then to minimize the quaternion length of the frame assignment for curves and the quaternion area of the frame assignment for surfaces to achieve an optimal frame choice. This choice reduces to the parallel transport frame for simple cases. Our justification for choosing minimal quaternion length for curves is that there is a unique rotation in the plane of two neighboring tangents that takes each tangent direction to its next neighbor along a curve. This is the geodesic arc connecting the two frames in quaternion space, and is therefore the minimum distance between the quaternion points representing the two frames. The choice of minimal area for surface frames is more heuristic, basically a plausibility argument that the generalization of *minimal length* is *minimal area* (no doubt this could be made more rigorous).

By imposing other criteria, such as end-point derivative values and minimal bending energy (see Barr et al. [15] and Ramamoorthi and Barr [140]), the short straight line segments and polygons that result from the simplest minimization could be smoothed to become generalized splines passing through the required constraint rings. Because in practice our curve and surface samplings are arbitrarily dense, we will typically work directly with the unique quaternion rings giving the degrees of freedom at each sample point.

22.4   THE SPACE OF FRAMES

We are now ready to introduce the details of our key concept, the *space of possible frames*. Suppose at each sample point $\mathbf{x}(t)$ of a curve we are given a unit tangent vector, $\hat{\mathbf{T}}(t)$, computed by whatever method one likes (two-point sampling, five-point sampling, analytic, and so on). Then one can immediately write down a one-parameter family describing all possible choices of the normal plane orientation. This is simply the set of rotation matrices $R(\theta, \hat{\mathbf{T}}(t))$, or quaternions $q(\theta, \hat{\mathbf{T}}(t))$, that leave $\hat{\mathbf{T}}(t)$ fixed.

For surfaces, the analogous construction follows from determining the unit normal $\hat{\mathbf{N}}(u, v)$ at each point $\mathbf{x}(u, v)$ on the surface patch. The needed family of rotations $R(\theta, \hat{\mathbf{N}}(u, v))$, or quaternions $q(\theta, \hat{\mathbf{N}}(u, v))$, now leaves $\hat{\mathbf{N}}(u, v)$ fixed and parameterizes the space of possible *tangent* directions completing a frame definition at each point $\mathbf{x}(u, v)$.

However, there is one slight complication: the family of frames $R(\theta, \hat{\mathbf{v}})$ leaving $\hat{\mathbf{v}}$ fixed does not have $\hat{\mathbf{v}}$ as one column of the 3×3 rotation matrix, and thus does not actually describe the desired family of frames. Therefore, we proceed as follows.

We define $f(\theta, \hat{\mathbf{v}}) = (f_0, f_1, f_2, f_3)$ to be a quaternion describing the family of frames for which the direction $\hat{\mathbf{v}}$ is a preferred fixed axis of the frame, such as the tangent or normal vector. The orthonormal triad of three-vectors describing the desired frame is

$$F(\theta, \hat{\mathbf{v}}) = \begin{bmatrix} f_0^2 + f_1^2 - f_2^2 - f_3^2 & 2f_1 f_2 - 2f_0 f_3 & 2f_1 f_3 + 2f_0 f_2 \\ 2f_1 f_2 + 2f_0 f_3 & f_0^2 - f_1^2 + f_2^2 - f_3^2 & 2f_2 f_3 - 2f_0 f_1 \\ 2f_1 f_3 - 2f_0 f_2 & 2f_2 f_3 + 2f_0 f_1 & f_0^2 - f_1^2 - f_2^2 + f_3^2 \end{bmatrix},$$

$$(22.1)$$

where one column of our choice is picked to be $\hat{\mathbf{v}}$, the fixed direction.

The standard rotation matrix $R(\theta, \hat{\mathbf{v}})$ leaves $\hat{\mathbf{v}}$ fixed but does not have $\hat{\mathbf{v}}$ as one column of the 3×3 rotation matrix, and thus we have more work to do. To compute $f(\theta, \hat{\mathbf{v}})$, we need the following.

- A base reference frame $b(\hat{\mathbf{v}})$ that is guaranteed to have one column exactly aligned with a chosen vector $\hat{\mathbf{v}}$, which is either the tangent to a curve or the normal to a surface.
- A one-parameter family of rotations that leaves a fixed direction $\hat{\mathbf{v}}$ invariant.

The latter family of rotations is given simply by the standard quaternion

$$q(\theta, \hat{\mathbf{v}}) = \left(\cos \frac{\theta}{2}, \hat{\mathbf{v}} \sin \frac{\theta}{2} \right), \tag{22.2}$$

for $0 \leqslant \theta < 4\pi$. The base frame can be chosen as

$$b(\hat{\mathbf{T}}) = q\left(\arccos(\hat{\mathbf{x}} \cdot \hat{\mathbf{T}}), (\hat{\mathbf{x}} \times \hat{\mathbf{T}})/\|\hat{\mathbf{x}} \times \hat{\mathbf{T}}\| \right) \tag{22.3}$$

for a curve frame with the tangent $\hat{\mathbf{T}}$ in the first column,

$$\text{Curve Frame} = [\hat{\mathbf{T}} \quad \hat{\mathbf{N}}_1 \quad \hat{\mathbf{N}}_2],$$

and as

$$b(\hat{\mathbf{N}}) = q\left(\arccos(\hat{\mathbf{z}} \cdot \hat{\mathbf{N}}), (\hat{\mathbf{z}} \times \hat{\mathbf{N}})/\|\hat{\mathbf{z}} \times \hat{\mathbf{N}}\| \right) \tag{22.4}$$

for a surface frame with the normal $\hat{\mathbf{N}}$ in the last column:

$$\text{Surface Frame} = [\hat{\mathbf{T}}_1 \quad \hat{\mathbf{T}}_2 \quad \hat{\mathbf{N}}].$$

We have already introduced the frame $b(\hat{\mathbf{v}})$, which we will refer to as the *geodesic reference frame* because it tilts the reference vector (e.g., $\hat{\mathbf{x}}$, $\hat{\mathbf{y}}$, or $\hat{\mathbf{z}}$) along a geodesic arc until it is aligned with $\hat{\mathbf{v}}$ (see Figure 22.4). If $\hat{\mathbf{v}} = \hat{\mathbf{z}}$ (or $\hat{\mathbf{x}}$ or $\hat{\mathbf{y}}$), there is no problem, in that we simply take $b(\hat{\mathbf{v}})$ to be the quaternion $(1, \mathbf{0})$. If $\hat{\mathbf{v}} = -\hat{\mathbf{z}}$, we may choose any compatible quaternion (such as $(0, 0, 0, 1)$, and so on). We escape the classic difficulty of being unable to assign a global frame to all of $\mathbf{S}^2$ because we need a parameterization of *all possible* frames, not any one particular global frame. If one wants to use a reference frame that is not the identity frame, one must premultiply $b(\hat{\mathbf{v}})$ on the right by a quaternion rotating from the identity into that reference frame. This is important when constructing a nonstandard geodesic reference frame such as that required to smoothly describe a neighborhood of the southern hemisphere of $\mathbf{S}^2$.

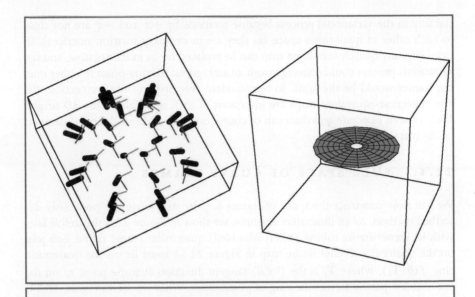

FIGURE 22.4 *Example of the geodesic reference frame: (a) On the northern hemisphere of a two-sphere, the geodesic reference frame tilts the $\hat{\mathbf{z}}$ axis of the North Pole's identity frame along the shortest arc to align with a specified reference direction. (b) The quaternion map has just a single possible plane for the tilt axes allowed by this procedure.*

We can thus write the full family of possible quaternion frames keeping $\hat{\mathbf{v}}$ as a fixed element of the frame triad to be the quaternion product

$$f(\theta, \hat{\mathbf{v}}) = q(\theta, \hat{\mathbf{v}}) \star b(\hat{\mathbf{v}}), \tag{22.5}$$

where $\star$ denotes quaternion multiplication and all possible frames are described twice (in that $0 \leqslant \theta < 4\pi$). To summarize, if we specify a frame axis $\hat{\mathbf{v}}$ to be fixed then the variable θ in $f(\theta, \hat{\mathbf{v}})$ serves to parameterize a ring in quaternion space, each point of which corresponds to a particular 3D frame and each frame of which has a diametrically opposite twin.

We argue that because optimization will typically be done in the full quaternion space the fact that two opposite-sign quaternions map to the same physical three-space rotation is not a detriment. In fact, it potentially permits an additional stability in the variational process because rotations by $+\pi$ and $-\pi$ are not close to each other in quaternion space (as they are in ordinary rotation matrices). In principle, any quaternion Gauss map can be replaced by its exact negative, and the variational process could converge from an ambiguous starting point to either one; the frames would be the same. In our standard (vector-part-only) projection, the two reflection-equivalent maps are inversions of each other about the 3D origin. Their unseen opposite q_0 values can of course cause an additional large separation of the maps in 4D space.

22.4.1 FULL SPACE OF CURVE FRAMES

We can now construct the space of frames step by step using the previously described method. As an illustrative example, we show in Figure 22.5 the trefoil knot with its Frenet-frame tubing and its (doubled) quaternion Frenet frame. Each point on the quaternion Frenet frame map in Figure 22.5b must lie on the quaternion ring $f(\theta, \hat{\mathbf{T}}_i)$, where $\hat{\mathbf{T}}_i$ is the (local) tangent direction at some point $\mathbf{x}_i$ on the curve and θ parameterizes the ring of possibilities. What we see in Figure 22.5 is one set of the values $\{\theta_i\}$ corresponding to the frame uniquely determined by the Frenet formulas. Next, we examine what happens when we release θ in $f(\theta, \hat{\mathbf{T}})$ from its Frenet values at the first few sample points of the curve.

Figure 22.6 shows the steps in the construction of the space of frames for the trefoil knot, beginning with a few tangent vectors and the quaternion basis frames corresponding to quaternions that tilt the reference axis into this tangent direction. The circular curve of quaternions representing the space of normal frames is drawn for each tangent. Each basis frame touches this curve once, modulo quaternion doubling. Then the family of these circular curves sweeps out a cylindrical two-manifold, the full space of invariant frames for a 3D curve.

This full space, shown in Figure 22.7, has several nontrivial properties. One is that, given one circular ring of frames, a neighboring ring that is a parallel-transported version of the first ring is a so-called "Clifford parallel" of the first ring, meaning that the distance from any point on one ring to the nearest point on the second ring is the same. This is nontrivial to visualize and is a feature of the 4D space we are working in. Another property is that the intervals between rings in the quaternion space directly indicate the curvature. This comes about because the magnitude of $\hat{\mathbf{T}}'$ is related to the parallel transport transition between any two sample points, given by Equation 20.12. Because the parallel transport frames are legal frames, and because the starting frame is arbitrary, each full ring is a parallel

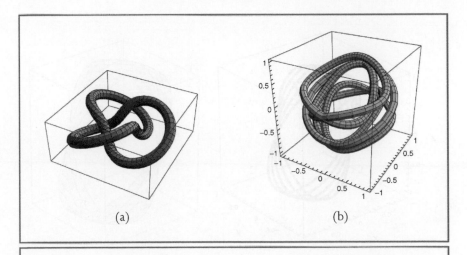

(a)   (b)

FIGURE 22.5   (a) A trefoil torus knot. (b) Its quaternion Frenet frame projected to 3D. For
this trefoil knot, the frame does not close on itself in quaternion space unless the curve is traversed
twice, corresponding to the double-valued "mirror" image of the rotation space that can occur in
the quaternion representation. Observe the longer segments in (b). These correspond to the three
high-torsion segments observable in (a).

transport of its predecessor, with the angular distance of the transition rotation
providing a measure of the curvature relative to the sampling interval.

22.4.2   FULL SPACE OF SURFACE MAPS

The full space of frames for a surface patch is even more complex to visualize,
because it is a hypercylindrical three-manifold formed by the direct product of
patches of surface area with the rings of possible frames through each surface point.

As a very simple case of a surface, consider the patch shown in Figure 22.2a.
The coordinate system used does not provide a unique tangent frame, and thus one
cannot immediately determine a logical frame choice.

Figure 22.8 shows spaces of possible frames for the four corners as four rings
of quaternion values compatible with the normals at the patch corners. Recall that
parallel transporting the initial frame along two different routes (Figures 22.2b and

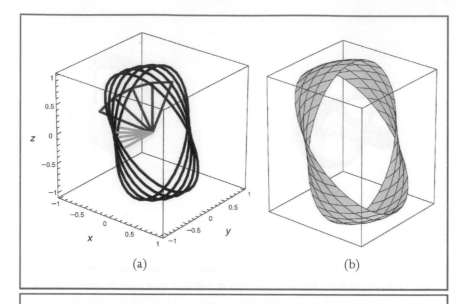

(a)   (b)

FIGURE 22.6 (a) *The first several pieces of the construction of the invariant quaternion space for the frames of the trefoil knot. The red fan of vectors shows the first several elements of the tangent map, represented as vectors from the origin to the surface of the two-sphere and connected by a line. Each green vector points from the origin to the geodesic reference element of the quaternion space* $q(\arccos(\hat{\mathbf{T}} \cdot \hat{\mathbf{x}}), \hat{\mathbf{x}} \times \hat{\mathbf{T}}/\|\hat{\mathbf{x}} \times \hat{\mathbf{T}}\|)$ *guaranteed to produce a frame with the tangent* $\hat{\mathbf{T}}$. *The black curves are the first several elements of the one-parameter space of quaternions representing all possible quaternion frames with the tangent* $\hat{\mathbf{T}}$. *(b) This piece of the space of possible frames is represented as a continuous surface, where a circle on the surface corresponds to the space of frames for one point on the curve. All quaternions are projected to 3D using only the vector part.*

22.2c) produces incompatible frames at the final corner. We represent this situation in Figure 22.8 by drawing the routes in quaternion space between the initial frame (the degenerate circle appearing as a central vertical line) and the final frame. The mismatch between the two final frames is illustrated by the fact that the two paths meet at different points on the final ring specifying the frame freedom for the bottom corner's frame.

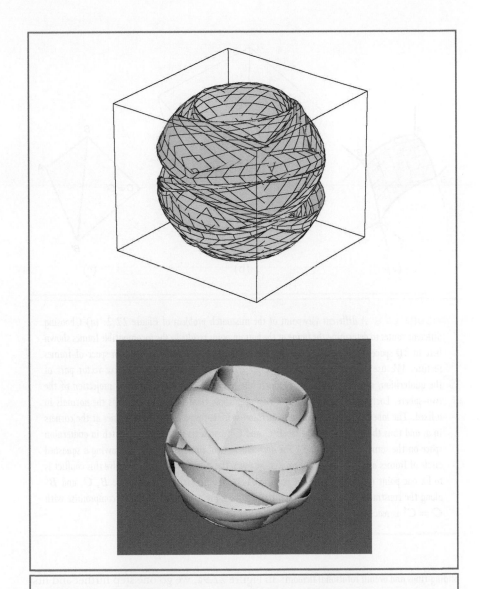

FIGURE 22.7 *Full surface of the invariant quaternion frame space for the frames of the trefoil knot. All quaternions are projected to 3D using only the vector part.*

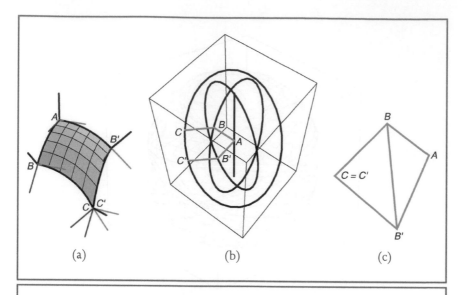

(a)   (b)   (c)

FIGURE 22.8 *A different viewpoint of the mismatch problem of Figure 22.2. (a) Choosing different routes to determine the frame at the bottom point results in the incompatible frames shown here in 3D space. (b) The same information is presented here in the quaternion space-of-frames picture. We use throughout a quaternion projection that shows only the three-vector part of the quaternion, dropping q_0. This is much like projecting away z in a polar projection of the two-sphere. Each heavy black curve is a ring of possible frame choices that keeps the normals in a fixed. The labels mark the point in quaternion space corresponding to the frames at the corners in a, and thus the gap between the labels C and C' represents the frame mismatch in quaternion space on the same constraint ring. (The apparent vertical line is the result of drawing a squashed circle of frames at vertex A in this projection.) (c) The method proposed to resolve this conflict is to fix one point (say A), divide the polygon $ABCB'$ into triangles, and slide B, C, and B' along the constraint rings until the total triangle areas are minimized and some compromise with $C = C'$ is reached.*

Sliding rings and overall rotational freedom: In Figure 22.9a, we go one step further and first show how the quaternion Gauss map of an entire patch is situated relative to the ring space. Keeping one corner fixed and sliding the rest of the frames around the circular rings takes us to distinct families of frames, which obviously have different

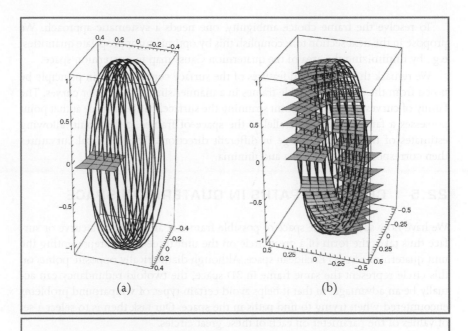

(a)   (b)

FIGURE 22.9 (a) *A more complete picture of the space of frames for this surface patch. The surface shown is a sparse quaternion frame choice for the surface, and we show a subset of the rings of constraints. Each ring passes through one quaternion point on the frame map, the point specifying the current frame choice. Variations must keep each vertex on its ring. (b) An equivalent set of frames is formed by applying a rotation to the entire set of frames. All points follow their own ring of constraints to keep the same normal. These pictures represent the three-manifold in quaternion space swept out by the possible variations.*

areas in the quaternion space. Finally, in Figure 22.9b we keep the fundamental space of frames the same but exercise the freedom to choose the single parameter describing the basis for the overall orientation. Rotating the basis sweeps out both the three-manifold describing the space of frames for this patch and the family of equivalent frames differing by an insignificant orientation change in the basis vector.

To resolve the frame choice ambiguity, one needs a systematic approach. We propose in the next section to accomplish this by optimizing appropriate quantities, e.g., by minimizing the area of the quaternion Gauss map in quaternion space.

We remark that the general features of the surface curvature can in principle be noted from the space of possible frames in a manner similar to that for curves. The family of curves through any point spanning the surface's tangent space at that point possesses a family of rings parallel to the space of frames at the point, allowing estimates of the rates of change in different directions. The principal curvatures then correspond to the maxima and minima.

22.5   CHOOSING PATHS IN QUATERNION SPACE

We have now seen that the space of possible frames at any point of a curve or surface thus takes the form of a great circle on the unit three-sphere representing the unit quaternions in 4D Euclidean space. Although diametrically opposite points on this circle represent the same frame in 3D space, the twofold redundancy can actually be an advantage, in that it helps avoid certain types of wraparound problems encountered when trying to find paths in the space. Our task then is to select a set of values of the parameter on each of these great circles.

The advantage of looking at this entire problem in the space of quaternions is that one can clearly compare the intrinsic properties of the various choices by examining such properties as length and smoothness in the three-sphere. We note the following issues.

- *Frame–frame distance*: Suppose we are given two neighboring tangents, $\hat{\mathbf{T}}_1$ and $\hat{\mathbf{T}}_2$, and two corresponding candidate frame choices parameterized by θ_1 and θ_2. What is the "distance" in frame space between these? The simplest way to see how we should define the distance is by observing that by Euler's fundamental theorem there is a single rotation matrix $R(\theta, \hat{\mathbf{n}})$ or quaternion $q(\theta, \hat{\mathbf{n}})$ that takes one frame to the other. If $R_1(\theta_1, \hat{\mathbf{T}}_1)$ and $R_2(\theta_2, \hat{\mathbf{T}}_2)$ are the two frames, one can write $R = (R_2 \cdot (R_1)^{-1})$ (or $q = (q_2 \star (q_1)^{-1})$) and solve for θ and $\hat{\mathbf{n}}$. Clearly, the value of θ gives a sensible measure of the closeness of the two frames.
- *Quaternion distance*: We remark that essentially the same procedure is required to obtain the parameters of R directly or to find the value of the equivalent quaternion. If we work in quaternion space, we compute $q_1(\theta_1, \hat{\mathbf{T}}_1)$ and $q_2(\theta_2, \hat{\mathbf{T}}_2)$ and then find rather more straightforwardly an equivalent result

by noting that the zeroth component of $q = q_2 \star (q_1)^{-1}$ is identical to the rotation-invariant scalar product of the two quaternions, $q_1 \cdot q_2$, and thus provides the needed angle at once:

$$\theta = 2\arccos(q_1 \cdot q_2).$$

- *Approximation by Euclidean distance:* Using the methods previously discussed, one can in principle compute precise arc-length distances among frames when dealing with fine tessellations of smoothly varying geometric objects. In this case, it may be sufficient for numerical purposes to estimate frame-to-frame distances using the Euclidean distance in 4D Euclidean space, in that the chord of an arc approximates the arc length well for small angles.

22.5.1 OPTIMAL PATH CHOICE STRATEGIES

Why would one want to choose one particular set of values of the frame parameters over another? The most obvious reason is to keep a tubing from making wild twists such as those that occur for the Frenet frame of a curve with inflection points. In general, one can imagine wanting to minimize the total twisting, the aggregate angular acceleration, and so on subject to a variety of boundary conditions. A bewildering variety of energy functions to minimize can be found in the literature (e.g., see Brakke [22]). The following summarize a selection of such criteria for choosing a space of frames, with the caveat that one certainly may think of others!

- *Minimal length and area:* The most obvious criterion is to minimize the total turning angle experienced by the curve frames. Fixing the frames at the ends of a curve may be required by periodicity or external conditions, and thus a good solution would be one that minimizes the sum total of the turning angles needed to get from the starting to the ending frame. The length to minimize is simply the sum of the angles rotated between successive frame choices (as noted previously), either exact or approximate. Similar arguments apply to the area of a surface's quaternion Gauss map.
- *Parallel transport along geodesics:* Given a particular initial frame, and no further boundary constraints, one may also choose the frame that uses the minimum local distance to get between each neighboring frame. Because the parallel transport algorithm corresponding to the Bishop frame uses precisely the smallest possible rotation to get from one frame to the next, this gives the minimal free path that could be computed frame by frame. On a surface,

the resulting paths are essentially geodesics, but (as noted in Figure 22.2) there is no obvious analog of a global parallel transport approach to surface framing.

- *Minimal acceleration:* Barr, Currin, Gabriel, and Hughes [15] proposed a direct generalization of the no-acceleration criterion of cubic Euclidean splines for quaternion curves constrained to the three-sphere. The basic concept was to globally minimize the squared tangential acceleration experienced by a curve of unit quaternions. Although the main application of that paper was animation, the basic principles can be adopted and used to numerically compute optimal frames for curves and surfaces in our context as well.

- *Keyframe splines and constraints:* If for some reason one must pass through or near certain specified frames with possible derivative constraints, a direct spline construction in the quaternion space may actually be preferred (e.g., see Kim et al. [111], Nielson [132], Schlag [145], Shoemake [149,152], and Chapter 25). Most splines can be viewed in some sense as solving an optimization problem with various constraints and conditions, and thus the keyframe problem essentially reverts again to an optimization.

22.5.2 GENERAL REMARKS ON OPTIMIZATION IN QUATERNION SPACE

For both curves and surfaces, there is a single degree of freedom in the frame choice at each point where we can determine the tangent or normal direction, respectively. This degree of freedom corresponds to a relatively common sliding ring constraint that occurs in such minimization problems. General packages for solving systems with constraints are mentioned in Barr et al. [15], who chose MINOS [131]. For our own experiments, we have chosen Brakke's Surface Evolver package [22], which has a very simple interface for handling parametric constraint conditions and can be used for a wide variety of general optimization problems. See Appendix G for more details on the use of Surface Evolver. Alternatively, one can in principle construct one's own custom optimization packages.

Numerical optimization remains a bit of an art, requiring patience and resourcefulness on the part of the investigator. We found, for example, that curve optimization was relatively more stable than surface optimization because single curve outliers add huge amounts to the length, whereas single surface points stuck in a faraway crevice may contribute only a tiny amount to the area of a large surface. Although Surface Evolver in principle handles spherical distances, we used the default

4D Euclidean distance measure as an approximation. This generally corresponded well to explicit area calculations using solid angle performed on the same data sets. However, we did find that extremely random initial conditions (unrealistic for most applications) could produce isolated points stuck in local minima diametrically across quaternion space, at $q \to -q$, from where they should be. This type of problem can be largely avoided simply by running a consistency pre-processor to force nearby neighbors to be on the same side of the three-sphere. Another useful technique is to organize the data into hierarchies and optimize from the coarse scale down to the fine scale. In other cases, when things seem unreasonably stuck a manual "simulated annealing" procedure like that afforded by Surface Evolver's `jiggle` option often helps.

22.6   EXAMPLES

We now present some examples of frame choices computed by minimizing the length of the total path among sliding ring constraints for selected curves, and the total area spanned by analogous sliding rings for surfaces. One interesting result is that there appear to be families of distinct minima. If the initial data for a periodic surface, for example, are set up to return directly to the same point in quaternion space after one period, one has two disjoint surfaces (one the $q \to (-q)$ image of the other). If the data do not naturally repeat after one cycle, they must after two, because there are only two quaternion values that map to the same frame. The family of frame surfaces containing their own reflected images has a minimum distinct from the disjoint family.

22.6.1   MINIMAL QUATERNION FRAMES FOR SPACE CURVES

The helix provides a good initial example of the procedure we have formulated. We know that we can always find an initial framing of a curve based on the geodesic reference algorithm. However, suppose we wish to impose minimal length in quaternion space on the framing we select, and we do not know whether this frame is optimal with respect to that measure. Then, as illustrated in Figure 22.10, we can send the ring constraints on the possible quaternion frames at each sample point to our chosen optimizer and let it automatically find the optimal framing. The results and energies for several stages of this evolution are shown in the figure.

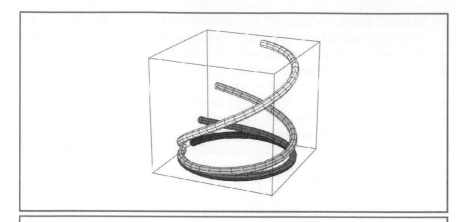

FIGURE 22.10 *Starting from the geodesic reference quaternion frame for a single turn of the helix (the very dark gray circle), the optimization produces these intermediate steps while minimizing the total quaternion curve length subject to the constraints in the space of frames. The final result is the white curve, which to several decimal points is identical to the parallel transport quaternion frame for the same helix. The numerical energies of the curves, from dark to light in color, are 3.03, 2.91, 2.82, and 2.66 for the parallel transport frame. The individual tubings used to display these curves are created using the parallel transport frame for each curve.*

The final configuration is indistinguishable from the parallel transport frame, confirming experimentally our theoretical expectation that parallel transport produces the minimal possible twisting.

In Figure 22.1, we introduced the question of finding an optimal framing of a particular $(3, 5)$ torus knot whose almost-optimal parallel transport framing was not periodic. In Figure 22.11, we show the solution to this problem achieved by clamping the initial and final quaternion frames to coincide, and then letting the optimizer pick the shortest quaternion path for all other frames. It would be possible, as in the case of the $(2, 3)$ torus knot framing shown in Figure 22.5, to have different conditions produce a framing solution containing its own reflected image rather than having a distinct reflected image (as is the case in Figure 22.11).

The types of solutions we find are remarkable in that they should be essentially the same for all reparameterizations of the curve. Regardless of the spacing of the sampling, the continuous surface of possible frames is geometrically the same in

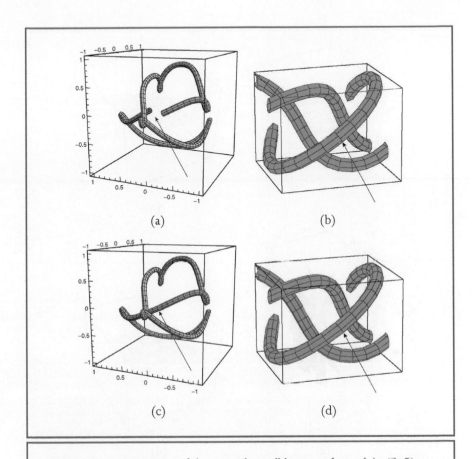

(a)

(b)

(c)

(d)

FIGURE 22.11 *Optimization of the nonperiodic parallel transport frame of the* $(3, 5)$ *torus knot introduced in Figure 22.1 to produce a nearby periodic framing.* (a) *The original quaternion parallel transport frame used to produce the tubing in Figures 22.1b and 22.1c.* (b) *The frame mismatch, repeated for completeness.* (c) *The result of fixing the final frame to coincide with the initial frame, leaving the other frames free to move on the constraint rings and minimizing the resulting total length in quaternion space. The length of the original curve was* 13.777, *and that of the final was* 13.700—*not a large difference, but noticeable enough in the tube and the quaternion space plot.* (d) *Close-up of the corresponding framing of the knot in ordinary 3D space, showing that the mismatch problem has been successfully resolved. This tube can now be textured, because the frames match exactly.*

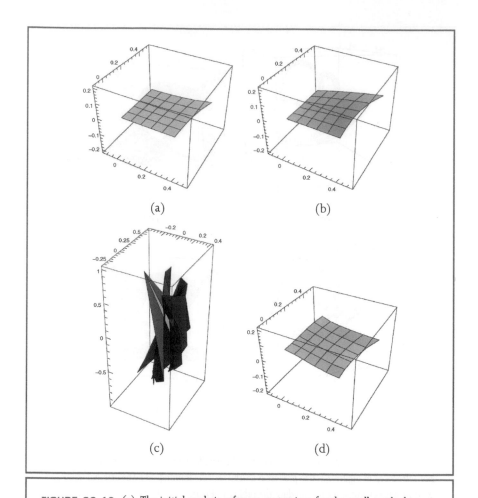

FIGURE 22.12 (a) *The initial geodesic reference quaternions for the small patch shown in Figure 22.2.* (b) *Initial quaternions from parallel transporting the vertex frame down one edge, and then across line by line.* (c) *A random starting configuration with the single same fixed corner point as in a and b, and a range of* $-\pi$ *to* $+\pi$ *relative to the geodesic reference frame.* (d) *The result of minimization of the quaternion area is the same for all starting configurations. The relative areas are* 0.147, 0.154, 0.296, *and* 0.141, *respectively. Thus, the geodesic reference is very close to optimal.*

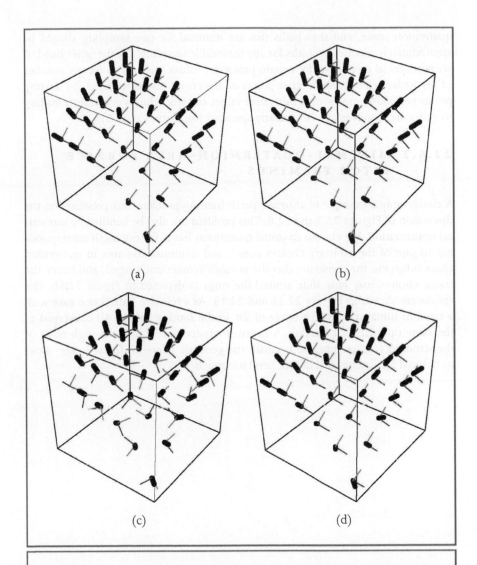

(a)   (b)

(c)   (d)

FIGURE 22.13 The 3D frame configurations corresponding to the quaternion fields shown in Figure 22.12. (a) The geodesic reference frame. (b) Two-step parallel transport frame. (c) Random frames. (d) The frame configuration resulting from minimizing area in quaternion space.

quaternion space, and thus paths that are minimal for one sampling should be approximately identical to paths for any reasonable sampling. On the other hand, if we *want* special conditions for certain parameter values it is easy to fix any number of particular orientations at other points on the curve, just as we fixed the starting points previously mentioned. Derivative values and smoothness constraints leading to generalized splines can be similarly specified (see Barr et al. [15]).

22.6.2 MINIMAL-QUATERNION-AREA SURFACE PATCH FRAMINGS

A classic simple example of a surface patch framing problem was presented in the discussion of Figures 22.2 and 22.8. This problem can also be handled by numerical optimization. We choose an initial quaternion frame for the mesh corresponding to one of the arbitrary choices noted, and minimize the area in quaternion space subject to the constraint that the normals remain unchanged, and hence the frame choices may only slide around the rings as depicted in Figure 22.8b. The results are shown in Figures 22.12 and 22.13. As a test, we started one case with a random initial state with a range of 2π in the starting values. All converged to the same optimal final framing. A heuristic observation is that although none of the standard guesses appeared optimal, the geodesic reference frame is very close to optimal for patches that do not bend too much.

Quaternion Volumes

23

We have now treated in detail two fundamental families of frames: the 1D family of frames corresponding to points on a curve carrying a continuous frame, with one axis following the curve's tangent vector and the 2D family of frames attached to a surface, with one axis following the normal vector perpendicular to the surface at each point. By transforming each of these families of frames to quaternion form, we achieved a *map* from each point in the manifold to a manifold of the same dimension in quaternion space. Because the metric properties of quaternion space permit quantitative evaluation of proximity among frames, the geometry of these maps reveals both local and global properties of the frame assignments to the original manifolds that cannot be readily seen any other way.

In this chapter we complete our investigation of quaternion maps by extending our treatment to the last unexamined domain, *quaternion volumes*. Quaternion volumes can in principle arise in the following ways.

- *As fields attached to discrete sampled points in a spatial volume:* In this domain, one can imagine a regular or irregular lattice of points connected by edges to form a tessellated volume. Given some unspecified mechanism associating a frame and its quaternion value to each lattice point, the 3D sample spatial lattice generates a corresponding quaternion lattice, as shown schematically in Figure 23.1.
- *Continuous generation of frames:* A set of frames need not be associated directly to discrete points in space. Indeed, instead of associating quaternion curves to

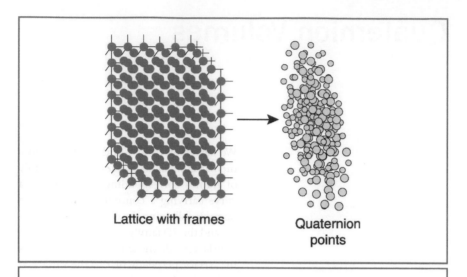

Lattice with frames

Quaternion
points

FIGURE 23.1 *A set of coordinate frames associated in some way with each point of a 3D lattice (typically a sampled volumetric distribution) is associated with a quaternion lattice generated by the quaternion frame map.*

moving frames on differentiable space curves we could have simply had a rigid object fixed at a single point and been given a one-parameter process that continuously changed the object's orientation from an initial state to a final state. A quaternion surface could have been generated by a continuous two-parameter process with specified boundaries and regions, not necessarily simply connected. The 1D and 2D cases have practical examples corresponding to restricted degrees of freedom in the motion of a single robotic or biological joint. The *volumetric quaternion map* corresponds to the quaternion volume generated by an unrestricted local orientation exploration (e.g., see Herda et al. [89,91,92]). Typical scenarios are presented schematically in Figure 23.2.

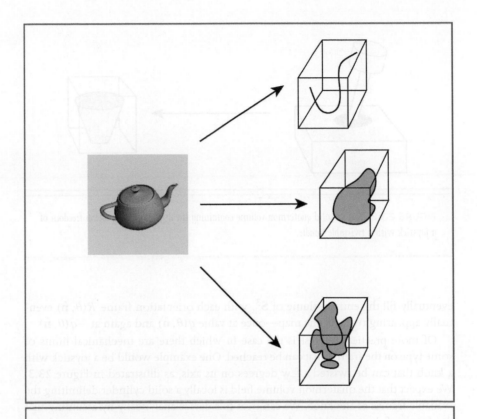

FIGURE 23.2 *Generating quaternion frame maps from 1D, 2D, and 3D sequences of frames.*

23.1 THREE-DEGREE-OF-FREEDOM ORIENTATION DOMAINS

To provide a motivating context, we will now focus our investigation of quaternion volume maps primarily on the final example in our introduction, the generation of a set of orientations by the reorientation of an abstract frame. If such a frame has truly no restrictions, and can achieve every possible orientation, the result is not very interesting. As all possible orientations are explored, the quaternion map will

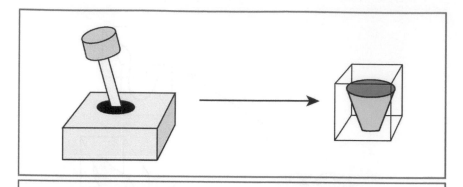

FIGURE 23.3 *Plot of solid quaternion volume containing the degrees of orientation freedom of a joystick with a twistable handle.*

eventually fill the entire volume of $\mathbf{S}^3$, with each orientation frame $R(\theta, \hat{\mathbf{n}})$ eventually appearing twice in the map—once at value $q(\theta, \hat{\mathbf{n}})$ and again at $-q(\theta, \hat{\mathbf{n}})$.

Of more practical interest is the case in which there are mechanical limits of some type on the frames that can be reached. One example would be a joystick with a knob that can be twisted a few degrees on its axis, as illustrated in Figure 23.3. We expect that the quaternion volume field is locally a solid cylinder delimiting the available frames of the knob. We can treat this device analytically by assuming that the frame of the joystick's knob can be represented as a tilting operation relative to the base frame of the joystick housing. This orientation can be written as the following tilt relative to the z axis, as illustrated in Figure 23.4:

$$\hat{\mathbf{s}}(\alpha, \beta) = (\cos\alpha \sin\beta, \sin\alpha \sin\beta, \cos\beta).$$

The entire transformation is then a tilt back up from the current joystick axis direction to the z axis, a spin about the current z axis by the polar angle γ, and a return tilt. The resulting matrix is simply the concatenation of the transformations

$$N(\alpha, \beta, \gamma) = R(\alpha, \hat{\mathbf{z}}) R(\beta, \hat{\mathbf{y}}) R(\gamma, \hat{\mathbf{z}}) R(-\beta, \hat{\mathbf{y}}) R(-\alpha, \hat{\mathbf{z}}). \qquad (23.1)$$

To produce the geometry restrictions of the joystick, we take

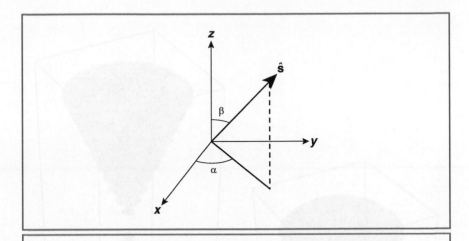

FIGURE 23.4 *The angles α and β specify the tilt of $\hat{s}$ to mark the orientation of the joystick.*

$$0 \leqslant \alpha \leqslant 2\pi,$$

$$0 \leqslant \beta < b,$$

$$-c \leqslant \gamma \leqslant c,$$

which corresponds to limiting the tilt in β to a maximum value $\beta = b$, permitting any axis orientation projection onto the xy plane whatsoever, and restricting the knob twist to $\pm c$. Mapping $N(\alpha, \beta, \gamma)$ into quaternion space with $N(0, 0, 0) =$ Identity Frame, we find the quaternion

$$v(\alpha, \beta, \gamma) = q(\alpha, \hat{\mathbf{z}}) \star q(\beta, \hat{\mathbf{y}}) \star q(\gamma, \hat{\mathbf{z}}) \star q(-\beta, \hat{\mathbf{y}}) \star q(-\alpha, \hat{\mathbf{z}})$$

$$= \left(\cos \frac{\gamma}{2}, \cos \alpha \sin \beta \sin \frac{\gamma}{2}, \sin \alpha \sin \beta \sin \frac{\gamma}{2}, \cos \beta \sin \frac{\gamma}{2} \right).$$

We now restrict the values of b and c suitably, and plot the three-vector part of $v(\alpha, \beta, \gamma)$ in our usual visualization tool, thus producing the explicit analytic volume map visualization exhibited in Figure 23.5. We emphasize that this type of information can now be *combined* with, e.g., human interface measurements about the limits of the user's comfort level to produce a complete visualization of product design constraints for such a device.

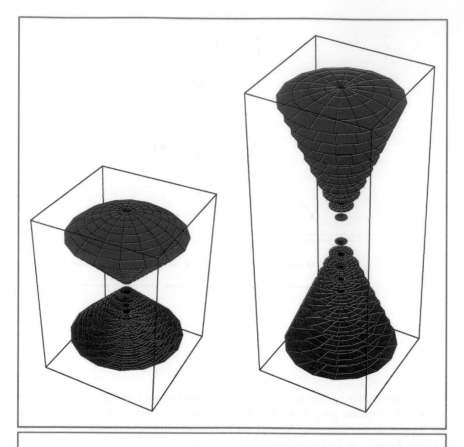

FIGURE 23.5 *This solid cone describes the joystick access space as a quaternion volume. Two ranges of the tilt limit of the joystick are given: $b = \pi/2$ and $b = \pi/4$.*

23.2   APPLICATION TO THE SHOULDER JOINT

Now that we see how a quaternion three-manifold can be used in practice for orientation maps that can be computed analytically, we are ready to study an application that is more complex. The human shoulder joint [89,91,92] is an excellent prototype of a generator for a frame set that does not have a simple analytic model.

In fact, one of the sources of key interest in this system is that individual human anatomy and individual robotic ball-and-socket joints can be quite distinct in their limits. The quaternion volume map is perfectly suited for several different analyses.

- *Comparison of range signatures*: The quaternion map for the allowed ranges for an individual joint can be as distinct in shape as a fingerprint. Because the metric properties allow one to make meaningful comparisons of limit differences, one can *compare* two individuals or mark the stages of development of a *single* individual (e.g., in the course of a conditioning regimen or recovery from an injury). Two maps (before and after treatment of an injury) can precisely quantify the restoration of joint functionality, for example.

- *Projection to allowed orientation domains*: The quaternion volume map of an experimentally determined range of motion for a three-degree-of-freedom joint can have an arbitrarily complex boundary surface. This surface must of course be compact, but it need not be convex or even simply connected. This surface defines the allowed domain for plausible models of the given joint, and more interestingly the domain to which any *proposed* positions of the joint must be clamped. For example, if a visual detector performs a measurement on the joint and proposes an interpretation of its 3D orientation, the quaternion point may be outside the experimental volume and thus illegal. To determine the optimal allowed reinterpretation of the joint orientation, one would locate the nearest point (in the quaternion metric) on the surface, and clamp the interpretation to that point.

- *Processed enhancements of allowed orientation domains*: Given an experimentally determined three-degree-of-freedom joint domain represented as a quaternion volume and its surface, it is possible to create new structures that enhance the application of determining allowed domains and dealing with the fact that measurements of this type tend to be error prone. One natural approach is to enhance the volume itself by running a 3D medial axis transformation (e.g., see Russ [143]) on the volume. In this way, one can define a family of isosurfaces (with the measured volume corresponding to the base isosurface value) and normalize the field outside the base value to give an instantaneous lookup of the signed distance to the base surface for any given sample orientation. Following the gradient of this distance field sends one directly to the nearest legal point.

23.3   DATA ACQUISITION AND THE DOUBLE-COVERING PROBLEM

23.3.1   SEQUENTIAL DATA

Suppose we are acquiring orientation data for a three-degree-of-freedom joint in a sequential manner. Each frame will be captured and converted in some way into a 3×3 matrix. This matrix is in turn transformed (using one of the standard

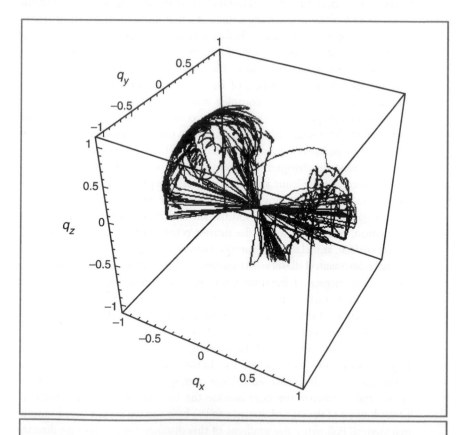

FIGURE 23.6 *Quaternion shoulder joint data before correction for doubling.*

algorithms) into quaternion parameters, which of necessity have a sign ambiguity: any given frame can appear after quaternion mapping as either q or $-q$. Thus, one can in principle wind up with a continuous sequence of closely spaced frames connected by line segments (Figure 23.6)—a collection of connected points in which almost identical 3D frames have quaternion images that are connected by line segments 2π radians apart (technically π radians in the quaternion logarithm, or $-1 \rightarrow +1$ in the unit sphere itself).

23.3.2   THE SEQUENTIAL NEAREST-NEIGHBOR ALGORITHM

To make a sequence of points such as that shown in Figure 23.6 into something more sensible, we simply force each point to be as close to its neighbor as possible. This is achieved using the following steps of an algorithm, which we call the force-close-1D-neighbors algorithm:

1   Assume a quaternion point p has been chosen for the *previous* frame in the sequence.
2   Transform the *current* frame into a quaternion q.
3   Check the sign of $p \cdot q = \cos\theta(p, q)$.
4   If the sign is positive, adjoin q to the list of points.
5   If the sign is negative, adjoin $-q$ to the list of points.
6   If the dot product vanishes, the point spacing is unreasonable. Resample, replacing q by a new point halfway back to p.

In this way, we can obtain a continuous family of points such as that shown in Figure 23.7. In terms of a point set, we have

$$\{p_i | p_i \cdot p_{i-1} > 0\}.$$

A mathematical curve (such as a spline) could be fitted to these points, generating a continuous model $p(t)$, which could then be used for various types of data analysis on the actual time sequence of the joint orientation motion.

23.3.3   THE SURFACE-BASED NEAREST-NEIGHBOR ALGORITHM

The next situation is the case in which one allows, for example, the *direction* of a limb to vary—keeping its axial twist fixed or restricted in some way—yielding a

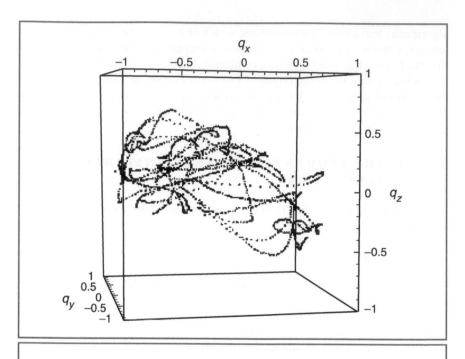

FIGURE 23.7 *A continuous sequence of quaternion shoulder orientations with neighbors forced to be in the same hemisphere of quaternion space as their predecessors.*

two-degree-of-freedom system. This system will then be described by a quaternion *surface* when the frames are mapped to the three-sphere. Once again, frames that are quite nearby can in principle wind up across the diameter of $\mathbf{S}^3$ in quaternion values. If there is any definite lattice or 2D array order of the points, we use the following 2D variant of the force-close-1D-neighbors algorithm.

1   Assume a quaternion 2D corner point p has been chosen.
2   Assume there is a known sequence of samples along an edge, and a known sequence of edges that, when connected, tessellate the surface.
3   Perform the force-close-1D-neighbors algorithm on the first edge.

4 Take p to be the first point on the first edge, and q to be the first point on the next edge. Check the sign of $p \cdot q = \cos \theta(p, q)$, and select q or $-q$ so that this is positive in the manner of the 1D algorithm.

5 Run the 1D algorithm on the edge starting at q.

6 Repeat until the entire surface-tessellation lattice is reconciled.

There are some anomalous cases that can in principle come up. For example, if the sequence of points on one edge obeys the local-positive-dot-product criterion, it is still conceivable that two adjacent points on neighboring edge sequences could fail to satisfy $p \cdot q > 0$.

23.3.4 THE VOLUME-BASED NEAREST-NEIGHBOR ALGORITHM

Finally, given a volumetric collection of quaternion points one needs to pick a particular anchor point and then walk through the entire set to force close 3D neighbors. In a randomly ordered set, it is still possible to migrate to an inappropriate association unless one presorts clusters in some way.

No lattice order: If there is no lattice order, the brute-force ordering can be had from making a $-q$ copy of each point q and then searching through the parameter space of possible "equators" specified by the family of $\mathbf{S}^2$s defined by the intersection of the hyperplane

$$\hat{\mathbf{n}} \cdot \mathbf{x} = 0,$$

with $\mathbf{S}^3$, which is itself simply parameterized by the three parameters placing the unit four-vector $\hat{\mathbf{n}}$ in $\mathbf{S}^3$. One seeks the cleanest partition (with the most points well separated from the chosen equator). For given applications, more specific optimization methods may be appropriate.

Existing lattice order: If there is a lattice order, we repeat the previous procedures, one level higher, as follows.

1 Assume a quaternion 3D corner point p has been chosen.

2 Assume there is a known sequence of samples along an edge, a known sequence of edges that, when connected, tessellate the surface, and known surfaces that sequentially tessellate the volume.

3 Perform the force-close-2D-neighbors algorithm on the first surface.

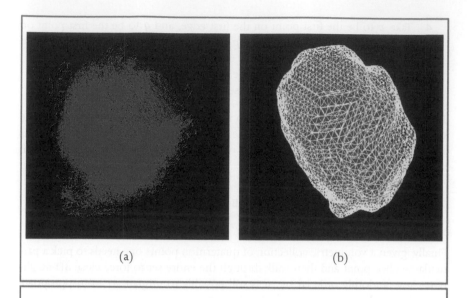

(a)   (b)

FIGURE 23.8   (a) *A dense sample of shoulder orientation data in quaternion space.* (b) *Implicit surface model fitted to the data.*

4   Take p to be the first point on the first surface, and q to be the first point on the next surface. Check the sign of $p \cdot q = \cos \theta(p, q)$, and select q or $-q$ so that this is positive in the manner of the 1D and 2D algorithms.

5   Run the 2D algorithm on the edge, starting at q.

6   Repeat until the entire volume-tessellation lattice is reconciled.

23.4   APPLICATION DATA

Given a collection of samples of volumetric orientation-frame data and a process (such as one of those previously described) for refining the data's quaternion presentation, we can finally visualize the data and develop an application for it. Our example is chosen from a set of Vicom-generated shoulder data generated by Herda (e.g., see [89,91,92]). Figure 23.8a shows the processed projection of a dense sample of data for the range of shoulder motion of a single subject. The force-close-3D-neighbors process has grouped the collection of points into a compact quaternion volume region, and all double-covering and $q \rightarrow -q$ problems have been eliminated. Figure 23.8b shows the result of fitting an implicit surface model to the salient boundaries of the quaternion volume. With this implicit surface, one can now perform a selection of applications, including the following.

- *Legality checking*: One can check the range of motion of a new sample (e.g., a body position generated by an animation program) and determine if it conforms to the legal range of a real human motion. If not, the hypothesized sample is projected to the premeasured realistic orientation by mapping it to the nearest legal point.
- *Physiological progress metric*: Suppose a patient is undergoing treatment for a joint injury or problem. A range-of-motion data sample before treatment can be compared to the sample at later times, and a quantitative measure of progress in physiologically accessible motion range can be carried out.

Quaternion Maps of Streamlines

24

We now apply quaternion frame methods to the study of data containing a large number of related curves. We assume that we are given a family of analytic or experimentally sampled curves, with sufficient information to extract smooth first derivatives (and perhaps higher derivatives) from the data. A representative application domain is a flow field that has been processed to generate streamlines by integrating the paths of test particles. Standard methods are typically limited to visual representations showing the 3D spatial positions of the test particle traces. The methods we shall study indicate that the data contain a wealth of additional information we can exploit using intrinsic geometric features and quaternion maps.

24.1   VISUALIZATION METHODS

The moving-frame field of a set of streamlines is potentially a rich source of detailed information about the data. However, the nine-component frame is unsuitable for direct superposition on dense data due to the high clutter resulting when its three orthogonal three-vectors are displayed; direct use of the frame is only practical at very sparse intervals, which prevents the viewer from grasping at a glance important structural details and changes. Displays based on 3D angular coordinates are potentially useful, but lack metric uniformity [3]. We believe the quaternion frame is potentially a more informative and flexible basis for visualizing collections of frames.

271

We can study a single curve in a flow field using exactly the same methods we did in Chapter 20. Given each curve and a method of computing its tangent vectors, we can compute the Frenet frame, along with the corresponding curvature and torsion, by the standard formulas. Gray [31,60] in particular has advocated the use of curvature and torsion-based color mapping to emphasize the geometric properties of curves. Because this information is essentially trivial to obtain simultaneously with the Frenet frame, we can easily supply encodings of the curvature and torsion as scalar fields mapped onto the streamlines. If the Frenet frame has anomalies, we can choose a starting frame and compute the parallel transport (Bishop) frame at each point. Once we have calculated the 3D moving frames, we can determine the corresponding quaternion frames, retaining the information about curvature and torsion for each point on the curve. We may then exploit these quantities to supplement the ordinary 3D frame information and to expose the intrinsic properties of the family of curves. Because of the metric properties of the quaternion representation, we also have the ability to visually note the similarities and differences of families of neighboring curves. Although we focus in this chapter on space curves, we remark that collections of frames of isolated points, frames on stream surfaces [99], and volumetric frame fields could also be represented using a similar mapping into quaternion space.

24.1.1 DIRECT PLOT OF QUATERNION FRAME FIELDS

We now emphasize the crucial observation: *For each 3D space curve, the moving quaternion frames define a completely new 4D space curve lying on the unit three-sphere embedded in 4D Euclidean space.* These curves can have *entirely different geometry* from the original space curve, in that distinct points on a curve correspond to distinct orientations. Families of space curves with exactly the same shape will map to the *same* quaternion curve, whereas quaternion curves that fall away from their neighbors will stand out distinctly in the $\mathbf{S}^3$ plot. Regions of vanishing curvature will show up as discontinuous gaps in the otherwise continuous quaternion Frenet-frame field curves, but will be well behaved in the quaternion parallel-transport frame fields. Straight 3D lines will of course map to single points in quaternion space, which is a distinctive feature of the display.

24.1.2 SIMILARITY MEASURES FOR QUATERNION FRAMES

We also remind ourselves that quaternion frames carry with them a natural geometry that may be exploited to compute meaningful similarity measures. Rather than use the Euclidean distance in 4D Euclidean space $\mathbb{R}^4$, one should use a distance based on the four-vector scalar product of unit quaternions

$$d(q, p) = |q \cdot p| = |q_0 p_0 + q_1 p_1 + q_2 p_2 + q_3 p_3|,$$

or the corresponding angle

$$\theta(q, p) = \arccos(d(q, p)),$$

which is the length of the geodesic arc connecting the two 4D unit vectors and is the natural distance measure on $\mathbf{S}^3$. Choosing such a distance measure results in a quantity that is invariant under 4D rotations, is invariant under quaternion multiplication, and is insensitive to the sign ambiguity in the quaternion representation for a given frame. Thus, we may quantitatively measure the similarity of any two 3D frames. This is a natural way to compare either successive frames on a single streamline or pairs of frames on different streamlines.

24.1.3 EXPLOITING OR IGNORING DOUBLE POINTS

The unique feature of quaternion representations of orientation frames is that they are doubled. If we have a single smooth curve, it technically does not matter which of the two points in $\mathbf{S}^3$ is chosen as a starting point because the others follow by enforcing small distances or continuously integrating small transformations. A collection of points with a uniform orientation as an initial condition will similarly evolve in tandem, and one normally need make only a single choice to see the pattern.

However, it is possible for a frame to rotate a full 2π radians back to its initial orientation, and then be on the opposite side of $\mathbf{S}^3$, or for a collection of streamlines to have a wide range of starting orientations that preclude a locally consistent method for choosing a particular quaternion q over its conjugate neighbor $-q$. We then have the following alternatives.

- *Include a reflected copy* of every quaternion field in the display. This doubles the data density but ensures that no two frame fields that are similar will appear diametrically opposite. The metric properties of similar curves will be easy to detect. In addition, 4D rotations will do no damage to the continuity of fields that are rotated to the outer surface and pass from the northern to the southern hyperhemisphere. If 4D depth is represented via color coding, for example, a point that rotates up to the surface of the displayed solid ball will smoothly pass to the surface and then pass back toward the center while its color changes from positive to negative depth coding.

- *Keep only one copy*, effectively replacing q with $-q$ if it is not in the default viewing hyperhemisphere. This has the effect that each data point is unique, but that curve frames very near diametrically opposite points on the $\mathbf{S}^2$ surface of the solid ball representing the north hyperhemisphere will be close in orientation but far away in the projection. In addition, when 4D rotations are applied curves that reach the $\mathbf{S}^2$ surface of the solid ball will jump to the diametrically opposite surface instead of passing smoothly "around" the edge to the southern hyperhemisphere.

24.2 3D FLOW DATA VISUALIZATIONS

We now examine some typical examples of streamline data, and see how the simultaneous application of our visualization methods to large families of curves helps emphasize important features and trends. We render each curve in the data set using the following alternative visualization modes.

- *Curve and curvature*: Use a 3D Euclidean space representation of the actual curve, pseudocolored by the curvature value. This singles out rapidly bending portions of the curve.
- *Curve and torsion*: Use a 3D Euclidean space picture, pseudocolored by torsion value. This singles out rapidly twisting portions of the curve.
- *Quaternion Frenet-frame map*: Plot the four-vector quaternion Frenet-frame fields in the three-sphere using one of our standard quaternion visualization methods. In the vector-only display method, pseudocolor can be used to remind us of the value of the hidden fourth q_0 component, even though it is in principle redundant.
- *Quaternion parallel-transport frame map*: Plot the four-vector quaternion parallel-transport frame field in the three-sphere using one of our standard quaternion visualization methods. The redundant fourth component can again be emphasized by color encoding.

24.2.1 AVS STREAMLINE EXAMPLE

Our first example is shown in Figure 24.1. These data are from a standard public AVS-generated streamline data set. The flow is obstructed somewhere in the center, causing sudden jumps of the streamlines and their quaternion maps in certain regions.

24.2.2 DEFORMING SOLID EXAMPLE

Our second example is derived from the simulation of a physical deformation process. If we imagine a solid wall of deformable spaghetti strands running from the top to the bottom of the wall, Figure 24.2 shows the result of twisting around the vertical axis and tracing the deformations of the spaghetti strands. The result is a symmetric but surprisingly complicated set of streamlines and quaternion maps.

24.3   BRUSHING: CLUSTERS AND INVERSE CLUSTERS

One of the most interesting properties of the quaternion frame method is the appearance of clusters of similar frame fields in the three-sphere display. Two reciprocal tools for exploring these properties immediately suggest themselves. In Figure 24.3, we illustrate the effect of grabbing a cluster of streamlines that are spatially close in 3D space and then highlighting their counterparts in the 4D quaternion field space, thus allowing the separate study of their moving frame properties. This technique distinguishes curves that are similar in 3D space but have drastically different frame characteristics.

Figure 24.4, in contrast, shows the result of selecting a cluster of curves with similar frame-field properties and then highlighting the original streamlines back in the 3D space display. This method assists in the location of similar curves that could not easily be singled out in the original densely populated spatial display. A variety of approaches can be used to design such tools for the exploration of streamlines.

24.4   ADVANCED VISUALIZATION APPROACHES

The quaternion frame curves displayed in Figures 24.1 and 24.2 are, strictly speaking, 2D projections of two overlaid 3D solid balls corresponding to the "front" and

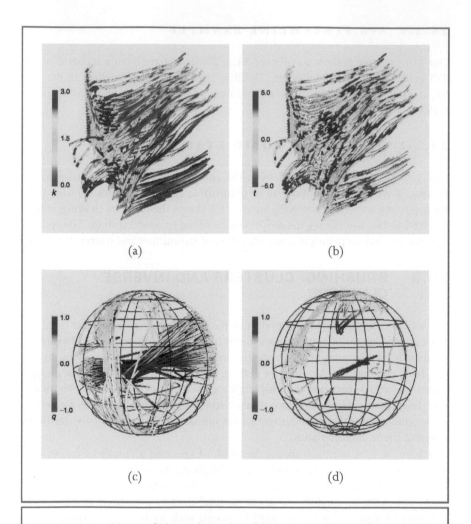

FIGURE 24.1 (a) *Vector field streamlines, color coded by curvature.* (b) *Vector field streamlines, color coded by torsion.* (c) *The corresponding quaternion field paths for the Frenet frames.* (d) *The corresponding quaternion field paths for the parallel transport frames. The color code is keyed to the value of the quaternion component q_0 that is collapsed in the projection from 4D to 3D. (AVS data set.)*

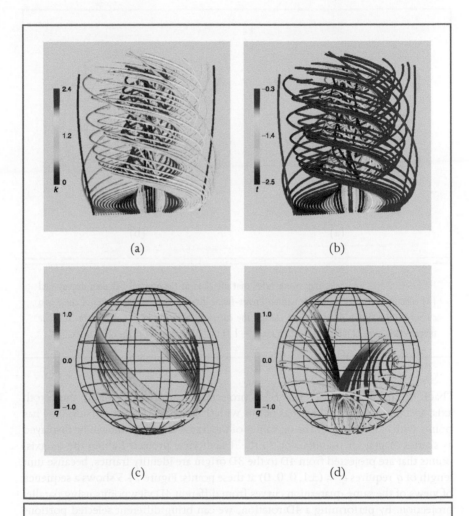

FIGURE 24.2 (a) Deformed volume, color coded by curvature. (b) Deformed volume, color coded by torsion. (c) The corresponding quaternion field paths for the Frenet frames. (d) The corresponding quaternion field paths for the parallel transport frames. The color code is keyed to the value of the quaternion component q_0 that is collapsed in the projection from 4D to 3D.

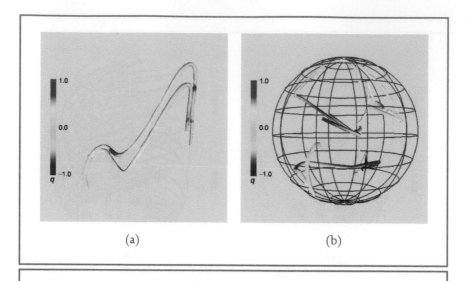

(a)   (b)

FIGURE 24.3 (a) *Selecting stream fields that are close in the original 3D data display and* (b) *echoing them in the 4D quaternion Frenet-frame display. The moving frames of these two curves are drastically different, even though the curves appear superficially similar in 3D. The unseen component of 4D depth, with a range* -1.0 *to* 1.0, *is mapped to the color index.*

"back" hemispheres of $\mathbf{S}^3$. This $\mathbf{S}^3$ is projected from 4D to 3D along the zeroth axis, and thus the "front" ball has points with $0 \leqslant q_0 \leqslant +1$ and the "back" ball has points with $-1 \leqslant q_0 < 0$. The q_0 values of the frame at each point can be displayed as shades of gray or pseudocolor. In the default view projected along the q_0 axis, points that are projected from 4D to the 3D origin are identity frames, because unit length of q requires $q = (\pm 1, 0, 0, 0)$ at these points. Figure 24.5 shows a sequence of views of the same quaternion curves from different 4D viewpoints using parallel projection. By performing a 4D rotation, we can bring different selected portions of the streamline quaternion map data set into the less-distorted area around the origin of the solid ball. Using the polar sphere projection methods of Chapter 17, we can create the alternative viewpoints shown in Figure 24.6, which show the additional contrast in structure sizes resulting from a 4D perspective projection.

The simplest viewing strategy plots wide lines that may be viewed in stereo or using motion parallax. A more expensive viewing strategy requires projecting a line

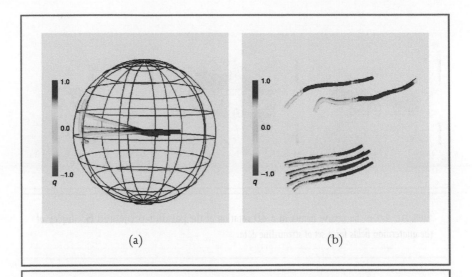

(a)   (b)

FIGURE 24.4 (a) Selecting stream fields that are close in the 4D quaternion Frenet-frame display and (b) echoing them in the original 3D data display, thus showing the locations of similar curves that could not be easily singled out in the original 3D spatial display. The unseen component of 4D depth, with a range −1.0 to 1.0, is mapped to the color index.

or solid from the 4D quaternion space and reconstructing an ideal tube in real time for each projected streamline. The parallel transport techniques introduced earlier are extremely relevant to this task, and may be applied to the tubing problem as well (see Bloomenthal [21] and Hanson and Ma [79]).

24.4.1 3D ROTATIONS OF QUATERNION DISPLAYS

Using, for example, the 3D rolling ball interface, we can generate quaternion representations of 3D rotations of the form $q = (\cos\frac{\theta}{2}, \hat{\mathbf{n}}\sin\frac{\theta}{2})$ and transform the entire quaternion display by quaternion multiplication (i.e., by changing each point to $p' = q \star p$). This effectively displaces the 3D identity frame in quaternion space from $(1, 0, 0, 0)$ to q. This may be useful when trying to compare curves whose properties differ by a rigid 3D rotation (a common occurrence in the parallel transport frame due to the arbitrariness of the initial condition).

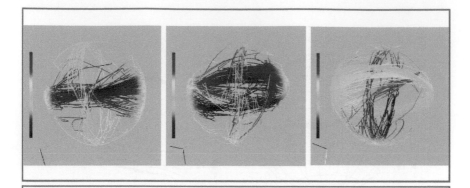

FIGURE 24.5 *Successive frames in a 4D rotation of the parallel-projected* $\mathbf{S}^3$ *display of the quaternion fields for a set of streamline data.*

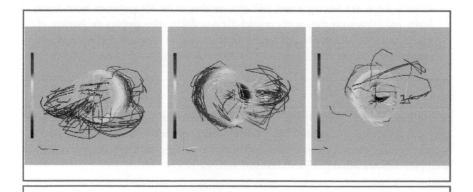

FIGURE 24.6 *Successive frames in a 4D rotation of the polar-projected* $\mathbf{S}^3$ *display of the quaternion fields for a set of streamline data.*

Other refinements might include selecting and rotating single streamlines in the quaternion field display to make interactive comparisons with other streamlines differing only by rigid rotations. One might also use automated tools to select rotationally similar structures based on minimizing the 4D scalar product between quaternion field points as a measure of similarity.

24.4.2   PROBING QUATERNION FRAMES WITH 4D LIGHT

We next explore techniques developed for the exploitation of probes that are based on 4D lighting models (e.g., see Hanson and Cross [72], Hanson and Heng [73, 76]). In these approaches, the critical element is the observation that 4D light can be used to selectively emphasize geometric structure provided we can find a way (such as thickening curves or surfaces until they become true three-manifolds) to define a unique 4D normal vector that has a well-defined scalar product with the 4D light. When that objective is achieved, we can interactively employ a moving 4D light and a generalization of the standard illumination equations to produce images that selectively expose new structural details.

Given a quaternion field, we may simply select a 4D unit vector L to represent a "light direction" and employ a standard lighting model such as $I(t) = L \cdot q(t)$ to select individual components of the quaternion fields for display using pseudocolor coding for the intensity.

Figure 24.7 shows a streamline data set rendered by computing a pseudocolor index at each point using the 4D lighting formula and varying the directions of the four-vector L.

True 4D illumination: For completeness, we remark that quaternion curves in 4D may also be displayed in an entirely different mode by thickening them—using the method of Hanson and Heng [73,76]—to form three-manifolds and replacing $q(t)$ in the 4D lighting formula and its specular analogs by the 4D normal vector for each volume element or vertex. Furthermore, the massive expense of volume rendering the resulting solid tubes comprising the 4D projection to 3D can be avoided by extending the "bear-hair" algorithm to 4D curves [14,72,108] and rendering the tubes in the limit of vanishing radius.

4D light orientation control: Direct manipulation of 3D orientation using a 2D mouse is typically handled using a rolling ball [66] or virtual sphere [30] method to give the user a feeling of physical control. This philosophy extends well to 4D orientation control [34,69], giving a practical approach to interacting with the visualization approaches exploiting 4D lighting.

A 3D unit vector has only two degrees of freedom, and is thus determined by picking a point within a unit circle to determine the direction uniquely up to

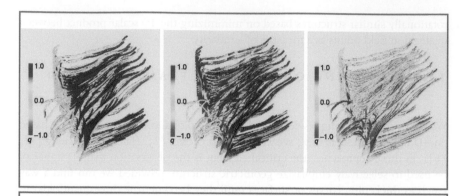

FIGURE 24.7 *Color coding a streamline data set using an interactively moving 4D "light" as a probe to isolate similar components of the quaternion fields associated to each point of each curve.*

the sign of its view-direction component. The analogous control system for 4D lighting is based on a similar observation: Because the 4D normal vector has only three independent degrees of freedom, choosing an *interior point* in a solid sphere determines the vector uniquely up to the sign of its component in the unseen fourth dimension (the "4D view-direction component"). Figure 24.7 shows via a series of snapshots an example of this interactive interface at work. One can optionally print or display the components of the 4D light vector at any particular moment.

Quaternion Interpolation

25

In this chapter we pursue the details of parametric quaternion interpolation directly on $\mathbf{S}^3$. The main issue is the problem of creating smooth transitions among orientation frames while retaining the convenience of parametric forms familiar from Euclidean polynomial parameterizations of curves and surfaces. We will also briefly discuss alternate methods of logarithmic-space interpolation and direct variational optimization of spherical curves.

The methods of quaternion animation splines were introduced to the graphics community originally by Shoemake [149]. This chapter provides an overview of the techniques of constructing splines with various desirable continuity properties, following the method of Schlag [145] applied to quaternion Bezier, Catmull–Rom, and uniform B-splines. Some problems (such as those involving derivative computation) may be resolved by the exponential map approach of Kim et al. (e.g., see [59,111]), which is briefly noted, along with the variational methods of Barr et al. [15,140]. Alternative approaches, such as the rational quaternion spline method of Jüttler [104,105], are not treated in detail. We begin with some basic technology of Euclidean splines to lay the mathematical groundwork, and then apply that intuition to the development of several useful families of quaternion splines.

25.1   CONCEPTS OF EUCLIDEAN LINEAR INTERPOLATION

In a Euclidean space of any dimension, the linear interpolation defining a parameterized straight line between two points $\mathbf{x}_0 = (x_0, y_0, z_0, \ldots)$ and $\mathbf{x}_1 = (x_1, y_1, z_1, \ldots)$ is given by

$$\mathbf{x}(t) = (1 - t)\mathbf{x}_0 + t\mathbf{x}_1$$

$$= \mathbf{x}_0 + t(\mathbf{x}_1 - \mathbf{x}_0). \tag{25.1}$$

An important feature of Euclidean interpolation is that each dimension is treated independently. The first form is essentially a linear solution of the barycentric coordinate form

$$\mathbf{x}(u_0, u_1) = u_0\mathbf{x}_0 + u_1\mathbf{x}_1,$$

where $u_0 + u_1 = 1$ is the barycentric constraint in one dimension (shown in Figure 25.1). Note that u_0 is *opposite* the vertex $\mathbf{x}_0$, and thus $u_0 = 1$ gives full weight to the point $\mathbf{x}_0$. Choosing

$$u_0 = 1 - t,$$

$$u_1 = t \tag{25.2}$$

gives the first parameterization, which is essentially the lowest-order Bernstein polynomial basis. The second form of Equation 25.1, which starts with the con-

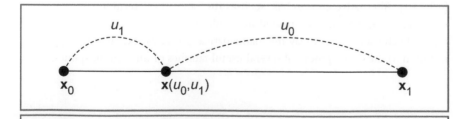

FIGURE 25.1 *The 1D barycentric weight partition for an interpolated straight line.*

stant $\mathbf{x}_0$, has the feature that it exposes the derivative of the curve,

$$\frac{d\mathbf{x}(t)}{dt} = \dot{\mathbf{x}}(t) = (\mathbf{x}_1 - \mathbf{x}_0),$$

which is the coefficient of t; that is, the constant slope of the straight line $\mathbf{x}(t)$.

25.1.1   CONSTRUCTING HIGHER–ORDER POLYNOMIAL SPLINES

Euclidean polynomial splines of higher order can be derived iteratively using the de Casteljau construction, which takes a set of anchor points $\{\mathbf{x}_0, \mathbf{x}_1, \mathbf{x}_2, \ldots, \mathbf{x}_n\}$ and the linear interpolants

$$\mathbf{x}_{01}(t) = (1 - t)\mathbf{x}_0 + t\mathbf{x}_1,$$

$$\mathbf{x}_{12}(t) = (1 - t)\mathbf{x}_1 + t\mathbf{x}_2,$$

$$\vdots$$

$$\mathbf{x}_{ij}(t) = (1 - t)\mathbf{x}_i + t\mathbf{x}_j$$

and linearly interpolates each pair of these

$$\mathbf{x}_{012}(t) = (1 - t)\mathbf{x}_{01}(t) + t\mathbf{x}_{12}(t),$$

$$\vdots$$

$$\mathbf{x}_{i,i+1,i+2}(t) = (1 - t)\mathbf{x}_{i,i+1}(t) + t\mathbf{x}_{i+1,i+2}(t).$$

Repeating the process (as indicated schematically in Figure 25.2) generates a polynomial of order n,

$$\mathbf{x}_{01\ldots n}(t) = \sum_k c_k \mathbf{x}_k (1 - t)^{n-k} t^k, \tag{25.3}$$

which is the Bezier spline of order n.

25.1.2   MATCHING

To string together several sets of Bezier splines, it is necessary to match all derivatives at the end point (here, $\mathbf{x}_n$) with those at the starting point of the next set

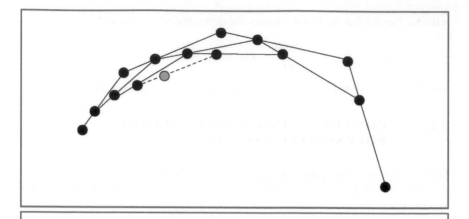

FIGURE 25.2 *Nested linear splines in the de Casteljau framework produce arbitrary polynomial spline curves.*

(say, $\mathbf{y}_0$). This is feasible but tedious, and thus one typically looks instead for splines with sliding sets of "windows" in the set of anchor points to automatically handle the continuity of one or more derivatives.

The Catmull–Rom spline (sometimes called a cardinal spline) produces a cubic spline that passes exactly through a set of anchor point *values* and matches *first derivatives* by using a sliding window of four anchor points (see the simple cases shown in Figure 25.3). The uniform cubic B-spline matches *both* the first and second derivatives of each point in the sliding window, but has insufficient degrees of freedom to force the curve to pass through the anchor points as the Catmull–Rom does; see Figure 25.4.

For the cubic spline, the three standard interpolations (corresponding, respectively, to the Bezier spline, Catmull–Rom spline, and uniform B-spline) are as follows.

$$Bz(t) = (1-t)^3 \mathbf{x}_0 + 3t(1-t)^2 \mathbf{x}_1 + 3t^2(1-t)\mathbf{x}_2 + t^3 \mathbf{x}_3,$$

$$Cr(t) = -\frac{1}{2}t(1-t)^2 \mathbf{x}_0 + \frac{1}{2}(1-t)\left(2 + 2t + 3t^2\right)\mathbf{x}_1$$

$$+ \frac{1}{2}t\left(1 + 4t - 3t^2\right)\mathbf{x}_2 + -\frac{1}{2}(1-t)t^2 \mathbf{x}_3,$$

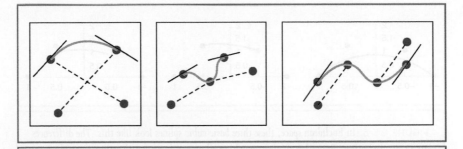

FIGURE 25.3 *Examples of the behavior of the Euclidean Catmull–Rom spline.*

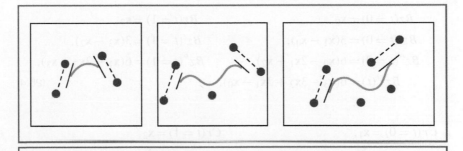

FIGURE 25.4 *Examples of the behavior of the cubic uniform B-spline.*

$$Ub(t) = \frac{1}{6}(1-t)^3 \mathbf{x}_0 + \frac{1}{6}\left(4 - 6t^2 + 3t^3\right)\mathbf{x}_1$$

$$+ \frac{1}{6}\left(1 + 3t + 3t^2 - 3t^3\right)\mathbf{x}_2 + \frac{1}{6}t^3 \mathbf{x}_3.$$

In Euclidean space, side-by-side comparison of the three cubic splines (shown in Figure 25.5) clearly exposes their fundamental characteristics. Their derivatives give the directions of their slopes as follows.

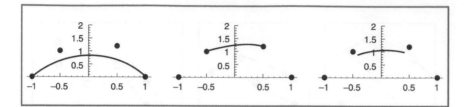

FIGURE 25.5 In Euclidean space, these three basic cubic splines look like this. The differences are in the derivatives: Bezier has to start matching all over at every fourth point, Catmull–Rom matches the first derivative, and B-spline is the "Cadillac," matching all derivatives but matching no control points.

$$
\begin{aligned}
Bz(t=0) &= \mathbf{x}_0, & Bz(t=1) &= \mathbf{x}_3, \\
Bz'(t=0) &= 3(\mathbf{x}_1 - \mathbf{x}_0), & Bz'(t=1) &= 3(\mathbf{x}_3 - \mathbf{x}_2), \\
Bz''(t=0) &= 6(\mathbf{x}_0 - 2\mathbf{x}_1 + \mathbf{x}_2), & Bz''(t=1) &= 6(\mathbf{x}_1 - 2\mathbf{x}_2 + \mathbf{x}_3), \\
Bz'''(t) &= 6(\mathbf{x}_3 - 3\mathbf{x}_2 + 3\mathbf{x}_1 - \mathbf{x}_0), & &\text{[25.4]}
\end{aligned}
$$

$$
\begin{aligned}
Cr(t=0) &= \mathbf{x}_1, & Cr(t=1) &= \mathbf{x}_2, \\
Cr'(t=0) &= \frac{1}{2}(\mathbf{x}_2 - \mathbf{x}_0), & Cr'(t=1) &= \frac{1}{2}(\mathbf{x}_3 - \mathbf{x}_1), \\
Cr''(t=0) &= (2\mathbf{x}_0 - 5\mathbf{x}_1 + 4\mathbf{x}_2 - \mathbf{x}_3), & Cr''(t=1) &= (-\mathbf{x}_0 + \mathbf{x}_1 - 5\mathbf{x}_2 + 2\mathbf{x}_3), \\
Cr'''(t) &= 3(\mathbf{x}_3 - 3\mathbf{x}_2 + 3\mathbf{x}_1 - \mathbf{x}_0), & &\text{[25.5]}
\end{aligned}
$$

$$
\begin{aligned}
Ub(t=0) &= \frac{1}{6}(\mathbf{x}_0 + 4\mathbf{x}_1 + \mathbf{x}_2), & Ub(t=1) &= \frac{1}{6}(\mathbf{x}_1 + 4\mathbf{x}_2 + \mathbf{x}_3), \\
Ub'(t=0) &= \frac{1}{2}(\mathbf{x}_2 - \mathbf{x}_0), & Ub'(t=1) &= \frac{1}{2}(\mathbf{x}_3 - \mathbf{x}_1), \\
Ub''(t=0) &= (\mathbf{x}_0 - 2\mathbf{x}_1 + \mathbf{x}_2), & Ub''(t=1) &= (\mathbf{x}_1 - 2\mathbf{x}_2 + \mathbf{x}_3), \\
Ub'''(t) &= (\mathbf{x}_3 - 3\mathbf{x}_2 + 3\mathbf{x}_1 - \mathbf{x}_0). & &\text{[25.6]}
\end{aligned}
$$

25.1.3 SCHLAG'S METHOD

There are several ways of rewriting these expressions to achieve a uniform representation. One is the iterated linear interpolation method of Schlag [145], which solves for the linear combination $a + bt$ at each level of the de Casteljau algorithm that actually transforms the spline into each of the forms above simply by substituting the appropriate expression $(a + bt)$ into each level of the linear interpolation. Schlag defines the linear interpolator

$$L(a, b; t) = (1 - t)a + tb,$$ (25.7)

writes

$$S(x_1, x_2, x_3, x_4; t) = L\big(L\big(L(x_1, x_2; f_{12}(t)),\ L(x_2, x_3; f_{23}(t)); f_{123}(t)\big),$$
$$L\big(L(x_2, x_3; f_{23}(t)),\ L(x_3, x_4; f_{34}(t)); f_{234}(t)\big);$$
$$f(t)\big),$$

and solves for the forms of the set of $f(t)$'s for each cubic spline:

- Bezier: The de Casteljau algorithm for the Bezier spline is the trivial case $f(t) = t$.

$$
\begin{array}{ccccc}
f_{12} = t & & f_{23} = t & & f_{34} = t \\
& f_{123} = t & & f_{234} = t & \\
& & f = t & &
\end{array}
$$

- Catmull–Rom:

$$
\begin{array}{ccccc}
f_{12} = t + 1 & & f_{23} = t & & f_{34} = t - 1 \\
& f_{123} = \frac{(t+1)}{2} & & f_{234} = \frac{t}{2} & \\
& & f = t & &
\end{array}
$$

- Uniform B-spline:

$$f_{12} = \frac{(t+2)}{3} \qquad\qquad f_{23} = \frac{(t+1)}{3} \qquad\qquad f_{34} = \frac{t}{3}$$
$$f_{123} = \frac{(t+1)}{2} \qquad\qquad\qquad f_{234} = \frac{t}{2}$$
$$f = t$$

This procedure can obviously be repeated for any additional level of nested interpolations $L(a, b; f(t))$ to generate any desired polynomial order for any spline.

25.1.4 CONTROL–POINT METHOD

An alternative approach is to ask: "What control points substituted into the de
Casteljau algorithm produce the Catmull–Rom and uniform B-spline?" Because
the control points always appear linearly, there must exist some combination that
achieves this. If we simply substitute

$$Cr(\mathbf{x}_0, \mathbf{x}_1, \mathbf{x}_2, \mathbf{x}_3; t) = Bz(\mathbf{u}_0, \mathbf{u}_1, \mathbf{u}_2, \mathbf{u}_3; t),$$
$$Ub(\mathbf{x}_0, \mathbf{x}_1, \mathbf{x}_2, \mathbf{x}_3; t) = Bz(\mathbf{v}_0, \mathbf{v}_1, \mathbf{v}_2, \mathbf{v}_3; t),$$

we find the solution

$$\mathbf{u}_0^{cr} = \mathbf{x}_1,$$
$$\mathbf{u}_1^{cr} = \mathbf{x}_1 + \frac{1}{6}(\mathbf{x}_2 - \mathbf{x}_0),$$
$$\mathbf{u}_2^{cr} = \mathbf{x}_2 + \frac{1}{6}(\mathbf{x}_1 - \mathbf{x}_3),$$
$$\mathbf{u}_3^{cr} = \mathbf{x}_2,$$
$$\mathbf{x}_0^{cr} = \mathbf{u}_3 + 6(\mathbf{u}_0 - \mathbf{u}_1),$$
$$\mathbf{x}_1^{cr} = \mathbf{u}_0,$$
$$\mathbf{x}_2^{cr} = \mathbf{u}_3,$$
$$\mathbf{x}_2^{cr} = \mathbf{u}_0 + 6(\mathbf{u}_3 - \mathbf{u}_2)$$

for the Catmull–Rom and

$$\mathbf{u}_0^{ub} = \frac{1}{6}(\mathbf{x}_0 + 4\mathbf{x}_1 + \mathbf{x}_2),$$

$$\mathbf{u}_1^{ub} = \frac{1}{3}(2\mathbf{x}_1 + \mathbf{x}_2),$$

$$\mathbf{u}_2^{ub} = \frac{1}{3}(\mathbf{x}_1 + 2\mathbf{x}_2),$$

$$\mathbf{u}_3^{ub} = \frac{1}{6}(\mathbf{x}_1 + 4\mathbf{x}_2 + \mathbf{x}_3),$$

$$\mathbf{x}_0^{ub} = 6\mathbf{u}_0 - 7\mathbf{u}_1 + 2\mathbf{u}_2,$$

$$\mathbf{x}_1^{ub} = 2\mathbf{u}_1 - \mathbf{u}_2,$$

$$\mathbf{x}_2^{ub} = -\mathbf{u}_1 + 2\mathbf{u}_2,$$

$$\mathbf{x}_2^{ub} = 2\mathbf{u}_1 - 7\mathbf{u}_2 + 6\mathbf{u}_3$$

for the uniform B-spline.

If we examine the Catmull–Rom solution in Figure 25.6, we see that the interior Bezier control points have become slope-controlling combinations:

$$\mathbf{u}_1^{cr} - \mathbf{u}_0^{cr} = \frac{1}{6}(\mathbf{x}_2 - \mathbf{x}_0), \qquad \mathbf{u}_3^{cr} - \mathbf{u}_2^{cr} = \frac{1}{6}(\mathbf{x}_3 - \mathbf{x}_1). \qquad (25.8)$$

Similarly, for the uniform B-spline we have

$$\mathbf{u}_1^{ub} - \mathbf{u}_0^{ub} = \frac{1}{6}(\mathbf{x}_2 - \mathbf{x}_0), \qquad \mathbf{u}_3^{ub} - \mathbf{u}_2^{ub} = \frac{1}{6}(\mathbf{x}_3 - \mathbf{x}_1). \qquad (25.9)$$

These observations are of some importance to us for several reasons. First, we shall see that the Schlag procedure generalizes perfectly from Euclidean linear interpolators to unit-length-preserving spherical interpolators, which is what we require for 3D orientation frame interpolation. Second, we will see that the alternative slope-controlling transformation also has a direct relationship to an alternative view of quaternion anchor points when we discover how *linear algebra* has exact analogs in *spherical algebra*.

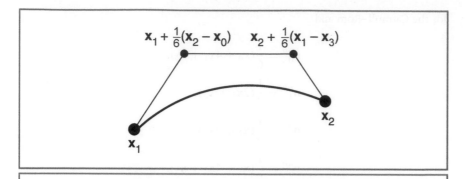

FIGURE 25.6 *Diagram of the values to use in the Bezier iterated spline formula with* $\mathbf{u}_0(\mathbf{x})$, $\mathbf{u}_1(\mathbf{x})$, $\mathbf{u}_2(\mathbf{x})$, $\mathbf{u}_3(\mathbf{x})$ *to get the same result as the Catmull–Rom formula with anchor points* $\mathbf{x}_0$, $\mathbf{x}_1$, $\mathbf{x}_2$, $\mathbf{x}_3$.

25.2 THE DOUBLE QUAD

There is one further transformation among the linear splines (named the "Squad" by Shoemake [149]) that is of interest because it reduces the required number of linear interpolations and is therefore more efficient. Suppose we take a set of four Bezier control points $\{\mathbf{x}_0, \mathbf{x}_1, \mathbf{x}_2, \mathbf{x}_3\}$ and form first a pair of "bottom" and "top" linear splines (as in Figure 25.7):

$$\mathbf{x}_{\mathrm{bot}}(t) = (1 - t)\mathbf{x}_0 + t\mathbf{x}_3,$$

$$\mathbf{x}_{\mathrm{top}}(t) = (1 - t)\mathbf{x}_1 + t\mathbf{x}_2.$$

Then, if we perform a third interpolation using a quadratic function $2t(1 - t)$—whose features are plotted in Figure 25.8—instead of our usual linear function of t, we find

$$\mathbf{x}_{\mathrm{squad}}(t) = L\big(\mathbf{x}_{\mathrm{bot}}(t), \mathbf{x}_{\mathrm{top}}(t); 2t(1 - t)\big)$$

$$= \big(1 - 2t + 2t^2\big)\mathbf{x}_{\mathrm{bot}}(t) + 2t(1 - t)\mathbf{x}_{\mathrm{top}}(t)$$

$$= \big(1 - 2t + 2t^2\big)(1 - t)\mathbf{x}_0 + 2t(1 - t)^2\mathbf{x}_1$$

$$+ 2t^2(1 - t)\mathbf{x}_2 + \big(1 - 2t + 2t^2\big)\mathbf{x}_3.$$

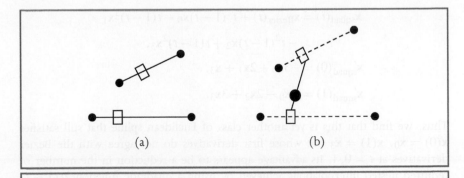

FIGURE 25.7 *The initial stage of the Squad interpolates separately along a "top" and a "bot-tom" straight line.*

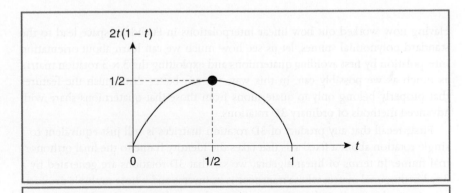

FIGURE 25.8 *The curve $2t(1-t)$ that goes symmetrically to zero at both ends of the range $0 \leqslant t \leqslant 1$, thereby interpolating efficiently with the correct limits.*

However, what is this with respect to our familiar cubic spline methods? It is in fact a completely different cubic curve, which reduces to the same end points as the standard Bezier curve. We can verify this by going through the algebra to subtract the Bezier formula, with the result

$$\mathbf{x}_{\text{squad}}(t) = \mathbf{x}_{\text{Bezier}}(t) + t^2(1-t)\mathbf{x}_0 - t(1-t)^2\mathbf{x}_1$$
$$\qquad\qquad - t^2(1-t)\mathbf{x}_2 + t(1-t)^2\mathbf{x}_3,$$
$$\mathbf{x}'_{\text{squad}}(0) = -3\mathbf{x}_0 + 2\mathbf{x}_1 + \mathbf{x}_3,$$
$$\mathbf{x}'_{\text{squad}}(1) = -\mathbf{x}_0 - 2\mathbf{x}_2 + 3\mathbf{x}_3.$$

Thus, we find that this is yet another class of Euclidean spline that still satisfies $\mathbf{x}(0) = \mathbf{x}_0$, $\mathbf{x}(1) = \mathbf{x}_3$ but whose first derivatives do not agree with the Bezier derivatives at $t = 0, 1$. Its advantage appears to be a reduction in the number of required nested interpolations achieved by using a quadratic function $2t(1-t)$ in the interpolator. However, it is unclear just what geometric motivation suggests that the "Squad" is a significant improvement over any of the standard splines.

25.3 DIRECT INTERPOLATION OF 3D ROTATIONS

Having now worked out how linear interpolations in Euclidean space lead to the standard polynomial splines, let us see how much we can learn about orientation interpolation by first avoiding quaternions and exploiting the 3×3 rotation matrix as much as we possibly can. In this way, we can better distinguish the features that properly belong only to quaternions from those that quaternions share with advanced methods of ordinary 3D rotations.

First, recall that any product of 3D rotation matrices is still just equivalent to a single rotation about a fixed axis that takes the identity frame to the final orthonormal frame. In terms of linear algebra, we say that 3D rotations are generated by a 3×3 orthogonal matrix whose transpose is its inverse and whose multiple products preserve orthogonality: the columns of an orthogonal matrix give the orientation vectors of the axes into which the identity frame is transformed. In terms of group theory, we say that any 3×3 3D rotation matrix is an element of the Lie group $\mathbf{SO}(3)$ and that matrix multiplication by definition preserves membership in the group.

The critical feature of an arbitrary rotation A (possibly resulting from any number of successive component rotations) is that, from Euler's theorem, the matrix has a single unique, real eigenvector $\hat{\mathbf{n}}$ and that there is an angle θ (unique up to 2π periodicity) that gives the rotation about that axis that goes from the identity matrix I to the new matrix A whose columns are the transformed coordinate frame axes. This matrix is by now well known to us:

$$R(\theta, \hat{\mathbf{n}}) = \text{Rotate Identity by } \theta \text{ about } \hat{\mathbf{n}} \text{ to give } A,$$

$$A = R(\theta, \hat{\mathbf{n}}) \cdot I. \tag{25.10}$$

We can see from Figure 25.9 that the rotation fixing the axis $\hat{\mathbf{n}}$ is the most direct route from the identity to the frame A. This operation can be thought of as rotating a platform or spinning a gyroscope whose axis is $\hat{\mathbf{n}}$ and whose equator is the unit circle (Figure 25.9) perpendicular to $\hat{\mathbf{n}}$ at the origin. Thus, the *smoothest angular interpolation* results when we simply replace θ with $t\theta$:

$$\text{Interp}(t) = R(t\theta, \hat{\mathbf{n}}),$$

$$\text{Interp}(0) = I,$$

$$\text{Interp}(1) = \Lambda.$$

25.3.1 RELATION TO QUATERNIONS

We already know that the quaternion

$$q(\theta, \hat{\mathbf{n}}) = \big(\cos(\theta/2), \hat{\mathbf{n}} \sin(\theta/2)\big)$$

is a point on $\mathbf{S}^3$, and that $q(t\theta, \hat{\mathbf{n}})$ parameterizes a great circle, the shortest-distance path on the hypersphere $\mathbf{S}^3$ as $t : 0 \to 1$. Furthermore, $q(t\theta, \hat{\mathbf{n}})$ coincides point for point (via the quadratic map) with the 3D interpolation matrix $R(t\theta, \hat{\mathbf{n}})$. Therefore, we conclude that for single-axis rotations $R(t\theta, \hat{\mathbf{n}})$ is going to work as well as the quaternion $q(t\theta, \hat{\mathbf{n}})$. The only important difference for this special case is that *relative distances* between two frames are measured in units of $t\theta/2$ radians in quaternion space, instead of $t\theta$ radians in ordinary space (if we had never heard of quaternions). The concrete evidence of an essential difference between the two spaces, however, must be kept in mind—as the belt trick (Chapter 12) showed us quite directly.

Note: The geometric properties of the hypersphere $\mathbf{S}^3$ are precisely described by Riemannian geometry when we define a standard metric on $\mathbf{S}^3$. Topologically, $\mathbf{S}^3$ is a *simply connected space*, which is the deeper mathematical reason the belt trick (Chapter 12) works the way it does. On $\mathbf{S}^3$, Riemannian geometry precisely defines the

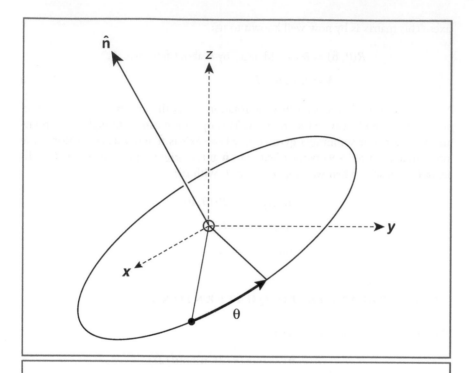

FIGURE 25.9 *Spinning plane perpendicular to the Euler eigenvector.*

path with shortest distance (or geodesic) in terms of the metric, and this can be verified to correspond to the path parameterized by t in $q(t\theta, \hat{\mathbf{n}})$ as $t : 0 \to 1$. The set of points $q(t\theta, \hat{\mathbf{n}})$ coincide point for point, via the quadratic map, with the 3D interpolation matrix $R(t\theta, \hat{\mathbf{n}})$. However, R is a 3×3 orthogonal matrix—a member of the group $\mathbf{SO}(3)$ and the topological space $\mathbf{RP}^3$ (real projective space), which is *not* a simply-connected metric space.

25.3.2 METHOD FOR ARBITRARY ORIGIN

As one can deduce from methods we have already seen, interpolating between nonidentity frames is qualitatively similar to a Euclidean translation, except that we

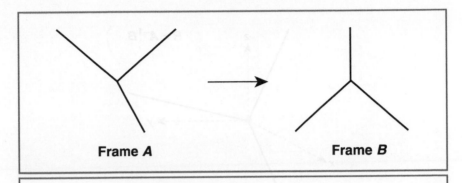

Frame A   **Frame B**

FIGURE 25.10 *Set up the task of going from frame A to frame B.*

apply a group transformation to the identity frame (the identity element of the rotation group). Given two frames A and B (as shown schematically in Figure 25.10) the shortest-angular-distance rotation matrix $M(t\theta, \hat{\mathbf{n}})$, taking $A \to B$ with

$$M(0, \hat{\mathbf{n}}) = A,$$

$$M(\theta, \hat{\mathbf{n}}) = B, \tag{25.11}$$

can be found starting from

$$R = A^{-1}B = A^T B. \tag{25.12}$$

(See Figure 25.11.) The eigenvector of R, $\hat{\mathbf{n}}$ such that

$$R \cdot \hat{\mathbf{n}} = \hat{\mathbf{n}},$$

can be found from linear algebra or directly from the elements of $(R - R^T)$. The angle θ, as usual, comes from the trace, $\operatorname{tr}(A^{-1}B) = 2\cos(\theta) - 1$. (See also Chapters 6 and 16.) Thus, we immediately find all that is necessary to directly construct

$$R(\theta, \hat{\mathbf{n}}) = A^{-1}B \tag{25.13}$$

and the great-circle interpolation $R(t\theta, \hat{\mathbf{n}})$ from the identity to $A^{-1}B$. Prepending the initial condition from Equation 25.11 to displace the identity to the frame A,

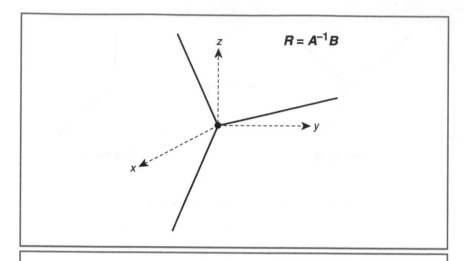

FIGURE 25.11 *The structure of* $(A^{-1}B)$.

we have

$$M(t\theta, \hat{\mathbf{n}}) = A \cdot R(t\theta, \hat{\mathbf{n}}) \qquad \text{(25.14)}$$

as the general solution (see Figure 25.12). Hence, we basically find the difference between the two rotations, extract the axis, and use this effective SLERP to perform an incremental rotation.

25.3.3 EXPONENTIAL VERSION

Many authors find it useful to use the behavior of matrix exponents and logarithms to express Equation 25.14 in an alternative fashion. Writing

$$\log R(t\theta, \hat{\mathbf{n}}) = t\theta\hat{\mathbf{n}} \cdot \mathbf{L} \qquad \text{(25.15)}$$

so that

$$R = e^{t\theta\hat{\mathbf{n}} \cdot \mathbf{L}} = R(\theta, \hat{\mathbf{n}})^t,$$

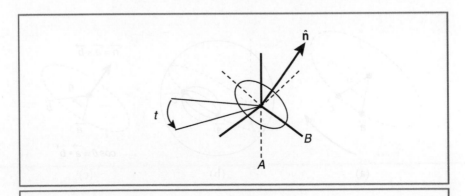

FIGURE 25.12 *Continuous transition from A to B using the parameter t. Note that the dotted lines are the frame A in Figure 25.11 and the heavy lines are the frame B in Figure 25.11.*

we arrive at

$$M(t\theta, \hat{\mathbf{n}}) = A\left(A^{-1}B\right)^t = A\, e^{(t \log(A^{-1}B))} \tag{25.16}$$

as an alternative means of looking at the great-circle interpolation in terms of 3×3 rotation matrices. However, its simplicity is deceptive. Although Equation 25.16 is formally useful (e.g., for derivatives), in fact one must always compute $\hat{\mathbf{n}}$ by hand from $A^{-1}B$ and return to the matrix $R(t\theta, \hat{\mathbf{n}})$ to perform the actual transformation. The quaternion form of the formula embodying the correct distance measure on $\mathbf{S}^3$ can be immediately deduced as

$$m(t\theta, \hat{\mathbf{n}}) = q_A \star \left(q_A^{-1} \star q_B\right)^t = q_A \star q(t\theta, \hat{\mathbf{n}}_{AB}). \tag{25.17}$$

25.3.4 SPECIAL VECTOR–VECTOR CASE

An extremely common special case involves the task of smoothly rotating one vector to another in their common plane. Because the final result is as always a fixed-eigenvector interpolation, this is equivalent to the frame-frame interpolation problem but with a different phrasing of the initial data. Figure 25.13a shows the conditions of the task, which requires the construction of a rotation matrix that takes the three-vector direction $\hat{\mathbf{a}}$ to the direction $\hat{\mathbf{b}}$. Any $\hat{\mathbf{n}}$ in the plane bisecting the arc

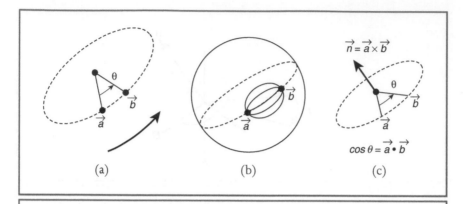

FIGURE 25.13 (a) Diagram of the fixed axis rotating $\hat{\mathbf{a}}$ to $\hat{\mathbf{b}}$. (b) Choices of circles that connect a to b on the sphere. (c) Minimal-length solution with $\hat{\mathbf{n}} = \hat{\mathbf{a}} \times \hat{\mathbf{b}}/|\hat{\mathbf{a}} \times \hat{\mathbf{b}}|$ and $\cos\theta = \hat{\mathbf{a}} \cdot \hat{\mathbf{b}}$.

of the great circle from $\hat{\mathbf{a}} \to \hat{\mathbf{b}}$ on $\mathbf{S}^2$ will have a θ that accomplishes this, as shown schematically in Figure 25.13b. However, only one unique $\hat{\mathbf{n}}$ (up to a sign) and its θ will give the minimal arc length $\theta = \cos^{-1}(\hat{\mathbf{a}} \cdot \hat{\mathbf{b}})$ in 3D, and $\theta/2 = (1/2)\cos^{-1}(\hat{\mathbf{a}} \cdot \hat{\mathbf{b}})$ in quaternion space.

The rotation parameters of the unique minimal-arc-length rotation, as shown in Figure 25.13c, can be immediately read off as

$$\cos\theta = \hat{\mathbf{a}} \cdot \hat{\mathbf{b}}, \qquad \hat{\mathbf{n}} = \frac{\hat{\mathbf{a}} \times \hat{\mathbf{b}}}{|\hat{\mathbf{a}} \times \hat{\mathbf{b}}|}, \tag{25.18}$$

and thus

$$R(\hat{\mathbf{a}} \to \hat{\mathbf{b}}) = R\left(\cos^{-1}(\hat{\mathbf{a}} \cdot \hat{\mathbf{b}}), \frac{\hat{\mathbf{a}} \times \hat{\mathbf{b}}}{|\hat{\mathbf{a}} \times \hat{\mathbf{b}}|}\right). \tag{25.19}$$

This is guaranteed of course to leave the plane defined by $\hat{\mathbf{a}}$ and $\hat{\mathbf{b}}$ fixed (in that $\hat{\mathbf{a}} \cdot \hat{\mathbf{n}} = \hat{\mathbf{b}} \cdot \hat{\mathbf{n}} = 0$), and thus all linear combinations of $\hat{\mathbf{a}}$ and $\hat{\mathbf{b}}$, forming all possible points in the plane, are perpendicular to $\hat{\mathbf{n}}$ as well. See Appendix F for an elegant quaternion version of this transformation.

25.3.5 MULTIPLE-LEVEL INTERPOLATION MATRICES

Now the question that should be uppermost in our minds is "what next?" In our Euclidean-space introductory sections, we constructed polynomial splines from multiple anchor points using the de Casteljau algorithm, and then showed that all polynomial splines could be rephrased in this way by using suitably adjusted linear combinations of the anchor points. Even though we may be aware of the quaternion-based approach to orientation splines, what if we plunge ahead and work directly with 3D frames to see how far we can go? To take an explicit example, consider three frames A, B, C and two interpolations, as illustrated in Figures 25.14 and 25.15:

$$T_{AB}(t) = A\left(A^{-1}B\right)^t = A\,R(t\theta_{AB}, \hat{\mathbf{n}}_{AB}),$$

$$T_{BC}(t) = B\left(B^{-1}C\right)^t = B\,R(t\theta_{BC}, \hat{\mathbf{n}}_{BC}).$$

Here, we define

$$\cos\theta_{AB} = \text{angle}\left(A^{-1}B\right) \text{ from trace formula},$$

$$\hat{\mathbf{n}}_{AB} = \text{normalized eigenvector of }\left(A^{-1}B\right), \text{etc.},$$

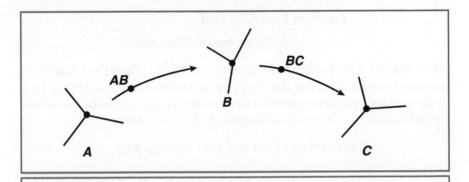

FIGURE 25.14 Transition from frame A to frame B to frame C.

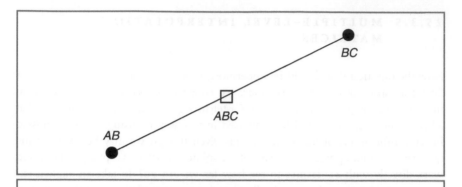

FIGURE 25.15 *Given intermediate frames AB and BC, we can form the equivalent of a quadratic Bezier spline by forming the joint interpolator ABC.*

or equivalently but less usefully

$$\theta_{AB}\hat{\mathbf{n}}_{AB} \cdot \mathbf{L} = \log A^{-1}B, \qquad (25.20)$$

where $\mathbf{L}$ is the set of 3×3 antisymmetric matrices we used in Chapter 18 to define rotations by matrix exponentiation.

If we now attempt the analog of a quadratic Bezier spline by interpolating from T_{AB} to T_{AC}, we would find

$$T_{ABC}(t) = T_{AB}(t)\left[T_{AB}^{-1}T_{BC}\right]^{t}$$

$$= A\, R(t\theta_{AB}, \hat{\mathbf{n}}_{AB})\, R(t\theta_{ABC}, \hat{\mathbf{n}}_{ABC}),$$

where $\hat{\mathbf{n}}_{ABC}(t)$ is the t-dependent eigenvector of $T_{AB}^{-1}(t) \cdot T_{BC}(t)$ and θ_{ABC} is the amount of rotation about that axis, obtainable as usual from the trace of $[T_{AB}^{-1}T_{BC}]$. To compare the equivalent quaternion form, we let (q_A, q_B, q_C) be the quaternions equivalent to the 3×3 orthogonal matrices A, B, C and examine

$$q_{AB}(t) = q_A \star (\bar{q}_A \star q_B)^t = q_A \star q(t\theta_{AB}, \hat{\mathbf{n}}_{AB}), \qquad (25.21)$$

where we trivially extract the needed quantities from the quaternion product

$$\bar{q}_A \star q_B = \left(\cos\frac{\theta_{AB}}{2}, \hat{\mathbf{n}}_{AB}\sin\frac{\theta_{AB}}{2}\right)$$

or (again, equivalently but with only formal utility)

$$\log \bar{q}_A \star q_B = \left(0, \frac{1}{2}\theta_{AB}\hat{\mathbf{n}}_{AB}\right).$$

Thus,

$$q_{ABC}(t) = q_{AB}(t) \star [\bar{q}_{AB} \star q_{BC}]^t$$

$$= q_{AB}(t) \star q\left(t\theta_{ABC}(t), \hat{\mathbf{n}}_{ABC}(t)\right),$$

where all quaternion quantities by now are well known and have clear analogs in 3×3 orthogonal matrices.

25.3.6 EQUIVALENCE OF QUATERNION AND MATRIX FORMS

It is obvious that the frames corresponding to points on any single geodesic arc generated by $T_{AB}(t)$ are equivalent to those generated by applying the quadratic form to $q_{AB}(t)$. It is also obvious that for some specific t_0 (fixing a particular pair of frames T_{AB} and T_{BC}) the arcs

$$R_{ABC}(t, t_0) = T_{AB}(t_0)\left[T_{AB}^{-1}(t_0)T_{BC}(t_0)\right]^t, \tag{25.22}$$

$$q_{ABC}(t, t_0) = q_{AB}(t_0) \star \left[\bar{q}_{AB}(t_0) \star q_{BC}(t_0)\right]^t \tag{25.23}$$

must also correspond point by point as t varies (see Figure 25.16). This is because

$$\log\left[T_{AB}^{-1}(t_0)T_{BC}(t_0)\right] = \theta_{ABC}(\hat{\mathbf{n}}_{ABC} \cdot \hat{\mathbf{L}}),$$

$$\log\left[\bar{q}_{AB}(t_0) \star q_{BC}(t_0)\right] = \left(0, \frac{1}{2}\theta_{ABC}\hat{\mathbf{n}}_{ABC}\right) \tag{25.24}$$

involve the identical fixed-eigenvector rotation parameters, $(\theta_{ABC}, \hat{\mathbf{n}}_{ABC})$. (Remark: This separation of t_0 is the first step of an orientation "blossom" construction.) By induction, repeated nesting of Equation 25.22 or Equation 25.23 corresponding to the Euclidean de Casteljau construction will agree point-by-point.

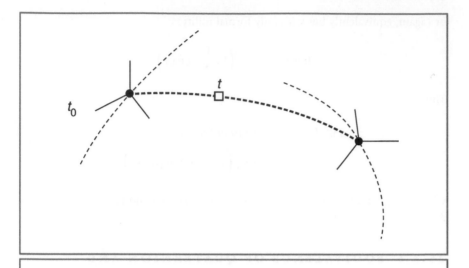

FIGURE 25.16 *Showing the interpolation point* t, *interpolating between two intermediate points on different curves fixed at* t_0.

25.4   QUATERNION SPLINES

We now have clearly in mind what it means to make a smooth, constant-angular-velocity rotation interpolation from one frame to another. The geodesic or minimum-angle transition from any frame A to any other frame B is produced by

$$M_{AB}(t) = A[A^{-1}B]^t = A\,R(t\theta_{AB}, \hat{\mathbf{n}}_{AB}), \tag{25.25}$$

and t parameterizes a constant-angular-velocity rotation. The quaternion

$$q_{AB}(t) = q_A \star [\bar{q}_A \star q_B]^t = q_A \star q\left(t\theta_{AB}, \hat{\mathbf{n}}_{AB}\right) \tag{25.26}$$

is completely equivalent because the standard quadratic map will exactly reproduce the matrix $M_{AB}(t)$. However, the quaternions q_A and q_B are also topological points on the compact metric space $\mathbf{S}^3$, which gives us the novel opportunity to explore a purely geometric viewpoint for orientation visualization.

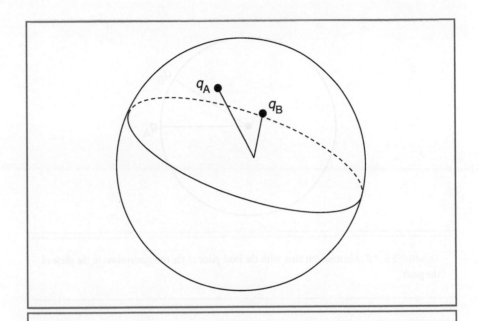

FIGURE 25.17 *Projection of three-sphere* $\mathbf{S}^3$ *with two quaternion points plotted as rays from the origin.*

In the geometric approach to orientation splines, we consider the two points (q_A, q_B) as arbitrary unit four-vectors, and plot them (as in Figure 25.17) using one of our sphere visualization methods. We can always place the two unit vectors in a 2D plane with an orientation of our choice (as shown in Figure 25.18) because, regardless of the dimension, two points on a unit sphere combine with the origin to define a simple 2D plane. The Gram–Schmidt process allows us to immediately project out an axis $q_\perp$ in the plane containing only the component of q_B perpendicular to q_A:

$$\tilde{q} = q_B - q_A(q_A \cdot q_B),$$

$$\|\tilde{q}\|^2 = \tilde{q} \cdot \tilde{q} = 1 - (q_A \cdot q_B)^2. \tag{25.27}$$

Defining the measure of the angular difference between q_A and q_B in quaternion space to be

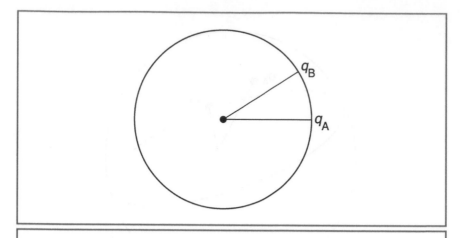

FIGURE 25.18 Flattened-out view with the local plane of the two quaternions in the plane of the paper.

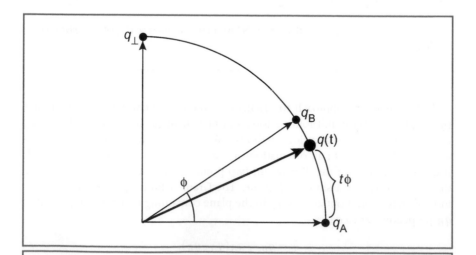

FIGURE 25.19 Gram–Schmidt orthonormal frame form of the unit-norm-preserving interpolation on a sphere.

$$\cos \phi = q_A \cdot q_B,$$

we find $\|\tilde{q}\| = \sqrt{1 - \cos^2 \phi} = |\sin \phi|$, and thus

$$q_\perp = \frac{\tilde{q}}{\|\tilde{q}\|} = \frac{1}{|\sin \phi|} [q_B - q_A \cos \phi]. \qquad (25.28)$$

The standard constant-angular-velocity interpolation from q_A to q_B in the fully geometrical Gram–Schmidt framework is thus recast into the local orthonormal basis of Figure 25.19 in the usual way:

$$q(t) = q_A \cos(t\phi) + q_\perp \sin(t\phi). \qquad (25.29)$$

A little trigonometry turns this into the nontraditional unit-norm-preserving SLERP

$$q_{\text{SLERP}}(t) = \text{SLERP}(t; q_A, q_B)$$

$$= q_A \left[\frac{\sin((1-t)\phi)}{\sin \phi} \right] + q_B \left[\frac{\sin(t\phi)}{\sin \phi} \right], \qquad (25.30)$$

where (as usual) $q_A \cdot q_B = \cos \phi$, $q(0) = q_A$ and $q(1) = q_B$, as expected.

Equation 25.30 is precisely the same as Equation 25.26, except that 25.30 adopts a geometric point of view as opposed to the algebraic viewpoint expressed in 25.26. We can quickly convince ourselves of the equivalence by transforming the system to a special coordinate frame with $q_A = (1, 0, 0, 0) = $ Identity and $q_B = (\cos(\theta/2), \hat{\mathbf{x}} \sin(\theta/2))$, so $\cos \phi = \cos \theta / 2$. (We can do this with no loss of generality due to the fact that only a 2D subplane is actually involved.) Thus, whereas

$$q_{AB}(t) = (\cos t\theta/2, \hat{\mathbf{x}} \sin t\theta/2),$$

we can see that

$$q_{\text{SLERP}}(t) = \left(\frac{\sin(\theta/2 - t\theta/2) + \sin(t\theta/2) \cos(\theta/2)}{\sin(\theta/2)}, \hat{\mathbf{x}} \frac{\sin(\theta/2) \sin(t\theta/2)}{\sin(\theta/2)} \right)$$

$$= \left(\frac{\sin(\theta/2) \cos(t\theta/2)}{\sin(\theta/2)}, \hat{\mathbf{x}} \sin(t\theta/2) \right)$$

$$= (\cos(t\theta/2), \hat{\mathbf{x}} \sin(t\theta/2)),$$

and the two will be identical in any frame.

The point is that, although we have shown that there exist equivalent de Casteljau constructions for 3×3 orthogonal matrices and quaternions, the geometrical viewpoint of the global behavior of the interpolation exists only for quaternions. We can examine each segment of the matrix interpolation locally using the appropriate fixed-eigenvector matrix, but we cannot meaningfully depict the entire curve and its metric properties, e.g., to see when a part of the curve passes near or through a previous orientation.

25.5   QUATERNION DE CASTELJAU SPLINES

We are now ready to create and visualize the quaternion analogs of the standard Bezier splines, as well as their counterparts such as Catmull–Rom splines and uniform B-splines, all in terms of the de Casteljau algorithm. In addition to the 3×3 matrix-based uniform-rotation interpolation primitive,

$$R_{AB}(t) = A\big[A^{-1}B\big]^t,$$

we have the following two equivalent quaternion-based uniform-rotation interpolation primitives, the algebraic and the geometric form,

$$q_{\text{algebraic}}(t) = q_A \star [\bar{q}_A \star q_B]^t,$$

$$q_{\text{geometric}}(t) = q_A \frac{\sin((1-t)\phi)}{\sin\phi} + q_B \frac{\sin t\phi}{\sin\phi}.$$

Although in principle any primitive can be used, the quaternion geometric form built from the Gram–Schmidt local 4D quaternion frame is much closer in both spirit and notation to the Euclidean spline framework. In fact, as pointed out by Schlag [145] the two systems are entirely isomorphic if we simply replace the Euclidean $\text{Lerp}(t; a, b)$ primitive with the norm-preserving spherical $\text{Slerp}(t; a, b)$ primitive. The nesting of SLERPs in the de Casteljau algorithm can in principle be repeated any number of times to get the analog of an nth-order spline, as illustrated in Figure 25.20.

Following this ansatz, we may write the three standard cubic-equivalent quaternion splines as

$$\begin{aligned}
I(q_1, q_2, q_3, q_4; t) = S\big(&S\big(S(q_1, q_2; f_{12}(t)), \ S(q_2, q_3; f_{23}(t)); \ f_{123}(t)\big), \\
&S\big(S(q_2, q_3; f_{23}(t)), \ S(q_3, q_4; f_{34}(t)); \ f_{234}(t)\big); \\
&\hspace{6cm} f(t)\big),
\end{aligned}$$

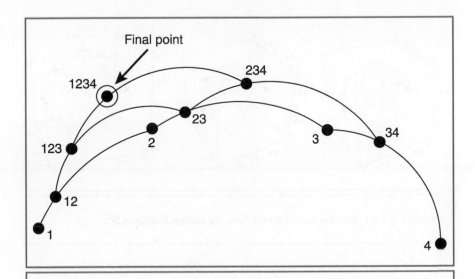

FIGURE 25.20 *Nested SLERPs produce the spherical analog of a spline of any order.*

where the weight functions are $f_{i\ldots j}(t) = t$ for Bezier quaternion curves,

$$f_{12} = t \qquad\qquad f_{23} = t \qquad\qquad f_{34} = t$$
$$f_{123} = t \qquad\qquad f_{234} = t$$
$$f = t,$$

and for Catmull–Rom splines

$$f_{12} = t + 1 \qquad\qquad f_{23} = t \qquad\qquad f_{34} = t - 1$$
$$f_{123} = \frac{(t+1)}{2} \qquad\qquad f_{234} = \frac{t}{2}$$
$$f = t,$$

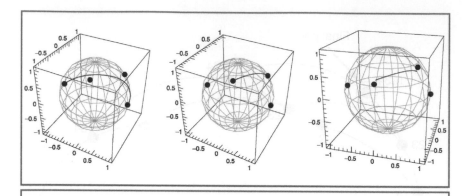

FIGURE 25.21 *Spherical Bezier, Catmull–Rom, and uniform B-spline on* $\mathbf{S}^2$.

and for the uniform B-spline

$$f_{12} = \frac{(t+2)}{3} \qquad\qquad f_{23} = \frac{(t+1)}{3} \qquad\qquad f_{34} = \frac{t}{3}$$

$$f_{123} = \frac{(t+1)}{2} \qquad\qquad f_{234} = \frac{t}{2}$$

$$f = t.$$

Because these formulas are valid for any set of points $\{q\}$ serving as anchor points on a sphere, we can first visualize a set of spherical spline paths of each type on $\mathbf{S}^2$ (the two-sphere in ordinary Euclidean 3D space), as shown in Figure 25.21. The curves are of course not strictly cubic, but are clearly cubic equivalents in that they have the exact expected behavior analogs with respect to the anchor points on the sphere. Figure 25.22 shows the analogous quaternion visualization using the $\mathbf{q}$ (vector-point-only) visualization, which places the identity frame at the apparent origin.

Matching angular slopes: In ordinary Euclidean spline applications, we are often modeling a *static shape*, and thus it is very important to match the *slope* of the tangent to the curve. For Euclidean splines, we can enforce slope matching by evaluating the curve derivative on each side of a given curve point. For orientations, the equiv-

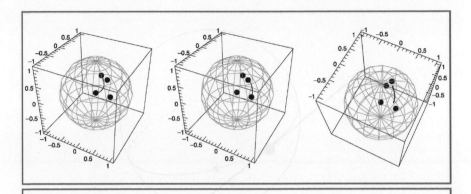

FIGURE 25.22 *Spherical Bezier, Catmull–Rom, and uniform B-spline on* $\mathbf{S}^3$ *using the* $\mathbf{q}$ *visualization projection.*

alent operation is to find the directions of the locally geodesic arcs at each curve point. If the rotation planes coincide, the orientation shape slopes will match.

The Catmull–Rom spline and uniform B-spline are automatically configured to match first derivatives at intermediate end points in moving windows of four anchor points in a series (groups of four are for a cubic; $n + 1$ anchors are needed for an nth-order spline). The Bezier must be matched by hand.

Figures 25.23 and 25.24 represent taking the derivative of a smooth curve on a sphere, as we must do to check derivative matching or to enforce it by hand in a Bezier sequence. Although the tangent line at a point on a sphere computed by taking the derivative lies outside the space containing the curve itself (unlike the Euclidean case), for our purposes it has qualitatively the same properties. For example, at a given point joining two curves, from each direction there will be a limiting value of the eigenvector $\hat{\mathbf{n}}$ of the rotation matrix. If the limiting values of $\hat{\mathbf{n}}$ do not match, as in Figure 25.25, the 2D planes of the local rotation will not match either, and there is a discontinuity. If they match as in Figure 25.26, the two rotations momentarily coincide and the slopes match.

In quaternion language, instead of looking at 3D rotations in the 2D plane perpendicular to momentary eigenvector to qualitatively check a slope match we examine

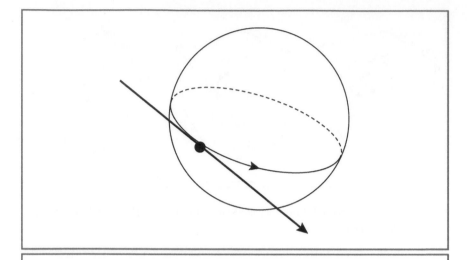

FIGURE 25.23 *The tangent vector to sphere determines the spherical curve derivatives to be matched.*

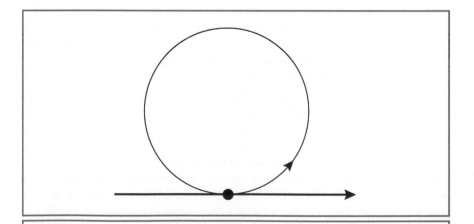

FIGURE 25.24 *The tangent vector to circle shows a clearer direct view of relationship of tangent to the curve itself at one point.*

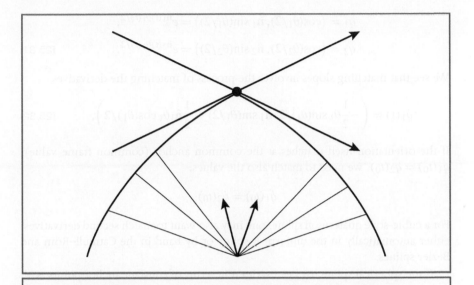

FIGURE 25.25 Mismatch of tangents (or derivatives) to two curves meeting at a point.

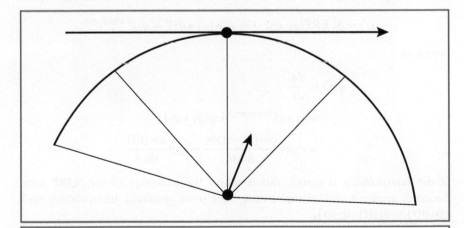

FIGURE 25.26 *Matching tangents show that the slopes of the curves match and no instantaneous angular velocity discontinuity will be observed.*

$$q_1 = \big(\cos(\theta_1/2), \hat{\mathbf{n}}_1 \sin(\theta_1/2)\big) = e^{\hat{\mathbf{n}}_1(t)\theta_1(t)/2},$$

$$q_2 = \big(\cos(\theta_2/2), \hat{\mathbf{n}}_2 \sin(\theta_2/2)\big) = e^{\hat{\mathbf{n}}_2(t)\theta_2(t)/2}. \qquad [25.31]$$

We see that matching slopes involves the process of matching the derivatives

$$\dot{q}_1(t) = \left(-\frac{1}{2}\dot{\theta}_1 \sin(\theta_1)/2, \, \dot{\hat{\mathbf{n}}}_1 \sin(\theta_1/2) + \frac{1}{2}\hat{\mathbf{n}}_1\dot{\theta}_1 \cos(\theta_1)/2\right). \qquad [25.32]$$

If the orientation itself matches at the common anchor (common frame value), $q_1(t_0) = q_2(t_0)$, we need to match also the values:

$$\dot{q}_1(t_0) = \dot{q}_2(t_0).$$

For a cubic-style quaternion spline, one may also want to match second derivatives, either automatically in the uniform B-spline or by hand in the Catmull–Rom and Bezier splines.

   If the spherical spline has been written in terms of constant anchor points with the conventional simple angular parameterization, we have significant simplification because any single component can be easily differentiated. For example, to find the slope or tangent direction of

$$q(t) = \text{SLERP}(q_1, q_2; t) = q_1 \star [\bar{q}_1 \star q_2]^t = q_1 e^{t \log(\bar{q}_1 q_2)}$$

we know

$$\dot{q}(t) = \frac{dq}{dt}$$

$$= q_1 \star e^{t \log(\bar{q}_1 \star q_2)} \log[\bar{q}_1 \star q_2]$$

$$= -q_1\theta \frac{\cos((1-t)\theta)}{\sin\theta} + q_2\theta \frac{\cos(t\theta)}{\sin\theta}.$$

If the interpolation is nested, this is where the advantage of the SLERP form becomes apparent. For example, in a three-point quadratic interpolation with $\cos\theta(t) = q_{12}(t) \cdot q_{23}(t)$,

$$q_{123}(t) = q_{12}(t)\frac{\sin((1-t)\theta(t))}{\sin\theta(t)} + q_{23}(t)\frac{\sin(t\theta(t))}{\sin\theta(t)} \qquad [25.33]$$

can be cleanly differentiated term by term.

25.6   EQUIVALENT ANCHOR POINTS

In our introductory review of Euclidean polynomial splines, we noted that every spline could be expressed as a Bezier spline with suitable combinations of anchor points. We can achieve an equivalent form for the Schlag nested-SLERP formulas that is somewhat more intuitive by adapting this procedure from translations to rotations.

The basic requirement is to transform a linear algebraic combination

$$x = a + sb \tag{25.34}$$

into a statement compatible with multiplicative group operations. We already know that group displacements from the identity matrix are equivalent to Euclidean displacements from the origin,

$$A = A \cdot I \leftrightarrow \mathbf{a} = \mathbf{a} + \mathbf{0}, \tag{25.35}$$

and thus that Euclidean translations with a plus sign are group multiplications and that those with a minus sign require the inverse matrix operation:

$$
\begin{aligned}
I &= A \cdot A^{-1} \cdot I, & \mathbf{0} &= \mathbf{a} - \mathbf{a} + \mathbf{0}, \\
A \cdot B &= A \cdot B \cdot I, & \mathbf{a} + \mathbf{b} &= \mathbf{a} + \mathbf{b} + \mathbf{0}, \\
B \cdot A &= B \cdot A \cdot I, & \mathbf{b} + \mathbf{a} &= \mathbf{b} + \mathbf{a} + \mathbf{0}, \\
A \cdot B &\neq B \cdot A, & \mathbf{a} + \mathbf{b} &= \mathbf{b} + \mathbf{a}.
\end{aligned}
\tag{25.36}
$$

Here we must choose an order for possibly noncommutative group operations.

The remaining operation, scalar multiplication, must correspond, say, to a factor of 2 being equivalent to twofold group multiplication, a factor of 3 to three-fold multiplication, and so on. The only consistent interpolation of scalar multiples is therefore

$$B^s = \underbrace{B \cdots B}_{s} \leftrightarrow s\mathbf{b}. \tag{25.37}$$

Thus, we find the translation between Euclidean linear algebra and group algebra must take the form

$$X = A \cdot B^s \leftrightarrow \mathbf{x} = \mathbf{a} + s\mathbf{b}, \tag{25.38}$$

up to a choice of matrix multiplication order. The same correspondence must hold for quaternions, whose algebra is isomorphic to a matrix algebra.

Bezier-equivalent polynomial quaternion splines: We can thus immediately translate the Catmull–Rom spline with control points (c_0, c_1, c_2, c_3), where in Euclidean space the corresponding Bezier anchor points are

$$\mathbf{x}_0 = \mathbf{c}_1,$$

$$\mathbf{x}_1 = \mathbf{c}_1 + \frac{1}{6}(\mathbf{c}_2 - \mathbf{c}_0),$$

$$\mathbf{x}_2 = \mathbf{c}_2 + \frac{1}{6}(\mathbf{c}_1 - \mathbf{c}_3),$$

$$\mathbf{x}_3 = \mathbf{c}_2, \tag{25.39}$$

to a nested SLERP with *all* simple $f(t) = t$ interpolating functions, but with the Catmull–Rom anchor quaternions $\mathbf{p}_i$ replaced (up to a choice of multiplication order) by Bezier control quaternions with the values

$$q_0 = p_1,$$

$$q_1 = p_1 \star (\bar{p}_0 \star p_2)^{1/6},$$

$$q_2 = p_2 \star (\bar{p}_3 \star p_1)^{1/6},$$

$$q_3 = p_2.$$

These equivalences are represented schematically in Figures 25.27 and 25.28. For uniform B-splines, the equivalent transformation (up to a choice of multiplication order) is

$$\mathbf{x}_0 = \frac{1}{6}(\mathbf{c}_0 + 4\mathbf{c}_1 + \mathbf{c}_2),$$

$$\mathbf{x}_1 = \frac{1}{3}(2\mathbf{c}_1 + \mathbf{c}_2),$$

$$\mathbf{x}_2 = \frac{1}{3}(\mathbf{c}_1 + 2\mathbf{c}_2),$$

$$\mathbf{x}_3 = \frac{1}{6}(\mathbf{c}_1 + 4\mathbf{c}_2 + \mathbf{c}_3),$$

$$q_0 = p_1 \star \left((\bar{p}_1)^2 \star p_0 p_2\right)^{1/6},$$

$$q_1 = p_1 \star (\bar{p}_1 \star p_2)^{1/3},$$

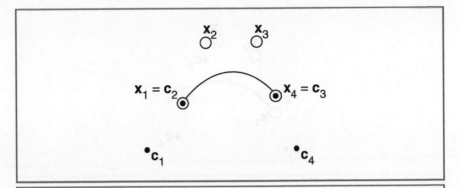

FIGURE 25.27 The 2D Catmull–Rom segment anchor points c_i are equivalent to the 2D Bezier anchor points x_i.

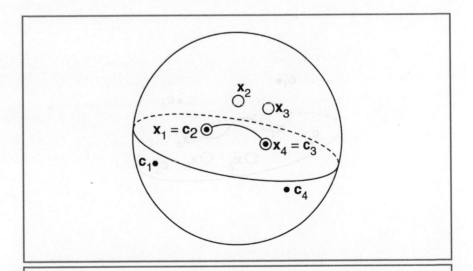

FIGURE 25.28 The 3D or 4D Catmull–Rom segment anchor points c_i are equivalent to the Bezier anchor points x_i.

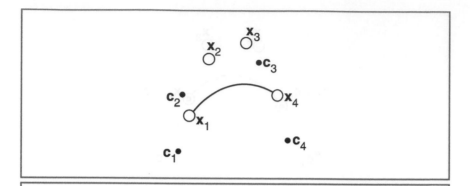

FIGURE 25.29 The 2D B-spline segment anchor points c_i are equivalent to the 2D Bezier anchor points x_i.

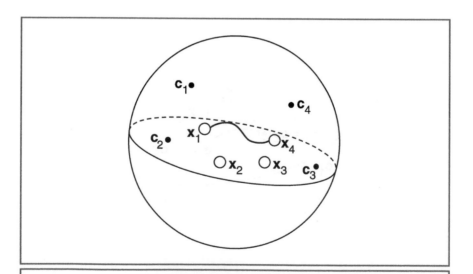

FIGURE 25.30 The 3D or 4D B-spline segment anchor points c_i are equivalent to the Bezier anchor points x_i.

$$q_2 = p_2 \star (\bar{p}_2 \star p_1)^{1/3},$$
$$q_3 = p_2 \star \left((\bar{p}_2)^2 \star p_1 \star p_3\right)^{1/6}.$$

These equivalences are represented schematically in Figures 25.29 and 25.30.

25.7   ANGULAR VELOCITY CONTROL

There is one final technical detail that adds complexity to the procedures for interpolating orientation frames. There are two aspects of derivative matching that may independently be of importance in practical applications.

- *Static curve slope and higher derivatives:* When modeling a *shape* with a Euclidean spline, the curve or surface parameterization or subdivision spacing is of little importance because the light-reflective properties of the static shape are what is relevant. Only the anchor point slope-matching features are visible in the reflectance, and if the normals (and possibly their derivatives) match all is well. Orientation interpolation, however, is rarely used to generate static light-reflective properties of object models, and thus the static shape of the quaternion curve is not a relevant model property.
- *Dynamic curve velocity and higher derivatives:* When modeling a *motion path* of either an object or a camera viewpoint, however, further parameters must be adjusted. The static slopes (and possibly higher derivatives) must still match, but in addition the *time steps* must be adjusted so that the dynamic object motion has the desired *spatial velocity* properties. For exactly the same mathematical curve in space, the apparent velocity of an object, moving on that curve can be wildly different, as illustrated in Figure 25.31, as can higher derivatives.

Although it is very straightforward (albeit tedious) to match the slopes of adjacent quaternion curve segments, this is basically *never enough*. Quaternion splines typically do not model shapes (like Euclidean splines do), but are used instead to model moving objects and cameras, complementing the Euclidean spline curves describing translations of objects and cameras in time.

Thus, what we must do is match the *sampling rate* in the curve distance to achieve the desired local angular velocity. Simply put, let there be three required orientations (q_1, q_2, q_3) on the same curve $q(t)$ defined by an arbitrary spherical spline

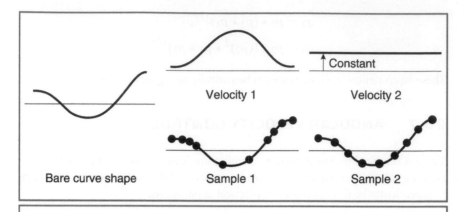

FIGURE 25.31 *Three curves comparing the spatial shape of a curve and two very different sampling intervals leading to drastically different object velocity profiles for exactly the same curve shape.*

(i.e., $q \cdot q = 1$). We take q_1 and q_3 to be given, for example in

$$q_1 = q(0),$$
$$q_3 = q(T),$$

and we want to find the value of t_0 such that $q_2 = q(t_0)$ describes an orientation that will produce a *uniform angular velocity*.

Mathematically, this requires simply that we match the quaternion arc length, and thus matching dot products works if the angles are $< \pi/2$, producing

$$q_1 \cdot q_2 \approx q_2 \cdot q_3.$$

The inverse cosines are the arc lengths shown in Figure 25.32:

$$d_1 = \cos^{-1}(q_1 \cdot q_2) \approx d_2 = \cos^{-1}(q_2 \cdot q_3). \tag{25.40}$$

Solving for the value of t_0 such that $q_2 = q(t_0)$ forces $d_1 = d_2$ is typically a numerical procedure. Starting from the fundamental principles of this simple example, we can effectively exploit the correctness of the quaternion distance measure to solve any variant of the problem, such as the following examples.

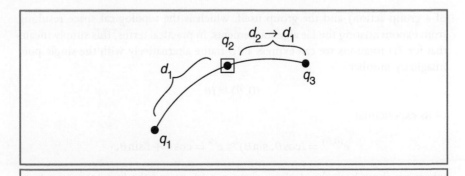

FIGURE 25.32 *Partitioning the spherical curve* (q_1, q_2, q_3) *so that* q_2 *results in uniform angular velocity.*

- *Constant angular velocity:* Using equal-arc-length quaternion curve subdivision.
- *Constant angular acceleration:* Using second-order differences.
- *Any desired model for the angular velocity curve:* Matching a heuristic function for the sample spacing.

25.8   EXPONENTIAL-MAP QUATERNION INTERPOLATION

It is annoying that in exchange for the desirable properties of quaternions as a way of parameterizing rotations we must work with and visualize $\mathbf{S}^3$. The hypersphere $\mathbf{S}^3$ is a beautiful space from the mathematical standpoint, but the most straightforward way of viewing it is as an embedding in four Euclidean dimensions instead of the familiar three. In addition, it is a compact three-manifold whereas human spatial intuition tends to falter beyond compact two-manifolds.

Many complaints have been voiced about the 4D nature of quaternions, and thus when people became aware of the 3D methods suggested by the paper of Kim, Kim, and Shin [111] there was great initial enthusiasm. In this section, we will briefly describe the exponential-map method and carefully delimit its strengths and weaknesses (e.g., see the treatment of Grassia [59] for further details).

The mathematical foundation of the exponential-map method is the elegant difference between a Lie algebra (describing the tangent space, or infinitesimal limit, of a group action) and the group itself, which is the topological space resulting from exponentiating the Lie algebra elements. In practical terms, this simply means that for 2D rotations we can express a 2D frame alternatively with the single pure imaginary number

$$(0, \theta) \approx i\theta$$

or its exponential

$$e^{(0,\theta)} = (\cos\theta, \sin\theta) \approx e^{i\theta} = \cos\theta + i\sin\theta,$$

where we show both the algebraic form $z = (x, y)$ and the imaginary-unit form $z = x + iy$ for each complex number in the formulas. Conversely, the Euclidean exponential-map form of the 2D rotation parameterization is

$$\log e^{(0,\theta)} = (0, \theta) \approx \log e^{i\theta} = i\theta.$$

Note that whereas the parameter space (θ) is 1D, the complex algebra onto which the exponential maps is 2D, but with the circle constraint $x^2 + y^2 = 1$. For 2D frames, the fact that $\theta \in \mathbb{R}$ maps periodically with period 2π but without singularities to the matrix

$$R_2 = \begin{bmatrix} \cos\theta & -\sin\theta \\ \sin\theta & \cos\theta \end{bmatrix} \tag{25.41}$$

means that the exponential map has perfectly acceptable properties.

For 3D frames, once again we can relate the 3D parameter space to the infinitesimal transformation parameter, which we write as $(0, \mathbf{n}/2) = \log q$. Now with the notation conventions $|\mathbf{n}| = \theta$ and $\hat{\mathbf{n}} = \mathbf{n}/|\mathbf{n}|$ we have

$$q = e^{\frac{1}{2}(0,\mathbf{n})} = \left(\cos\left(\frac{1}{2}\theta\right), \hat{\mathbf{n}} \sin\left(\frac{1}{2}\theta\right) \right). \tag{25.42}$$

Purists will note that as $\theta \to 0$ some expressions become undefined and thus sometimes we must factor out θ, giving an expression such as

$$\theta \hat{\mathbf{n}} \left(\frac{\sin\theta}{\theta} \right)$$

to maintain formal regularity.

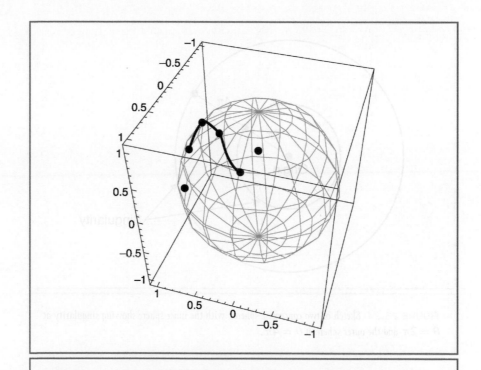

FIGURE 25.33 *Example of a many-control-point Catmull–Rom spline carried out using the exponential method in the 3D log space, then plotting (with our usual 3D projection) the full quaternion curve resulting from the exponentiation.*

Thus, any point q on $\mathbf{S}^3$ can be represented as $\log q = (0, \mathbf{n}/2)$. Again, we see that whereas $\mathbf{n}$ is a 3D Euclidean vector $q = e^{(0, \frac{1}{2}\mathbf{n})}$ is the standard 4D embedding of the 3D object $\mathbf{S}^3$. In this Euclidean 3D space, we can establish Euclidean anchor points, and construct ordinary Euclidean splines. As illustrated in Figure 25.33 using a Catmull–Rom spline, the quaternion results are qualitatively similar to SLERPs if one stays within the constraints of the methods.

However, we now note the following families of advantageous and disadvantageous properties of the exponential map.

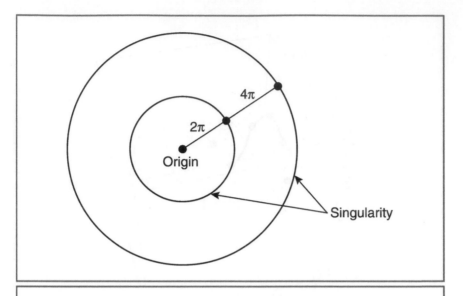

FIGURE 25.34 *Sketch of two concentric spheres, with the inner sphere showing singularity at $\theta = 2\pi$ and the outer sphere at $\theta = 4\pi$.*

- *Singularity at $\theta = 2\pi$*: When $|\mathbf{n}| = 2k\pi$, the exponential form maps all points into the origin, and thus there is no continuous route through the logarithmic orientation space if we have a path that lives on the sphere (as illustrated in Figure 25.34). Interpolation curves therefore are in danger of a type of "generalized gimbal lock," a situation in which a finite change in an orientation parameter produces no change whatever in the orientation frame and therefore differentiability is lost.

- *No algebra for combining rotations*: Quaternion multiplication moves from one orientation to another along a specific arc in $\mathbf{S}^3$. There is no 3D Euclidean algebra allowing insight into combined rotations in the exponential map. The only way to find the result is to write out the two exponentials as quaternions q_1 and q_2, perform quaternion multiplication to get $q_1 \star q_2$, and take the logarithm to find the new point $(0, \mathbf{n}_{12}/2) = \log(q_1 \star q_2)$ in the 3D Euclidean space.

- *No curve metric allowing distance comparisons*: Although the special case of a pure 2D (planar) rotation reduces to $\log e^{i\theta} = i\theta$, this is the only case in which distances among orientation states along curve can be straightforwardly computed in the exponential map.

Advantages of the exponential map: The exponential map does have advantages in several situations. In general, to exploit the exponential map one first needs to impose a restricted range. Conventionally, one stays within the half-sized ball of radius $r \leqslant \pi$ inside the singular surface at radius $r = 2\pi$ (e.g., see Grassia [59]). In this range, one can apply literally any 3D interpolation or spline from Euclidean geometry and can take derivatives and integrate differential equations without nonlinear constraints.

The apparent advantages of the exponential form—only three Euclidean degrees of freedom, no constraints on a higher-dimensional parameter space (compared to four with one constraint for quaternions, nine with six constraints for frame matrices), and the ability to use classical Euclidean splines—attracted great interest when proposed. However, one must be very careful to understand clearly that the exponential form has its mathematical origins in the Lie algebra, which (considering the following) achieves its simplicity at great cost.

- *No geometry*: The geometry appears only in the full Lie group (quaternions), in that the Lie algebra only knows about local properties of the group derived from its tangent space.
- *No global metric*: Only special cases, equivalent to 2D rotations passing through the origin, permit the measurement of comparative distances directly in the 3D exponential space.

Thus, for example, the nice 3D Euclidean spline forms possess illusory simplicity: the angular velocity is unknowable and uncontrollable without returning to the full $\mathbf{S}^3$ space one is trying to avoid. A Euclidean spline in the exponential—as in

$$\mathbf{n}(t) = \sum_{\text{anchors}} c_i \mathbf{x}(i) t^{n(i)} (1-t)^{m(i)}, \qquad (25.43)$$

where $q(t) = e^{(0, \mathbf{n}(t)/2)}$—will have frame-frame distances as a function of t that can only be measured using

$$d(t, t+dt) = \cos^{-1} e^{(0, \mathbf{n}(t)/2)} \cdot e^{(0, \mathbf{n}(t+dt)/2)}. \qquad (25.44)$$

If $|\mathbf{n}(t)| \approx 2\pi$, long distances in $\mathbf{n}$ coordinates could be vanishingly small in terms of frame-frame distance so that in the exponential formalism it is not possible to elegantly control the *rate* of frame change, which is critical to *all* orientation spline applications.

25.9   GLOBAL MINIMAL ACCELERATION METHOD

Spline curves in Euclidean space are derived originally from the observation that the eye is pleased by shapes that are naturally achieved by placing flexible pieces of wood in a framework of stakes to make certain points fixed. It was discovered that curves with minimal second derivatives were especially appealing. Mathematically, this observation is expressed by the fact that cubic splines—which have vanishing fourth derivative and constant third derivative—seem to be " just complex enough" to be aesthetically pleasing.

In this section, we will briefly outline a generalization of the minimal-second-derivative idea from Euclidean space to quaternion space that was originally introduced by Barr, Currin, Gabriel, and Hughes [15] to make more explicit the differential geometric spherical spline treatment of Kajiya and Gabriel [107]. Within this general framework, one can generate not only a more universal approach to the construction of smooth orientation changes but also a more systematic way to attack the requirement for constraining angular velocities.

25.9.1   WHY A CUBIC?

The basic concept of the approach is that one can understand our standard Euclidean-space splines by looking at a measure of the curvature energy based on the second derivative of an unspecified curve $\mathbf{x}(t)$:

$$E[\mathbf{x}] = \int_0^1 \mathbf{x}''(t) \cdot \mathbf{x}''(t)\, dt. \qquad (25.45)$$

If we seek the specific curve out of all possible curves $\mathbf{x}$ that minimizes this energy functional (a functional is an object that returns a number for each choice of a function), classical methods lead to the Euler–Lagrange equations, which for

Equation 25.45 are simply

$$\mathbf{x}''''(t) = \frac{d^4\mathbf{x}(t)}{dt^4} = 0.$$ (25.46)

In other words:

Cubic is Minimal

The minimal-bending-energy condition for $E[\mathbf{x}]$ is satisfied by $\mathbf{x}(t)$ with *constant third derivative*, or by a *cubic function of* t.

25.9.2 EXTENSION TO QUATERNION FORM

To carry out an analogous minimization in quaternion space, we can no longer find such a simple equation. We must instead pay attention to the spherical nature of the space and find the closest analog. Barr et al. argue that *minimal tangential acceleration* is the key generalization required to construct a spherical analog of Equation 25.45. Using the now-familiar Gram–Schmidt procedure (see also Chapter 10), we can directly compute

$$\alpha(t) = q''(t) - q(t)\big(q(t) \cdot q''(t)\big),$$ (25.47)

which is explicitly perpendicular to the quaternion's 4D radial unit vector $q(t)$,

$$\alpha(t) \cdot q(t) \equiv 0$$

(because $q \cdot q = 1$), and has the squared value

$$\alpha \cdot \alpha = q''(t) \cdot q''(t) - \big(q(t) \cdot q''(t)\big)^2.$$

The overall energy to be minimized is then

$$E[q] = \int_0^T \alpha(t) \cdot \alpha(t)\, dt,$$ (25.48)

where one can select any number of keyframes

$$q(t_i) = Q_i, \quad i = 1, 2, \ldots, N,$$

and the constraint $q \cdot q = 1$ must be enforced throughout. In addition, one may calculate or impose the desired angular velocities at each keyframe and impose those as additional conditions. The problem is now well-posed for numerical solution.

Quaternion Rotator Dynamics

26

In nature, rigid 3D extended objects follow two sets of laws that determine their motion in the course of interacting with their external environment.

- *Response to force:* The response to a force $\mathbf{F}$ acting on the center of mass is to accelerate or change the velocity $\mathbf{v} = \dot{\mathbf{x}}$ of the center-of-mass position $\mathbf{x}$. One can then in theory integrate the differential equations

$$\mathbf{F} = m\mathbf{a} = m\dot{\mathbf{v}} = \frac{d\mathbf{P}}{dt} \qquad [26.1]$$

and compute the path of the center of mass over time, as shown in Figure 26.1a.

- *Response to torque:* A rotating frame to which moments of inertia I and angular momentum $\mathbf{L} = I\boldsymbol{\omega}$ can be assigned responds to a torque $\mathbf{T}$ by changing its angular momentum in the direction of the torque according to the differential equation

$$\mathbf{T} = I\dot{\boldsymbol{\omega}} = \frac{d\mathbf{L}}{dt}. \qquad [26.2]$$

Thus, one can in principle integrate the differential equations and track the orientation of the frame defined by the angular momentum vector over time (see Figure 26.1b).

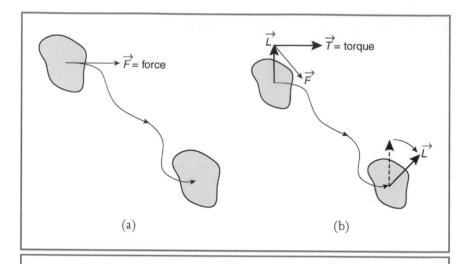

FIGURE 26.1 *Response of a rigid extended body to (a) external forces and (b) external torques.*

Because our subject is rotations, we will naturally focus on the response of a mechanical body's orientation frame to external torques. In particular, we will see how to exploit quaternions to explore the dynamic behavior of rotating body frames. This subject has been extensively treated in the physically based modeling literature (e.g., see Witkin and Baraff [174]), and was known to Hamilton and carefully explored by Tait [162].

26.1   STATIC FRAME

The first step in describing the dynamics of a rigid body is to establish a private coordinate system, the body frame, that typically corresponds to the directions of the principal moments of inertia (I_x, I_y, I_z). Spinning bodies possess stable rotation states about the axis of their maximal and minimal moments of inertia, and we will by convention choose the z axis to be the direction of the principal moment of inertia. Thus we can write the axes of the body frame in terms of a fixed external reference frame as

$$R = [\hat{\mathbf{x}} \quad \hat{\mathbf{y}} \quad \hat{\mathbf{z}}].$$

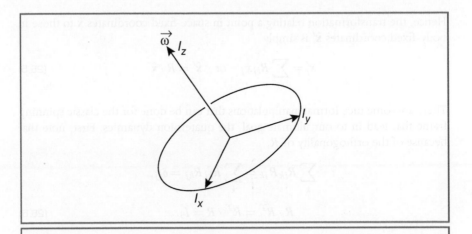

FIGURE 26.2 *Diagram of a spinning body with moment of inertia I_z by default along the axis of rotation.*

This is an orthogonal 3×3 matrix whose columns are the current directions of the principal axes. If the object is rotating, we will assume for simplicity that it has angular velocity $\boldsymbol{\omega}$ about the z axis, as shown schematically in Figure 26.2.

A classical approach in mechanics adjusts the coordinates so that the center of mass of the collection of points describing the composite rigid body is at the origin in both the *space-fixed* axes, with coordinates $\hat{\mathbf{x}}$, and the *body-fixed* axes, with coordinates $\hat{\mathbf{x}}'$. The space-fixed axes correspond to vectors indexed by the right-hand index of R_{ij},

$$\text{Space-fixed} = [\, R_{ix} \quad R_{iy} \quad R_{iz} \,] = [\hat{\mathbf{x}} \quad \hat{\mathbf{y}} \quad \hat{\mathbf{z}}], \qquad (26.3)$$

and the body-fixed axes to the left-hand index

$$\text{Body-fixed} = \begin{bmatrix} R_{xi} \\ R_{yi} \\ R_{zi} \end{bmatrix} = \begin{bmatrix} \hat{\mathbf{x}} \\ \hat{\mathbf{y}} \\ \hat{\mathbf{z}} \end{bmatrix}. \qquad (26.4)$$

Hence, the transformation relating a point in space-fixed coordinates $\hat{\mathbf{x}}$ to those in body-fixed coordinates $\hat{\mathbf{x}}'$ is simply

$$x_i' = \sum_{ij} R_{ij} x_j \quad \text{or} \quad \hat{\mathbf{x}}' = R \cdot \hat{\mathbf{x}}. \tag{26.5}$$

There are some nice formal manipulations that can be done for the classic spinning frame that lead in to our ultimate goal, the quaternion dynamics. First, note that because of the orthogonality of R,

$$\sum_k R_{ik} R_{jk} = \sum_k R_{ki} R_{kj} = \delta_{ij},$$

$$R \cdot R^T = R^T \cdot R = I_3. \tag{26.6}$$

We can make the assumption that each body-fixed point $\mathbf{x}'$ is a constant and take the derivative:

$$0 = \dot{R} \cdot \mathbf{x} + R \cdot \dot{\mathbf{x}}. \tag{26.7}$$

We can immediately find an elegant form for the angular velocity of the moving points by multiplying by R^{-1} to yield

$$\dot{\mathbf{x}} = -R^T \cdot \dot{R} \cdot \mathbf{x} = \dot{R}^T \cdot R \cdot \mathbf{x}, \tag{26.8}$$

where we used $d(R^T \cdot R)/dt = \dot{R}^T \cdot R + R^T \cdot \dot{R} = 0$. This allows us to define the antisymmetric matrix $\boldsymbol{\omega}$ with components

$$\omega_{ij} = \left(R^T \cdot \dot{R}\right)_{ij} = \sum_k R_{ki} \dot{R}_{kj}$$

$$= -\left(\dot{R}^T \cdot R\right)_{ij} = -\omega_{ji}.$$

Thus,

$$\dot{\mathbf{x}} = -\boldsymbol{\omega} \cdot \mathbf{x} \tag{26.9}$$

describes the rotating point $\mathbf{x}$ in the composite object, and (in any coordinate frame) we verify that the point stays at constant radius from the origin

$$\frac{1}{2} \frac{d|\mathbf{x}|^2}{dt} = \mathbf{x} \cdot \dot{\mathbf{x}} = -\mathbf{x} \cdot \boldsymbol{\omega} \cdot \mathbf{x} = 0 \tag{26.10}$$

due to $\omega_{ij} = -\omega_{ji}$.

Remark: It can sometimes be useful to know that

$$\sum_{ijk} \epsilon_{ijk} R^{ia} R^{jb} R^{kc} = \epsilon^{abc} \det R,$$

where $\det R = 1$ for orthogonal matrices and the Levi-Civita symbol, the totally antisymmetric pseudotensor ϵ_{ijk}, is treated in detail in the Appendix F. For the purposes of this section, it is sufficient to note that

$$\epsilon_{jkl} = 0 \qquad \text{if any two indices are equal,}$$

$$\epsilon_{jkl} = +1 \qquad \text{if the indices are in cyclic order,}$$

$$\epsilon_{jkl} = -1 \qquad \text{if the indices are in anticyclic order.}$$

Another conventional form uses the vector angular velocity referred to the space-fixed axes

$$\omega_i = +\frac{1}{2} \sum_{jk} \epsilon_{ijk} \omega_{jk},$$

$$\omega_{jk} = +\sum_l \epsilon_{jkl} \omega_l,$$

and thus

$$\dot{x}_i = -\sum_{jk} \epsilon_{ijk} \omega_k x_j$$

$$= +(\boldsymbol{\omega} \times \mathbf{x})_i,$$

$$\dot{\mathbf{x}} = \mathbf{v} = \boldsymbol{\omega} \times \mathbf{x}.$$

This reproduces the standard picture (shown in Figure 26.3) for the computation of the instantaneous velocity of any point $\mathbf{x}$ at a distance r from the rotation axis of a rigid body rotating at a constant angular velocity

$$\omega_i = \frac{1}{2} \sum_{jk} \epsilon_{ijk} \left(R^T \cdot \dot{R} \right)_{jk}. \tag{26.11}$$

By referring back to the fixed-eigenvector notation for a rotation, we see that in fact R must leave the direction $\boldsymbol{\omega}$ fixed. Thus, Figure 26.3 must be the consequence

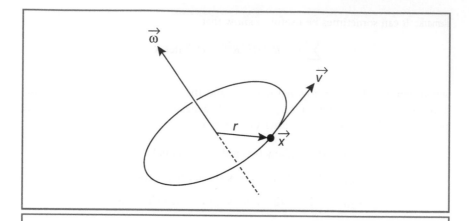

FIGURE 26.3 *A point* **x** *in a spinning rigid body has local velocity* $\mathbf{v} = \boldsymbol{\omega} \times \mathbf{x}$ *if* $\boldsymbol{\omega}$ *is the angular velocity.*

of a rotation matrix that we can now write exactly as

$$R(t) = R(\omega t, \hat{\boldsymbol{\omega}}) \cdot R_0, \qquad (26.12)$$

where $\hat{\boldsymbol{\omega}} = \boldsymbol{\omega}/|\boldsymbol{\omega}|$ and R_0 is the body-frame orientation at $t = 0$.

26.2   TORQUE

Continuing to neglect linearly accelerating forces, we now ask what happens when our constant-angular-velocity frame $R(t)$ is acted on by an external torque, which will normally add to the system's total energy. Defining the angular momentum

$$\mathbf{L} = I\boldsymbol{\omega}, \qquad (26.13)$$

where I is the assumed-constant moment of inertia, we recall that the change in angular momentum (the reaction to the torque) is equal to the applied torque itself:

$$\mathbf{T} = \dot{\mathbf{L}}. \qquad (26.14)$$

In terms of the frame matrix $R(t)$, we have

$$\frac{d\omega_{ij}}{dt} = \frac{d}{dt}\left(R^T \cdot \dot{R}\right)_{ij}$$

$$= \dot{R}^T \cdot \dot{R} + R^T \cdot \ddot{R}$$

$$= \dot{R}^T \cdot R \cdot R^T \cdot \dot{R} + R^T \cdot \ddot{R}$$

$$= -\sum_l \omega_{il}\omega_{lj} + \left(R^T \cdot \ddot{R}\right)_{ij}.$$

Another familiar way to relate the rotation matrix itself to the angular momentum is to think of $\boldsymbol{\omega} = (\omega_1, \omega_2, \omega_3)$ as components of an antisymmetric 3×3 matrix, in which case we can write

$$\dot{R} = R \cdot \Omega,$$

$$\Omega = \begin{bmatrix} 0 & \omega_3 & -\omega_2 \\ -\omega_3 & 0 & \omega_1 \\ \omega_2 & -\omega_1 & 0 \end{bmatrix}.$$

26.3   QUATERNION ANGULAR MOMENTUM

The basic formula for the angular velocity of a rigid body whose rotation is effected by a time-varying frame matrix $R(t)$ is

$$\omega(t)_{ij} = \left[R^T(t) \cdot \dot{R}(t)\right]_{ij}. \tag{26.15}$$

The basic differential equation follows at once as

$$\dot{R}_{ij} = \sum_k R_{ik}\omega_{kj}, \tag{26.16}$$

where the usual three-vector angular velocity is $[\boldsymbol{\omega}]_i = (1/2)\sum_{jk}\epsilon_{ijk}\omega_{jk}$. For constant angular velocity,

$$R(t) = R(\omega t, \hat{\boldsymbol{\omega}}) \cdot R_0. \tag{26.17}$$

Now, in fact, recalling that

$$R = \begin{bmatrix} q_0^2 + q_1^2 - q_2^2 - q_3^2 & 2q_1q_2 - 2q_0q_3 & 2q_1q_3 + 2q_0q_2 \\ 2q_1q_2 + 2q_0q_3 & q_0^2 - q_1^2 + q_2^2 - q_3^2 & 2q_2q_3 - 2q_0q_1 \\ 2q_1q_3 - 2q_0q_2 & 2q_2q_3 + 2q_0q_1 & q_0^2 - q_1^2 - q_2^2 + q_3^2 \end{bmatrix}, \tag{26.18}$$

and writing the columns of R as $[R_1, R_2, R_3]$, we may directly compute

$$\dot{R}_i = 2W_i \cdot \dot{q},$$

where

$$[W_1] = \begin{bmatrix} q_0 & q_1 & -q_2 & -q_3 \\ q_3 & q_2 & q_1 & q_0 \\ -q_2 & q_3 & -q_0 & q_1 \end{bmatrix}, \tag{26.19}$$

$$[W_2] = \begin{bmatrix} -q_3 & q_2 & q_1 & -q_0 \\ q_0 & -q_1 & q_2 & -q_3 \\ q_1 & q_0 & q_3 & q_2 \end{bmatrix}, \tag{26.20}$$

$$[W_3] = \begin{bmatrix} q_2 & q_3 & q_0 & q_1 \\ -q_1 & -q_0 & q_3 & q_2 \\ q_0 & -q_1 & -q_2 & q_3 \end{bmatrix}. \tag{26.21}$$

Because the rows of the W_i matrices form a mutually orthogonal system, they project the single quaternion frame system (see Equation 20.24)

$$\begin{bmatrix} \dot{q}_0 \\ \dot{q}_1 \\ \dot{q}_2 \\ \dot{q}_3 \end{bmatrix} = v(t) \frac{1}{2} \begin{bmatrix} 0 & -\omega_1 & -\omega_2 & -\omega_3 \\ +\omega_1 & 0 & +\omega_3 & -\omega_2 \\ +\omega_2 & -\omega_3 & 0 & +\omega_1 \\ +\omega_3 & +\omega_2 & -\omega_1 & \omega_0 \end{bmatrix} \cdot \begin{bmatrix} q_0 \\ q_1 \\ q_2 \\ q_3 \end{bmatrix}. \tag{26.22}$$

to each of the rows of the equivalent $\dot{R} = R \cdot \Omega$ system. Thus all of the forms are equivalent to the quaternion system

$$\dot{q} = \frac{1}{2} q \star (0, \boldsymbol{\omega}) \tag{26.23}$$

with unit-speed normalization.

Another alternative derivation is to directly differentiate

$$q(t) = \big(\cos(\omega t/2), \sin(\omega t/2)\hat{\omega}\big)$$

to get

$$\dot{q}(t) = \left(-\frac{\omega}{2}\sin\frac{\omega t}{2}, \frac{\omega}{2}\cos\frac{\omega t}{2}\hat{\omega}\right)$$

$$= \frac{\omega}{2} q \star (0, \hat{\omega})$$

$$= \frac{1}{2} q \star (0, \omega).$$

Note that the vector parts of q and $(0, \omega)$ are proportional, and so in fact the quaternion multiplication is in this case order-independent.

The differential equations $\dot{R} = R \cdot \Omega$ and $\dot{q} = \frac{1}{2} q \star (0, \omega)$ are both consistent with their respective constraints,

$$R \cdot R^T = I_3 \quad \Rightarrow \quad R \cdot \dot{R}^T + \dot{R} \cdot R^T = 0$$

$$q \cdot q = 1 \quad \Rightarrow \quad q \cdot \dot{q} = \frac{1}{2} q \cdot \left(q \star (0, \omega)\right) = 0.$$

Thus each will maintain its constraints in the course of numerical integration up to propagation of rounding error. Hence, the folklore that quaternions are unsuitable for numerical integration due to the $q \cdot q = 1$ constraint neglects the fact that the rigid body equations written in terms of 3×3 orthogonal matrices are even more prone to error propagation, with six constraints instead of one to preserve. Either can be periodically corrected by renormalization, but the single quaternion normalization is far easier to implement.

Finally, we can express the acceleration of the quaternion frame in terms of the angular acceleration $\dot{\omega} = \mathbf{T}/I$ as

$$\ddot{q} = \frac{1}{2} \dot{q} \star (0, \omega) + \frac{1}{2} q \star (0, \dot{\omega})$$

$$= \frac{1}{4} q \star (0, \omega) \star (0, \omega) + \frac{1}{2} q \star (0, \mathbf{T}/I).$$

The first term is the radial acceleration for constant orbital motion, and the second is the response to the external torque.

Concepts of the Rotation Group

27

In this Chapter we introduce some of the elementary constructs and ideas from the theory of representations of the 3D Euclidean rotation group. The most fundamental properties of representations of the rotation group ultimately have quaternionic origins, although we will not attempt to cover that connection in any depth here (for comprehensive accounts, see Edmonds [43] or Biedenharn and Louck [17,18]).

27.1 BRIEF INTRODUCTION TO GROUP REPRESENTATIONS

The mathematics of group theory—in particular of the rotation group—is an endless subject, and we will not attempt any rigorous treatment here (e.g., see Altmann [4], Edmonds [43], Gilmore [56], and van der Waerden [163]). However, from a practical computational point of view there are a few concepts we would like to present here that may ultimately be of general use in 3D graphics and modeling methods.

Therefore, we begin with a short review of some simple concepts of group theory. This will establish some of the language used in the following and provide some extremely useful analogies for understanding the ways in which spherical harmonics, a fundamental tool for analyzing systems involving 3D orientations, can and should be used.

We will assume that the reader has some familiarity with Fourier transforms, and we will use the basic properties of Fourier transforms as our main pedagogical tool. However, even before looking at Fourier transforms we need to look at ordinary Euclidean space. Suppose we have a basis for a Euclidean space that looks like

$$\{e\} = (\hat{\mathbf{e}}_1, \hat{\mathbf{e}}_2, \ldots, \hat{\mathbf{e}}_n).$$

Then, any vector $\mathbf{x}$ in this space can be expressed in terms of its projections onto the components of the basis

$$\mathbf{x} = x_1\hat{\mathbf{e}}_1 + x_2\hat{\mathbf{e}}_2 + \cdots + x_n\hat{\mathbf{e}}_n.$$

Our basic tool is the *dot product* between any two Euclidean vectors,

$$\mathbf{x} \cdot \mathbf{y} = x_1 y_1 + x_2 y_2 + \cdots + x_n y_n,$$

which is independent of the basis chosen. For *any particular basis*, the dot product allows each of the components in the expansion to be written as

$$x_i = \mathbf{x} \cdot \hat{\mathbf{e}}_i,$$

where we have made implicit use of the orthonormality of the basis,

$$\hat{\mathbf{e}}_i \cdot \hat{\mathbf{e}}_j = \delta_{ij}$$

and δ_{ij} is the Kronecker delta, which is 1 if $i = j$ and vanishes otherwise.

The basis vectors $\hat{\mathbf{e}}_i$ of Euclidean space have precise analogs in Fourier transform theory, with the slight complication that instead of using constant vectors the corresponding basis for expanding a function in a Fourier series is itself a function (but a function that "looks like" $\hat{\mathbf{e}}_i$ in terms of what it does). Using a Fourier expansion, we can write any function in two alternative ways: in terms of complex basis functions or in terms of real basis functions (see http://mathworld.wolfram.com/FourierTransform.html). In the full complex form, we find two dominant alternative forms in the literature: $F(k)$, $f(x)$ with normalization 1 in the signal processing literature and $\tilde{F}(k)$, $\tilde{f}(x)$ with normalization $1/\sqrt{2\pi}$ in the modern physics literature.

Object	Euclidean Vectors	Fourier Transforms	Spherical Harmonics
Group	Orthogonal group	Translation group	3D rotations $=\mathbf{SO}(3)$
Object	$\mathbf{x}$	$f(x)$	$f(\theta, \phi)$
Basis	$\hat{\mathbf{e}}_i$	e^{ikx}	$Y_{lm}(\theta, \phi)$
Projection	Dot ($\cdot$) product	$\int_{-\infty}^{+\infty} dx$	$\int_0^{2\pi} d\phi \int_{-1}^{+1} d\cos\theta$
Coefficients	x_i	$F(k)$	a_{lm}

TABLE 27.1   *Parallel structures between Euclidean vectors, Fourier transforms (translation group), and spherical harmonics (3D rotation group).*

$$F(k) = \int_{-\infty}^{+\infty} f(x) e^{-2\pi ikx} \, dx,$$

$$f(x) = \int_{-\infty}^{+\infty} F(k) e^{+2\pi ikx} \, dk,$$

$$\tilde{F}(k) = \frac{1}{\sqrt{2\pi}} \int_{-\infty}^{+\infty} \tilde{f}(x) e^{ikx} \, dx,$$

$$\tilde{f}(x) = \frac{1}{\sqrt{2\pi}} \int_{-\infty}^{+\infty} \tilde{F}(k) e^{-ikx} \, dk.$$

As summarized in Table 27.1, we can see that $f(x)$ is like the vector $\mathbf{x}$, integration acts like the dot product, e^{ikx} acts like the basis, and $F(k)$ is like the coefficients x_i that multiply each basis function in order to express the vector $f(x)$ in terms of its components. Finding the x_i by taking a dot product with a basis vector is like finding $F(k)$ by taking the integral with e^{ikx}.

27.1.1   COMPLEX VERSUS REAL

Because e^{ikx} is a complex function, the coefficients $F(k)$ may in general be complex (although if $f(x)$ starts out as real it will always stay real). However, when implementing explicit calculations or computer programs we sometimes find it necessary to work in terms of strictly real numbers. This is accomplished in the Fourier series by observing that every function has an even and an odd part:

$$f(x) = f_+(x) + f_-(x),$$

$$f_+(x) = \frac{1}{2}\big(f(x) + f(-x)\big),$$

$$f_-(x) = \frac{1}{2}\big(f(x) - f(-x)\big).$$

Because the Fourier basis also has an even and an odd part,

$$e^{ikx} = \cos(x) + i\sin(x),$$

$$\cos(x) = \frac{1}{2}\big(e^{ikx} + e^{-ikx}\big),$$

$$\sin(x) = \frac{1}{2i}\big(e^{ikx} - e^{-ikx}\big),$$

we can alternatively use two separate real bases (the cosine transform and the sine transform) to represent any function, in the form

$$F(k) = \int_{-\infty}^{+\infty} f_+(x)\cos(2\pi k x)\,dx - i\int_{-\infty}^{+\infty} f_-(x)\sin(2\pi k x)\,dx, \qquad [27.1]$$

$$\tilde{F}(k) = \frac{1}{\sqrt{2\pi}}\int_{-\infty}^{+\infty} \tilde{f}_+(x)\cos(kx)\,dx + \frac{i}{\sqrt{2\pi}}\int_{-\infty}^{+\infty} \tilde{f}_-(x)\sin(kx)\,dx. \quad [27.2]$$

27.1.2   WHAT IS A REPRESENTATION?

This is not the whole story, but in fact only the beginning. Fourier transforms have one more property that we will need in order to understand what is happening with spherical harmonics and how they represent shapes in space. This property comes from the remarkable fact that exponentials can be used to represent an important group—the group of translations in Euclidean space. What this means is the following: suppose we have a translation $T(a)$ in space that is implemented by the equation

$$x' = x + a.$$

Then there is a change in the function

$$f(x') = f(x + a) = \int_{-\infty}^{+\infty} F(k)e^{ik(x+a)}\, dk \tag{27.3}$$

$$= \int_{-\infty}^{+\infty} \left(e^{ika}F(k)\right)e^{ikx}\, dk \tag{27.4}$$

so that the *old basis* e^{ikx} forms a *new basis* that is multiplied by e^{ika}. We say that e^{ikx} is a basis for representations of the translation group and that the *action* of translation is embodied in multiplication by e^{ika}, and therefore by definition the function e^{ika} is a *representation* of the group of Euclidean translations.

We note that formally one more property is required; namely, that a *sequence* of translations has the correct properties,

$$T(a)T(b) = T(a + b),$$

which means that $(x + a) + b = x + (a + b)$. This property is satisfied by

$$e^{ika}e^{ikb} = e^{ik(a+b)},$$

and thus the exponential function intrinsically possesses the properties of the translation group under ordinary multiplication.

Note: Essentially the same algebra produces the N-dimensional Euclidean translations with $x \to (x_1, x_2, \ldots, x_N)$ and $k \to (k_1, k_2, \ldots, k_N)$, and $kx \to \mathbf{x} \cdot \mathbf{k}$.

Why is all of this important? It is important to understand these properties and relate them to Euclidean translations because we may take two pieces of data describing *the same object*, get two different Fourier series, and want to *prove* that they are the same object displaced by a distance a, which should of course be of no consequence! How can we do that?

To identify a unique object that is independent of arbitrary translations, we must find an intrinsic description that does not depend on whether the Fourier coefficient is $F(k)$ or $G(k) = (e^{ika}F(k))$. (Remember that $F(k)$ is analogous to the coefficient x_i of a coordinate direction $\hat{\mathbf{e}}_i$ for ordinary Euclidean vectors.) There are two ways to make this identification:

- Division: If we take $\phi(k) = F(k)/G(k)$, and find that $|\phi| = 1$, the two objects are equivalent up to a translation. However, this property will not generalize straightforwardly to spherical harmonics, which is our next concern.

- *Multiply by the complex conjugate:* The complex conjugate of e^{ika} is easily seen to be e^{-ika}, which corresponds to the representation of the *inverse translation*—the translation $T(-a)$ going back in the opposite direction in space. Thus, when we multiply two complex conjugate coefficients together we *undo* the translation and it cancels out. Therefore, if we compute the product with the inverse we get what we need:

$$G^*(k)G(k) = F^*(k)e^{-ika}e^{ika}F(k) = F^*(k)F(k).$$

For groups more complicated than the translation group, we will deal with inverses instead of complex conjugate exponentials.

27.2   BASIC PROPERTIES OF SPHERICAL HARMONICS

Having now established a language for representing data as vectors or as functions expanded in terms of basis vectors, we proceed to the technology of spherical harmonics. We begin with the fundamental properties of spherical harmonics, loosely following the material found at the web site http://mathworld.wolfram. com/SphericalHarmonic.html. More technical details, including a treatment of $\mathbf{SO}(3)$ representation theory, can be found in expositions such as that by Finley (see http://panda.unm.edu/Courses/finley/p495/handouts/ in the file rotations.pdf).

A *spherical harmonic* $Y_l^m(\theta, \phi)$ satisfies the angular part of the 3D Laplacian differential equation $\nabla^2 \phi(\mathbf{x}) = 0$, where we must pay particular attention to define the coordinate system as

$$\mathbf{x} = (r\cos\phi\sin\theta, \, r\sin\phi\sin\theta, \, r\cos\theta)$$

with $0 \leqslant \theta \leqslant \pi$ and $0 \leqslant \phi < 2\pi$ to be consistent with the standard conventions. Note that this differs from conventions we use elsewhere in our treatment of spheres. The spherical harmonics then take the form

$$Y_l^m(\theta, \phi) = \sqrt{\frac{2l+1}{4\pi} \frac{(l-m)!}{(l+m)!}} P_l^m(\cos\theta)e^{im\phi},$$

where $P_l^m(\cos\theta)$ is the associated Legendre polynomial.

One important property of Y_l^m is the relation between the complex conjugate and Y_l^{-m}:

$$\left(Y^*\right)_l^m(\theta, \phi) = (-1)^m Y_l^{-m}(\theta, \phi).$$

The analogs of the cosine and sine decomposition of the Fourier transform into real and imaginary parts for the spherical harmonics are sometimes written as $(Y^c)_l^m(\theta, \phi)$ and $(Y^s)_l^m(\theta, \phi)$ and defined simply as

$$(Y^c)_l^m(\theta, \phi) = \sqrt{\frac{2l+1}{4\pi} \frac{(l-m)!}{(l+m)!}} P_l^m(\cos\theta) \cos(m\phi),$$

$$(Y^s)_l^m(\theta, \phi) = \sqrt{\frac{2l+1}{4\pi} \frac{(l-m)!}{(l+m)!}} P_l^m(\cos\theta) \sin(m\phi).$$

With the standard chosen normalizations

$$\int_0^{2\pi} \int_0^{\pi} Y_l^m(\theta, \phi)(Y^*)_{l'}^{m'}(\theta, \phi) \, d\phi \, d\theta = \delta_{ll'}\delta_{mm'},$$

$$\int_0^{2\pi} \int_{-1}^{+1} Y_l^m(\theta, \phi)(Y^*)_{l'}^{m'}(\theta, \phi) \, d\phi \, d(\cos\theta) = \delta_{ll'}\delta_{mm'},$$

$$\int_0^{2\pi} \int_0^{\pi} (Y^c)_l^m(\theta, \phi)(Y^c)_{l'}^{m'}(\theta, \phi) \, d\phi \, d\theta = \delta_{ll'}\delta_{mm'},$$

$$\int_0^{2\pi} \int_0^{\pi} (Y^c)_l^m(\theta, \phi)(Y^s)_{l'}^{m'}(\theta, \phi) \, d\phi \, d\theta = 0,$$

$$\int_0^{2\pi} \int_0^{\pi} (Y^s)_l^m(\theta, \phi)(Y^c)_{l'}^{m'}(\theta, \phi) \, d\phi \, d\theta = 0,$$

$$\int_0^{2\pi} \int_0^{\pi} (Y^s)_l^m(\theta, \phi)(Y^s)_{l'}^{m'}(\theta, \phi) \, d\phi \, d\theta = \delta_{ll'}\delta_{mm'}$$

we may then perform the exact analog of the determination of the coefficients of a Euclidean vector and the Fourier coefficients by writing the representation of a surface in the form of a radial distance parameterized by the angular coordinates on the sphere:

$$\mathbf{x}(\theta, \phi) = f(\theta, \phi)(\sin\theta \cos\phi, \sin\theta \sin\phi, \cos\theta). \tag{27.5}$$

Although we will discuss several ways of expanding this shape function, the most straightforward is to make either an expansion in terms of complex coefficients using the standard spherical harmonics or to make an even-odd expansion using the real spherical harmonics analogous to the Fourier cosine–sine expansion:

$$f(\theta, \phi) = \sum_{l=0}^{\infty} \sum_{m=-l}^{+l} a_l^m Y_l^m(\theta, \phi),$$

$$f(\theta, \phi) = \sum_{l=0}^{\infty} \sum_{m=-l}^{+l} \left[c_l^m (Y^c)_l^m(\theta, \phi) + s_l^m (Y^s)_l^m(\theta, \phi) \right].$$

Projection of the function $f(\theta, \phi)$ onto any individual basic component should look familiar by now, and is given by

$$a_l^m = \int_0^{2\pi} d\phi \int_0^{\pi} \sin\theta \, d\theta \, (Y^*)_l^m(\theta, \phi) \, f(\theta, \phi)$$

$$= \int_0^{2\pi} d\phi \int_0^{\pi} \sin\theta \, d\theta (-1)^m Y_l^{-m}(\theta, \phi) \, f(\theta, \phi),$$

$$c_l^m = \int_0^{2\pi} d\phi \int_0^{\pi} \sin\theta \, d\theta \, (Y^c)_l^m(\theta, \phi) f(\theta, \phi),$$

$$s_l^m = \int_0^{2\pi} d\phi \int_0^{\pi} \sin\theta \, d\theta \, (Y^s)_l^m(\theta, \phi) \, f(\theta, \phi).$$

Table 27.1 shows how this fits into the overall context that we began with vectors and Fourier transforms.

27.2.1 REPRESENTATIONS AND ROTATION-INVARIANT PROPERTIES

Detailed descriptions of the spherical harmonic expansion and matching process can be found in a number of places. An example of a fairly complete treatment in the chemistry literature is that of Ritchie and Kemp [141].

We will first need the conventional expression for a specific rotation phrased in terms of the Euler angles,

Euler Angle Rotation $(\alpha, \beta, \gamma) = \mathbf{R}(\alpha, \hat{\mathbf{z}}) \cdot \mathbf{R}(\beta, \hat{\mathbf{y}}) \cdot \mathbf{R}(\gamma, \hat{\mathbf{z}})$

$$= \mathbf{R}(\gamma, \hat{\mathbf{c}}) \cdot \mathbf{R}(\beta, \hat{\mathbf{b}}) \cdot \mathbf{R}(\alpha, \hat{\mathbf{z}})$$

$$= \begin{bmatrix} \cos\alpha \cos\beta \cos\gamma - \sin\alpha \sin\gamma & -\cos\gamma \sin\alpha - \cos\alpha \cos\beta \sin\gamma & \cos\alpha \sin\beta \\ \cos\alpha \sin\gamma + \sin\alpha \cos\beta \cos\gamma & \cos\alpha \cos\gamma - \cos\beta \sin\alpha \sin\gamma & \sin\alpha \sin\beta \\ -\sin\beta \cos\gamma & \sin\beta \sin\gamma & \cos\beta \end{bmatrix},$$

where $\hat{\mathbf{c}} = (\cos\alpha \sin\beta, \sin\alpha \sin\beta, \cos\beta)$ and $\hat{\mathbf{b}} = (-\sin\alpha, \cos\alpha, 0)$. The quaternion corresponding to this sequence of rotations is

$$q_0 = \cos\frac{1}{2}(\alpha + \gamma)\cos\frac{1}{2}\beta,$$

$$q_1 = \sin\frac{1}{2}(\gamma - \alpha)\sin\frac{1}{2}\beta,$$

$$q_2 = \cos\frac{1}{2}(\gamma - \alpha)\sin\frac{1}{2}\beta,$$

$$q_3 = \sin\frac{1}{2}(\alpha + \gamma)\cos\frac{1}{2}\beta.$$

Thus, we can immediately produce the axis-angle form $\mathbf{R}(\rho, \hat{\mathbf{n}})$ from

$$q = (\cos\rho/2, \; \hat{\mathbf{n}}\sin\rho/2)$$

if we need it.

The spherical harmonics associated with each l form a $(2l + 1)$-dimensional basis for representations of the rotation group. When this basis is acted on by ordinary 3D rotations $\mathbf{R}(\rho, \hat{\mathbf{n}})$ (we write this action abstractly using the $\circ$ symbol), the result is a type of generalized matrix multiplication that acts separately on each value of l. This takes the form

$$R(\rho, \hat{\mathbf{n}}) \circ Y_l^m(\theta, \phi) = \sum_{m'} D_{m'm}^{(l)}(\alpha, \beta, \gamma) Y_l^{m'}(\theta, \phi), \qquad (27.6)$$

where (see Biedenharn and Louck [17])

$$D_{m'm}^{(l)}(\alpha, \beta, \gamma) = e^{-im'\alpha} d_{m'm}^{(l)}(\beta) e^{-im\gamma} \qquad (27.7)$$

and

$$d_{m'm}^{(l)}(\beta) = \left[\frac{(l+m')!\,(l-m')!}{(l+m)!\,(l-m)!}\right]^{1/2}$$

$$\sum_{k=\max(0,m-m')}^{\min(l-m',l+m)} \left[(-1)^{k+m'-m}\binom{l+m}{k}\binom{l-m}{l-m'-k}\right.$$

$$\left. \times (\cos\beta/2)^{2l+m-m'-2k}(\sin\beta/2)^{2k+m'-m}\right] \tag{27.8}$$

is the Wigner function. We finally make the connection to quaternions by noting that the entire representation formula (Equation 27.7) can be written more generally in terms of quaternion variables [17] as

$$D_{m'm}^{(j)}(q) = \left[(j+m')!(j-m')!(j+m)!(j-m)!\right]^{1/2}$$

$$\times \sum_{s} \frac{(q_0-iq_3)^{j+m-s}(-iq_1-q_2)^{m'-m+s}(-iq_1+q_2)^s(q_0+iq_3)^{j-m'-s}}{(j+m-s)!(m'-m+s)!s!(j-m'-s)!},$$

$$\tag{27.9}$$

where s ranges over the set of legal values determined by j, and j can in principle range over either half-integer representation labels or integer representation labels (we use l when restricting to integers only). This matrix is an *explicit* realization of the quaternion algebra and has the required property of a group representation, now exposed more clearly in terms of quaternions. The multiplication rule for $D_{m'm}^{(j)}(q)$ considered as a matrix $\mathbf{D}^j(q)$ is simply

$$\mathbf{D}^j(p)\cdot\mathbf{D}^j(q) = \mathbf{D}^j(p\star q). \tag{27.10}$$

27.2.2 PROPERTIES OF EXPANSION COEFFICIENTS UNDER ROTATIONS

If we let $f'(\theta,\phi)$ represent the *same* spatial distribution as $f(\theta,\phi)$ after a rigid rotation by $\mathbf{R}(\rho,\hat{\mathbf{n}})$,

$$f'(\theta,\phi) = \mathbf{R}(\rho,\hat{\mathbf{n}})\circ f(\theta,\phi), \tag{27.11}$$

then via Equation 27.6,

$$a'_{lm} = \sum_{m'} D^{(l)}_{mm'}\big(q(\rho, \hat{\mathbf{n}})\big) a_{lm'}.$$ (27.12)

This is the equivalent of the e^{ika} multiplication for Fourier transform coefficients, and basically tells us everything we need to know about how a radial shape model $f(\theta, \phi)$ behaves under rotations.

Spherical Riemannian Geometry

We have already mentioned many times that the principal advantage of using unit quaternions to represent 3D orientation frames is the fact that the topological space of unit quaternions is the three-sphere $\mathbf{S}^3$ and that $\mathbf{S}^3$ is a simply-connected manifold with a natural and elegant distance measure or metric. We have exploited this property in a variety of ways to facilitate the study of particular applications. In this chapter we complete this story by writing down the standard elements of the Riemannian geometry of $\mathbf{S}^3$, so that when heuristic methods provide insufficient power or insight, the rigorous mathematics is in principle at our disposal. (For extensive treatments, see for example Gray [61], Lee [118], or Waner [166], and especially the notes of Kajiya and Gabriel [107].)

28.1 INDUCED METRIC ON THE SPHERE

The basic element of Riemannian geometry is the concept of a distance on a manifold mediated by a function of the local coordinates called the *metric tensor*. The application of the metric tensor to manifolds, which may have arbitrarily complex global structure, is of necessity local, dealing in infinitesimals, and thus intimately involves differential geometry—the formal generalization of the familiar principles of calculus.

Therefore, the most fundamental object we encounter in this chapter is the infinitesimal distance element ds, which is defined in terms of a chosen local coordinate system $\mathbf{x} = (x^1, \ldots, x^n)$, the differentials in that coordinate system

$\mathbf{dx} = (dx^1, \ldots, dx^n)$, and the metric tensor $g_{ij}(\mathbf{x})$ as

$$ds^2 = \sum_{i=1}^{n} \sum_{j=1}^{n} dx^i g_{ij}(\mathbf{x}) dx^j = dx^i g_{ij} dx^j. \qquad (28.1)$$

In Equation 28.1, we adopt the Einstein summation convention for repeated tensor indices. It is a standard observation that the infinitesimal length element is invariant under reparameterization of the coordinate system, $\mathbf{x} \to \mathbf{x}'(\mathbf{x})$:

$$(ds')^2 = dx'^i g'_{ij} dx'^j = \frac{\partial x'^i}{\partial x^m} dx^m g'_{ij} \frac{dx'^j}{dx^n} dx^n$$

$$= dx^m g_{mn}(x) dx^n = (ds)^2. \qquad (28.2)$$

We now introduce the following fascinating pieces of folklore; each would of course be the subject of entire chapters in a more extensive treatment.

- *Manifolds generally require multiple local coordinate systems.* In fact, a Riemannian manifold is by definition a collection of local patches, like a bag of rags, each with its own coordinate system and corresponding expression of the metric tensor, together with a set of rules for sewing together the edges of the patches to make the required global structure.
- *Many metric tensors can describe one manifold.* The variables of the coordinate system can be changed almost arbitrarily, and many different metric tensors are possible without altering the manifold. However, there are quantities *derivable* from any of these metric tensors that are independent of any choice of coordinate system or metric tensor. These are the *topological invariants* of the manifold.
- *Using the metric tensor, one can study manifolds entirely in terms of their local internal coordinates.* A manifold of dimension n can be studied using n local coordinates. The commonsense picture of a manifold embedded in a higher-dimensional coordinate system is entirely superfluous in Riemannian geometry. In other words, although we typically picture the surface of a ball, a donut, or a pretzel embedded in 3D space, the surfaces themselves can be studied strictly in 2D without referring in any way to their 3D embeddings.
- *Nevertheless, such embeddings can be useful.* One of the conundrums of Riemannian geometry is that although the metric tensor is not unique, we must still somehow pick a useful one in order to perform explicit calculations. This

can be done in practice by exploiting the following facts: (1) any Riemannian manifold of dimension n can be embedded in a Euclidean space of dimension $2n$ or immersed (with possible self-intersections) in dimension $(2n - 1)$ [170,171], and (2) any embedding defines a metric, called the *induced metric*, that is in fact what we typically use in practice for simple manifolds (e.g., those described by algebraic equations).

- *There are at least three very different textbook approaches to Riemannian geometry.* (1) A mathematics text would typically define a *connection* ∇ that generalizes standard calculus to provide a directional derivative on vector fields. Then, for a given manifold M and a given metric g one takes the pair (M, g) defining a Riemannian manifold and selects the unique torsion-free connection derivable from the metric g. This is the *Levi-Civita connection*. (2) An alternative elegant formulation, the differential form approach, is essentially the dual to the directional derivative approach, replacing derivatives (tangent space) by differentials or exterior derivatives (cotangent space). Certain calculations can become substantially simpler using this technology for practical applications of Riemannian geometry [36,45,52]. (3) The tensor calculus approach, which does not take advantage of the coordinate-system-independent features of the first two approaches, requires an explicit choice of the coordinate system and is almost universally the choice of physicists, e.g., for the study of Einstein's theory of general relativity. (See, for example, Misner [125], Moller [127], or Weinberg [168].) This approach, which is less elegant but more visualizable, will be the primary method we use to study $\mathbf{S}^3$.

28.2   INDUCED METRICS OF SPHERES

We now expand on the concept of the induced metric and its application to spherical geometry. In general, if one has an appropriate vector-valued function $\mathbf{f(t)}$ mapping a parameter space $\mathbf{t}$ of dimension n into a larger Euclidean space of dimension $m > n$, $\mathbf{f}$ describes a surface of dimension n embedded in $\mathbb{R}^m$. Because we could actually take a piece of string and measure distances on the embedded surface, the embedding itself can be used to define a Riemannian metric. Obtaining an explicit parametric embedding of a surface is therefore a rather useful practical approach to determining a Riemannian metric for a surface whose properties we wish to study.

Explicit metrics can therefore be constructed for any sphere $\mathbf{S}^n$ by embedding it in Euclidean $(n + 1)$ space, $\mathbb{R}^{n+1}$, using the constant-radius algebraic equation

$$r^2 = \sum_{a=0}^{n} (x^a)^2 \qquad (28.3)$$

defining an n-sphere. In practice, we will normally take $r \equiv 1$. The induced metric for any set of n independent variables $\{t^i\}$ parameterizing a sphere is simply

$$g_{ij} = \sum_{a=0}^{n} \frac{\partial x^a(t)}{\partial t^i} \frac{\partial x^a(t)}{\partial t^j} = \frac{\partial \mathbf{x}(t)}{\partial t^i} \cdot \frac{\partial \mathbf{x}(t)}{\partial t^j}. \qquad (28.4)$$

Induced Metric Definition

Whatever the parameterization we choose for solving the sphere constraint (Equation 28.3), we will find that the corresponding metrics from Equation 28.4 will all have the same invariant properties describing a sphere in Riemannian geometry.

We will explore three different classes of parametric solutions to Equation 28.3: square-root-based solutions, trigonometric solutions, and the stereographic projection. These are defined as follows:

- *Square root*: The *square root solutions* simply solve for one dependent variable in terms of the remaining n independent variables, as in, for example,

$$x^0 = \pm \sqrt{1 - \sum_{i=1}^{n} (x^i)^2}.$$

  The patch covering the northern hemisphere corresponds to the plus $(+)$ sign, and the patch covering the southern hemisphere corresponds to the minus $(-)$ sign. Some treatments actually use all possible solutions, leading to a set of $2(n + 1)$ different patches—one pair of hemispheres for each coordinate axis in $\mathbb{R}^{n+1}$—rather than just the northern and southern hemispheres for the two solutions of the single chosen dependent variable x^0.
- *Trigonometric*: The trigonometric solutions are of the form

$$x^0 = \cos\theta_1,$$
$$x^1 = \sin\theta_1 \cos\theta_2,$$
$$x^2 = \sin\theta_1 \sin\theta_2 \cos\theta_3,$$

$$\vdots$$

$$x^{n-2} = \sin\theta_1 \sin\theta_2 \ldots \cos\theta_{n-1},$$

$$x^{n-1} = \sin\theta_1 \sin\theta_2 \ldots \sin\theta_{n-1}\cos\phi,$$

$$x^n = \sin\theta_1 \sin\theta_2 \ldots \sin\theta_{n-1}\sin\phi,$$

with the n independent angular variables having the ranges

$$0 \leqslant \theta_1, \quad \ldots, \quad \theta_{n-1} \leqslant \pi,$$

$$0 \leqslant \phi < 2\pi.$$

One can verify explicitly that with these n angular variables the $n+1$ functions $x^a(\{\theta\}, \phi)$ explicitly satisfy $\sum_{a=0}^{n}(x^a)^2 = 1$.

- *Stereographic projection:* Our third solution—the polar stereographic projection from the Euclidean plane $\mathbb{R}^n$—is singular at the North Pole, $x^0 = +1$ (or at the South Pole, $x^0 = -1$) but is otherwise nicely behaved. Letting $\mathbf{t} = t_1, \ldots, t_n$ be the n Euclidean variables and $\tau^2 = \mathbf{t} \cdot \mathbf{t}$, we have the following map from the Euclidean plane to the sphere embedded in $\mathbb{R}^{n+1}$ (excluding the North Pole):

$$x^0 = \frac{\tau^2 - 1}{\tau^2 + 1},$$

$$\mathbf{x} = \frac{2\mathbf{t}}{\tau^2 + 1}. \tag{28.5}$$

This again explicitly solves $(x^0)^2 + \mathbf{x} \cdot \mathbf{x} = 1$ using only n parameters. To describe a sphere excluding the South Pole, we switch the singularity at infinity from pole to pole by taking

$$x^0 = \frac{1 - \tau^2}{1 + \tau^2},$$

$$\mathbf{x} = \frac{2\mathbf{t}}{\tau^2 + 1}. \tag{28.6}$$

Note on hemisphere patches: The sphere must of course be described by at least two patches, because no single coordinate system can include a regular description of both the northern and southern hemispheres. The polar trigonometric system fails to be differentiable at either pole, $\theta_1 = 0$ or $\theta_1 = \pi$. In the following we work out the metric, connection, and invariants for just a single patch, noting that the three coordinate systems we examine cover all possible patches.

28.2.1 S^1 INDUCED METRICS

The one-sphere S^1 is a trivial case, but still shows us some useful properties. We can immediately compute (as follows) all three induced metrics from the parametric solutions of $x^2 + y^2 = 1$.

$$x = x, \qquad y = \sqrt{1 - x^2},$$

$$x = \cos\theta, \qquad y = \sin\theta,$$

$$x = \frac{2t}{1 + t^2}, \qquad y = \frac{1 - t^2}{1 + t^2},$$

and thus

$$ds^2 = (dx)^2\left(1 + \frac{x^2}{1 - x^2}\right) = \frac{dx^2}{1 - x^2} \rightarrow g_{xx} = \frac{1}{1 - x^2} \quad \left(g^{xx} = 1 - x^2\right),$$

$$ds^2 = (d\theta)^2\left(\sin^2\theta + \cos^2\theta\right) = (d\theta)^2 \rightarrow g_{\theta\theta} = g^{\theta\theta} = 1,$$

$$ds^2 = (dt)^2\left[\frac{4(1 - t^2)^2}{(1 + t^2)^4} + \frac{16t^2}{(1 + t^2)^4}\right] \rightarrow g_{tt} = \frac{4}{(1 + t^2)^2} \quad \left(g^{tt} = \frac{(1 + t^2)^2}{4}\right).$$

To check that the integrals are plausibly independent despite these radically different coordinate parameterizations, we verify (with the corresponding limits shown in Figure 28.1) that the integrals of the length elements ds over one quadrant are indeed identical:

$$\int_0^1 \frac{dx}{\sqrt{1 - x^2}} = \frac{\pi}{2},$$

$$\int_0^{\pi/2} d\theta = \frac{\pi}{2},$$

$$\int_1^\infty \frac{2dt}{(1 + t^2)} = \frac{\pi}{2}.$$

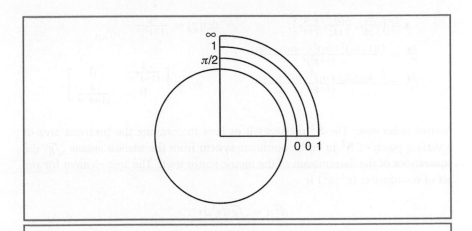

FIGURE 28.1 *The ranges of the alternate variable systems on* $\mathbf{S}^1$ *when we restrict integration to the first quadrant.*

28.2.2 $\mathbf{S}^2$ INDUCED METRICS

The ordinary two-sphere, $x^2 + y^2 + z^2 = 1$, is really the first nontrivial example, although we saw that even $\mathbf{S}^1$ had things to teach us about the reparameterization invariance of ds. The induced metrics on $\mathbf{S}^2$ are computed according to the by-now-familiar procedure:

$$\mathbf{x} = \left(x, y, z = \sqrt{1 - x^2 - y^2}\right), \qquad \det[g] = \tfrac{1}{1 - x^2 - y^2},$$

$$\tfrac{\partial \mathbf{x}}{\partial x} = \left(1, 0, -\tfrac{x}{z}\right),$$

$$\tfrac{\partial \mathbf{x}}{\partial y} = \left(0, 1, -\tfrac{y}{z}\right), \qquad g_{ij} = \begin{bmatrix} \frac{1-y^2}{1-x^2-y^2} & \frac{xy}{1-x^2-y^2} \\ \frac{xy}{1-x^2-y^2} & \frac{1-x^2}{1-x^2-y^2} \end{bmatrix},$$

$$\mathbf{x} = (\cos\theta, \sin\theta \cos\phi, \sin\theta \sin\phi), \qquad \det[g] = \sin^2\theta,$$

$$\tfrac{\partial \mathbf{x}}{\partial \theta} = (-\sin\theta, \cos\theta \cos\phi, \cos\theta \sin\phi),$$

$$\tfrac{\partial \mathbf{x}}{\partial \phi} = (0, -\sin\theta \sin\phi, \sin\theta \cos\phi), \qquad g_{ij} = \begin{bmatrix} 1 & 0 \\ 0 & \sin^2\theta \end{bmatrix},$$

$$\mathbf{x} = \left(\tfrac{2t_1}{1+t^2}, \tfrac{2t_2}{1+t^2}, \tfrac{1-t^2}{1+t^2}\right), \qquad \det[g] = \tfrac{16}{(1+t^2)^4},$$

$$\frac{\partial \mathbf{x}}{\partial t_1} = \frac{(2(1-(t_1)^2+(t_2)^2),-4t_1t_2,-4t_1)}{(1+t^2)^2},$$

$$\frac{\partial \mathbf{x}}{\partial t_2} = \frac{(-4t_1t_2,2(1+(t_1)^2-(t_2)^2),-4t_2)}{(1+t^2)^2}, \qquad g_{ij} = \begin{bmatrix} \tfrac{4}{(1+t^2)^2} & 0 \\ 0 & \tfrac{4}{(1+t^2)^2} \end{bmatrix}.$$

Invariant surface areas: The $\mathbf{S}^2$ metrics tell us how to compute the invariant area of a surface patch of $\mathbf{S}^2$ in each coordinate system from the *invariant measure* $\sqrt{g}$, the square root of the determinant of the metric tensor itself. The area element for any set of coordinates (s^1, s^2) is

$$d^2\mu = \sqrt{g}\, ds^1 ds^2.$$

Using the appropriate integration limits for the first octant of $\mathbf{S}^2$ (shown in Figure 28.2), we can verify that regardless of coordinate system we get the same value of the area for the curved surface provided we use the corresponding value of $\sqrt{g}$:

$$A = \iint \frac{dx\, dy}{\sqrt{1 - x^2 - y^2}} = \int_0^1 \frac{dr}{\sqrt{1 - r^2}} \int_0^{\pi/2} d\theta = \frac{\pi}{2},$$

$$A = \int_0^{\pi/2} d\phi \int_0^{\pi/2} \sin\theta\, d\theta = \frac{\pi}{2} \int_1^0 (-d\cos\theta) = \frac{\pi}{2},$$

$$A = \iint dt_1\, dt_2 \frac{4}{(1+t^2)^2} = \frac{\pi}{2}.$$

28.2.3 $\mathbf{S}^3$ INDUCED METRICS

Finally, the case of essential interest for us is $\mathbf{S}^3$. Here, we find

$$\mathbf{x} = \left(x, y, z, w = \sqrt{1 - x^2 - y^2 - z^2}\right),$$

$$\frac{\partial \mathbf{x}}{\partial x} = \left(1, 0, 0, -\tfrac{x}{w}\right),$$

$$\frac{\partial \mathbf{x}}{\partial y} = \left(0, 1, 0, -\tfrac{y}{w}\right),$$

$$\frac{\partial \mathbf{x}}{\partial z} = \left(0, 0, 1, -\tfrac{z}{w}\right),$$

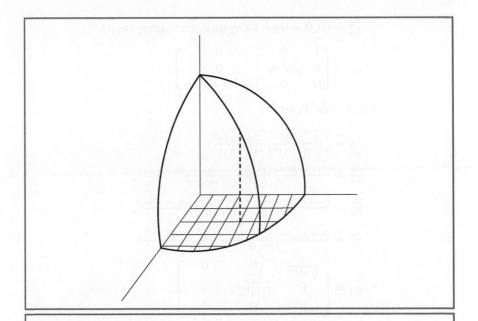

FIGURE 28.2 The first octant of $\mathbf{S}^2$, over which we can integrate using any of the available coordinate systems to get the same value of the area $(\pi/2)$ if we use the correct measure derived from the induced metric.

$$g_{ij} = \begin{bmatrix} 1 + \frac{x^2}{w^2} & \frac{xy}{w^2} & \frac{xz}{w^2} \\ \frac{xy}{w^2} & 1 + \frac{y^2}{w^2} & \frac{yz}{w^2} \\ \frac{xz}{w^2} & \frac{yz}{w^2} & 1 + \frac{z^2}{w^2} \end{bmatrix},$$

$$\det[g] = \frac{1}{1-x^2-y^2-z^2},$$

$$\mathbf{x} = (\cos\theta_1, \sin\theta_1\cos\theta_2, \sin\theta_1\sin\theta_2\cos\phi, \sin\theta_1\sin\theta_2\sin\phi),$$

$$\frac{\partial\mathbf{x}}{\partial\theta_1} = (-\sin\theta_1, \cos\theta_1\cos\theta_2, \cos\theta_1\sin\theta_2\cos\phi, \cos\theta_1\sin\theta_2\sin\phi),$$

$$\frac{\partial\mathbf{x}}{\partial\theta_2} = (0, -\sin\theta_1\sin\theta_2, \sin\theta_1\cos\theta_2\cos\phi, \sin\theta_1\cos\theta_2\sin\phi),$$

$$\frac{\partial \mathbf{x}}{\partial \phi} = (0, 0, -\sin\theta_1 \sin\theta_2 \sin\phi, \sin\theta_1 \sin\theta_2 \cos\phi),$$

$$g_{ij} = \begin{bmatrix} 1 & 0 & 0 \\ 0 & \sin^2\theta_1 & 0 \\ 0 & 0 & \sin^2\theta_1 \sin^2\theta_2 \end{bmatrix},$$

$$\det[g] = \sin^4\theta_1 \sin^2\theta_2,$$

$$\mathbf{x} = \left(\frac{2t_1}{1+t^2}, \frac{2t_2}{1+t^2}, \frac{2t_3}{1+t^2}, \frac{1-t^2}{1+t^2}\right),$$

$$\frac{\partial \mathbf{x}}{\partial t_1} = \frac{(2(1-(t_1)^2+(t_2)^2+(t_3)^2), -4t_1t_2, -4t_1t_3, -4t_1)}{(1+(t_1)^2+(t_2)^2+(t_3)^2)^2},$$

$$\frac{\partial \mathbf{x}}{\partial t_2} = \frac{(-4t_1t_2, 2(1+(t_1)^2-(t_2)^2+(t_3)^2), -4t_2t_3, -4t_2)}{(1+(t_1)^2+(t_2)^2+(t_3)^2)^2},$$

$$\frac{\partial \mathbf{x}}{\partial t_3} = \frac{(-4t_1t_3, -4t_2t_3, 2(1+(t_1)^2+(t_2)^2-(t_3)^2), -4t_3)}{(1+(t_1)^2+(t_2)^2+(t_3)^2)^2},$$

$$g_{ij} = \begin{bmatrix} \frac{4}{(1+t^2)^2} & 0 & 0 \\ 0 & \frac{4}{(1+t^2)^2} & 0 \\ 0 & 0 & \frac{4}{(1+t^2)^2} \end{bmatrix},$$

$$\det[g] = \frac{4^3}{(1+t^2)^6}.$$

28.2.4 TOROIDAL COORDINATES ON $\mathbf{S}^3$

The three-sphere has other traditional coordinate systems that can be useful from time to time. One of these comes from looking at $\mathbf{S}^3$ as a parameterized family of tori with major and minor radii contrived to satisfy $\mathbf{x} \cdot \mathbf{x} = 1$. These coordinates can conveniently be viewed as a pair of complex circles,

$$x + iy = \cos\theta e^{i\alpha}, \qquad z + iw = \sin\theta e^{i\beta},$$

or as the real coordinates

$$x = (\cos\theta \cos\alpha, \cos\theta \sin\alpha, \sin\theta \cos\beta, \sin\theta \sin\beta).$$

The induced metric in these coordinates then becomes

$$g_{ij} = \begin{bmatrix} 1 & 0 & 0 \\ 0 & \cos^2\theta & 0 \\ 0 & 0 & \sin^2\theta \end{bmatrix}. \qquad (28.7)$$

In many ways, this is the simplest of the metrics we can choose to describe the geometry of $\mathbf{S}^3$.

28.2.5 AXIS-ANGLE COORDINATES ON $\mathbf{S}^3$

When we wish to make a direct connection to the Euler eigenvector notation for a quaternion point on $\mathbf{S}^3$, we find it useful to rephrase the standard polar coordinate system as

$$q_0 = \cos(\theta/2),$$
$$\mathbf{q} = \sin(\theta/2)(\cos\alpha \sin\beta, \sin\alpha \sin\beta, \cos\beta), \tag{28.8}$$

giving us the $\mathbf{S}^3$ metric for the Euler eigenvector with coordinate order (θ, α, β):

$$g_{ij} = \begin{bmatrix} 1/4 & 0 & 0 \\ 0 & \sin^2\frac{\theta}{2} & 0 \\ 0 & 0 & \sin^2\frac{\theta}{2}\sin^2\beta \end{bmatrix}. \tag{28.9}$$

28.2.6 GENERAL FORM FOR THE SQUARE-ROOT INDUCED METRIC

For an arbitrary dimension, the induced metric of the sphere $\mathbf{S}^n$ can now easily be computed from the general square-root form. It is simply

$$g_{ij} = \delta_{ij} + \frac{x_i x_j}{1 - \mathbf{x}\cdot\mathbf{x}}, \tag{28.10}$$

$$g^{ij} = \delta^{ij} - x^i x^j, \tag{28.11}$$

where the Kronecker delta δ_{ij} (see Appendix F) is 1 when $i = j$ and 0 otherwise (that is, it is the identity matrix). One can also express the volume measure simply in terms of the square root of the determinant:

$$\det[g_{ij}] = \frac{1}{1 - \mathbf{x}\cdot\mathbf{x}}. \tag{28.12}$$

28.3   ELEMENTS OF RIEMANNIAN GEOMETRY

Given a metric $g_{ij}(x)$, in whatever coordinate system we choose, we can take further steps toward defining derivatives and moving paths on the sphere that are optimal (that is, those that follow the shortest-distance route). The unique torsion-free connection on a Riemannian manifold (M, g) is the Levi-Civita connection, given in terms of the metric by the Christoffel symbols,

$$\Gamma_{ij}^k = \frac{1}{2} g^{kl} (\partial_i g_{jl} + \partial_j g_{li} - \partial_l g_{ij}), \tag{28.13}$$

where, if g_{ij} is considered as a matrix, then $g^{ij} = (g_{ij})^{-1}$ is its inverse. Γ_{ij}^k allows us to define local directional derivatives of a basis vector ∂_j in the arbitrary direction ∂_i using

$$\nabla_{\partial_i} \partial_j = \Gamma_{ij}^k \partial_k. \tag{28.14}$$

The covariant derivative of a parameterized vector $\mathbf{V}(t)$ is then

$$\nabla_k V^i(x) = \frac{\partial V^i}{\partial x^k} + \Gamma_{jk}^i V^j, \tag{28.15}$$

and by the chain rule

$$\frac{DV^i(t)}{dt} = \nabla_k V^i(x) \frac{dx^k}{dt} = \frac{dV^i}{dt} + \Gamma_{jk}^i V^j \frac{dx^k}{dt}. \tag{28.16}$$

Curiously, whereas g_{ij} and x^k transform as vectors under coordinate transformations, Γ_{jk}^i does not, but it includes an extra term that is eventually cancelled in the Riemannian curvature:

$$\Gamma_{jk}^i = \Gamma_{bc}^{\prime a} \frac{\partial x^i}{\partial x^{\prime a}} \frac{\partial x^{\prime b}}{\partial x^j} \frac{\partial x^{\prime c}}{\partial x^k} + \frac{\partial x^i}{\partial x^{\prime e}} \frac{\partial^2 x^{\prime e}}{\partial x^j \partial x^k}. \tag{28.17}$$

The discrepancy is carefully contrived so that $D\mathbf{V}/dt$ transforms as a pure tensor:

$$\frac{DV^{\prime a}}{dt} = \frac{\partial x^{\prime a}}{\partial x^i} \frac{DV^i}{dt}. \tag{28.18}$$

Riemann curvature tensor, Ricci tensor, and scalar curvature: To discover intrinsic properties of a manifold, we need first to construct objects that are covariant under coordinate transformations and then to construct contractions of those objects with the metric tensor to form invariants. The Riemann curvature tensor, with four indices (no matter what the dimension of the manifold), is formed by manipulating first derivatives of the Christoffel symbols in such a way that all extra terms in Equation 28.17 cancel out, with the result (e.g., see Misner [125], Moller [127], and Weinberg [168])

$$R^i_{klm} = \frac{\partial \Gamma^i_{kl}}{\partial x^m} - \frac{\partial \Gamma^i_{km}}{\partial x^l} + \Gamma^i_{rm} \Gamma^r_{kl} - \Gamma^i_{rl} \Gamma^r_{km}$$

$$= -R^i_{kml}.$$

An alternative useful form can be shown from the definition of Γ^i_{jk} to be

$$R_{jklm} = g_{ji} R^i_{klm}$$

$$= \frac{1}{2} \left(\frac{\partial^2 g_{il}}{\partial x^k \partial x^m} + \frac{\partial^2 g_{km}}{\partial x^i \partial x^l} - \frac{\partial^2 g_{im}}{\partial x^k \partial x^l} - \frac{\partial^2 g_{kl}}{\partial x^i \partial x^m} \right)$$

$$+ g^{rs} (\Gamma_{r,il} \Gamma_{s,km} - \Gamma_{r,im} \Gamma_{s,kl}),$$

where

$$\Gamma_{i,jk} = g_{im} \Gamma^m_{jk} = \frac{1}{2} (\partial_k g_{ij} + \partial_j g_{ik} - \partial_i g_{jk}).$$

The Ricci tensor is formed by contracting two indices of the curvature with the metric tensor

$$R_{ij} = R^m{}_{imj} = g^{mn} R_{minj}, \qquad (28.19)$$

where, as always, the repeated indices are summed over the dimension of the space. The *scalar curvature*—which is, at last, an invariant property of the manifold regardless of the chosen coordinate system—is

$$R = g^{ij} R_{ij} = g^{ij} g^{mn} R_{minj}. \qquad (28.20)$$

28.4   RIEMANN CURVATURE OF SPHERES

Now that we have the machinery for all this, we can compute the Christoffel symbols and the curvatures for the spheres up to $\mathbf{S}^3$ in the various coordinate systems.

We will give representative values to illustrate the important features. When complete sets of functions are required, one should employ a symbolic algebra package to avoid algebraic mistakes.

28.4.1 $\mathbf{S}^1$

The circle $\mathbf{S}^1$ provides our first example, which is simple but not entirely trivial.

- Polar: Because $ds^2 = d\theta^2$, $\Gamma^\theta_{\theta\theta} = 0$, and $R^\theta_{\theta\theta\theta} = 0$ and the one-sphere is obviously flat.
- Square root: With the metric given by $g_{xx} = 1/(1-x^2)$, we find

$$\Gamma^x_{xx} = -\frac{x}{1-x^2}, \qquad R^x_{xxx} = 0. \tag{28.21}$$

- Projective: With the metric given by $g_{tt} = 4/(1+t^2)^2$, we find

$$\Gamma^t_{tt} = -\frac{2t}{1+t^2}, \qquad R^t_{ttt} = 0. \tag{28.22}$$

28.4.2 $\mathbf{S}^2$

For $\mathbf{S}^2$, the Christoffel symbols are no longer trivial but become a pair of 2×2 matrices. In polar coordinates, with variables (θ, ϕ),

$$g_{ab} = \begin{bmatrix} 1 & 0 \\ 0 & \sin^2\theta \end{bmatrix}, \qquad g^{ab} = \begin{bmatrix} 1 & 0 \\ 0 & \frac{1}{\sin^2\theta} \end{bmatrix},$$

and we find

$$\Gamma^\theta_{ab} = \begin{bmatrix} 0 & 0 \\ 0 & -\cos\theta\sin\theta \end{bmatrix}, \qquad \Gamma^\phi_{ab} = \begin{bmatrix} 0 & \cos\theta\sin\theta \\ \cos\theta\sin\theta & 0 \end{bmatrix}. \tag{28.23}$$

We omit the Riemann tensor in favor of the Ricci tensor:

$$R_{ab} = \begin{bmatrix} R_{\theta\theta} & R_{\theta\phi} \\ R_{\phi\theta} & R_{\phi\phi} \end{bmatrix} = \begin{bmatrix} -1 & 0 \\ 0 & -\sin^2\theta \end{bmatrix}.$$

Thus, the scalar curvature becomes

$$R = g^{\theta\theta} R_{\theta\theta} + g^{\phi\phi} R_{\phi\phi}$$

$$= -2.$$

Note: If we had used the induced metric for a sphere of arbitrary radius, $\mathbf{x} \cdot \mathbf{x} = r^2$, this would become

$$R = -\frac{2}{r^2}. \tag{28.24}$$

Observe that the sign convention for the Riemann curvature has a minus sign relative to the Gaussian curvature of classical differential geometry, and thus we have the expected result

$$K_{\text{Gaussian Curvature}} = -R = +\frac{2}{r^2} \tag{28.25}$$

and $\mathbf{S}^2$ has constant positive curvature. The resulting scalar curvature is the same in all other coordinate systems, which we leave as an exercise for the reader.

28.4.3 $\mathbf{S}^3$

Here, we work out the curvature for $\mathbf{S}^3$ in the Euler-eigenvector coordinates, doubling the θ variable and giving it range 2π instead of 4π to avoid inconvenient factors of 2. Thus, our $\mathbf{S}^3$ coordinates are

$$q = \left(\cos(\theta), \hat{\mathbf{n}} \sin(\theta) \right)$$

$$= (\cos\theta, \sin\theta \cos\alpha \sin\beta, \sin\theta \sin\alpha \sin\beta, \sin\theta \cos\beta),$$

which define our preferred parameterization of quaternion space for this computation. The metric, with order of the matrix indices (θ, α, β), is thus

$$g_{ab} = \begin{bmatrix} 1 & 0 & 0 \\ 0 & \sin^2\theta & 0 \\ 0 & 0 & \sin^2\theta \sin^2\beta \end{bmatrix}, \tag{28.26}$$

and thus $ds^2 = d\theta^2 + \sin^2\theta\, d\beta^2 + \sin^2\theta \sin^2\beta\, d\alpha^2$. We find Christoffel symbols that are now a triple of 3×3 matrices:

$$\Gamma^{\theta}_{ab} = \begin{bmatrix} 0 & 0 & 0 \\ 0 & -\cos\theta\sin\theta & 0 \\ 0 & 0 & \cos\theta\sin\theta\sin^2\beta \end{bmatrix},$$

$$\Gamma^{\alpha}_{ab} = \begin{bmatrix} 0 & 0 & \cot\theta \\ 0 & 0 & \cot\beta \\ \cot\theta & \cot\beta & 0 \end{bmatrix},$$

$$\Gamma^{\beta}_{ab} = \begin{bmatrix} 0 & \cot\theta & 0 \\ \cot\theta & 0 & 0 \\ 0 & 0 & -\cos\beta\sin\beta \end{bmatrix}.$$

Skipping again to the 3×3 Ricci tensor, we find

$$R_{ab} = \begin{bmatrix} -2 & 0 & 0 \\ 0 & -2\sin^2\theta & 0 \\ 0 & 0 & -2\sin^2\alpha\sin^2\theta \end{bmatrix}.$$

If we contract R_{ab} with the inverse metric

$$g^{ab} = \begin{bmatrix} 1 & 0 & 0 \\ 0 & \sin^{-2}\theta & 0 \\ 0 & 0 & \sin^{-2}\theta\sin^{-2}\alpha \end{bmatrix}, \tag{28.27}$$

we again find constant positive curvature

$$R = g^{ab}R_{ab} = -\frac{6}{r^2} = -6 \tag{28.28}$$

with the standard reversed sign convention for R.

28.5   GEODESICS AND PARALLEL TRANSPORT ON THE SPHERE

We are concerned with quaternion curves when we wish, for example, to describe a rotating rigid body or the continuous changes in orientation of a camera frame. Riemannian geometry provides a family of concepts that helps us to describe, analyze, and classify curves on the quaternion space $\mathbf{S}^3$.

If we choose any parameterization for a curve $q(t)$ on $\mathbf{S}^3$ (e.g., $q \cdot q = 1$ for all t), the metric can be used to define an essential concept: the *proper length* s. s is the integrated global value of the infinitesimal ds, provided we choose a local system $\mathbf{x}$ of three, not four, coordinates (e.g., (θ, α, β) or (x, y, z)):

$$s(t) = \int_0^t \sqrt{\dot{x}^i g_{ij} \dot{x}^j}\, dt.$$

To compute derivatives of $q(t)$ with respect to the more meaningful parameter s, there is a trick: we take

$$\frac{d\mathbf{x}}{ds} = \frac{d\mathbf{x}/dt}{ds/dt} = \frac{d\mathbf{x}}{dt}\left(g_{ij}\frac{dx^i}{dt}\frac{dx^j}{dt}\right)^{-1/2}.$$

The local curvature vector of the curve $x(s)$ is computable using the covariant derivative:

$$K^i = \frac{D}{ds}\left(\frac{dx^i}{ds}\right) = \frac{d^2x^i}{ds^2} + \Gamma^i_{jk}\frac{dx^j}{ds}\frac{dx^k}{ds}. \tag{28.29}$$

Note: Because the indices on vectors refer to local coordinates in $\mathbf{S}^3$, there are only three of them, corresponding to our chosen independent variables. These are not the Euclidean coordinates $q = (q_0, q_1, q_2, q_3)$ obeying $q \cdot q = 1$ but the three chosen independent local parameters, such as (θ, α, β).

A *geodesic* is a particular path with vanishing curvature vector, and thus is a solution to the differential equation

$$\frac{d^2x^i}{ds^2} + \Gamma^i_{jk}\frac{dx^j}{ds}\frac{dx^k}{ds} = 0, \tag{28.30}$$

or in terms of the arbitrary unnormalized curve parameter t,

$$\frac{d^2x^i}{dt^2}\frac{ds}{dt} - \frac{dx^i}{dt}\frac{d^2s}{dt^2} + \Gamma^i_{jk}\frac{dx^j}{dt}\frac{dx^k}{dt}\frac{ds}{dt} = 0. \tag{28.31}$$

It can be shown that the geodesic between two points has the minimal value of the total path length

$$s(a, b) = \int_a^b \sqrt{g_{ij}\frac{dx^i}{dt}\frac{dx^j}{dt}}\, dt. \tag{28.32}$$

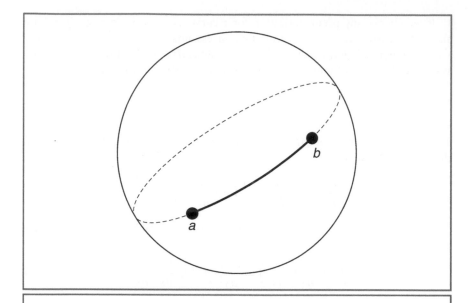

FIGURE 28.3 *The geodesic arc on a sphere equator passing through two points a and b. The minimal-length path is the solid arc, and the dashed arc is the maximal total path.*

On a sphere, the geodesics are arcs on great circles, which give both the minimal distance between two points and the maximal distance through the entire space passing through the two points (as indicated in Figure 28.3).

28.6   EMBEDDED-VECTOR VIEWPOINT OF THE GEODESICS

Although the traditional Riemannian geometry approach to describing the sphere in terms of intrinsic coordinates—with just n coordinate parameters describing points and curves on S^n—correctly gives all properties required, it is not easy to visualize. As an alternative, it is sometimes useful to return to the Euclidean embedding space in which we have focused our description of unit quaternions. The approach of Barr et al. [15] chooses this approach in contrast with that of

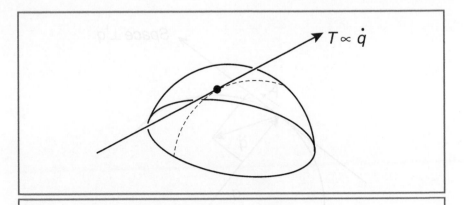

FIGURE 28.4 *The tangent vector T at a point on an embedded sphere.*

Kajiya and Gabriel [107], which adopted the standard local coordinate approach outlined in this chapter.

The elementary concept is to take the full unit-length four-vector $q(t)$ describing the quaternion curve, with $q \cdot q = 1$, and let the curve derivatives also be four-vectors, having directions that exist in $\mathbb{R}^4$, outside $\mathbf{S}^3$ itself. Because $d(q \cdot q)/dt = 2q \cdot \dot{q} = 0$, the first derivative

$$\dot{q} = cT$$

is necessarily in the tangent direction T perpendicular to the quaternion vector q from the origin to the sphere's surface.

The analog of the acceleration $d^2\mathbf{x}/(ds)^2$ of the curve, which we used to define the equations of a geodesic in intrinsic coordinates, now has components that need to be projected carefully back to the surface of the sphere. We use our by-now-standard Gram–Schmidt procedure to subtract the component of the acceleration $\ddot{q}(t)$ that is in the direction $q(t)$ itself. The result is the corrected tangential acceleration $A(t)$, constructed so that $A(t) \cdot q = 0$ and is therefore tangent to the spherical hypersurface. We see that

$$A(t) = \ddot{q}(t) - q(t)(q \cdot \ddot{q}) \qquad (28.33)$$

explicitly satisfies $A \cdot q = 0$. The relationship between $q, \dot{q}, \ddot{q}$, and A is shown graphically in Figures 28.4 and 28.5. If we are given only two boundary points,

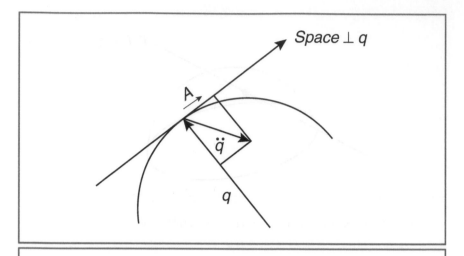

FIGURE 28.5 *Tangent and acceleration vectors computed at a point on a curve embedded in a sphere.*

$q(a)$ and $q(b)$, the path with vanishing tangential acceleration $A(t)$ will coincide with the great-circle geodesics in intrinsic coordinates.

In the more realistic case in which we require both keyframe control of a path and specified angular velocity, we see that specifying a minimal path is not enough because the derivatives can be discontinuous at the keyframes. We must also provide a principle for minimizing discontinuities in both path-shape and angular-distance sampling intervals. The path shape can be controlled either with the method of Barr et al.,

$$\text{minimize} \int_{t0}^{t_1} A \cdot A \, dt = \int_{t0}^{t_1} |\ddot{q} - q(q \cdot \ddot{q})|^2 \, dt, \qquad (28.34)$$

using extrinsic coordinates $q(t)$ or by using intrinsic sphere coordinates, minimizing

$$\int_{t0}^{t_1} |\ddot{\mathbf{x}}(t)|^2 \, dt. \qquad (28.35)$$

In each case, the locations of the anchor points as three-vectors $\{\mathbf{x}_i\}$ or four-vectors $\{q_i = q(\mathbf{x}_i)\}$ constrain the variation.

Finally, the value we have been denoting by $s(t)$, with $ds^2 = dx^i g_{ij}(x) dx^j$, is exactly the angular distance between sampled points on the quaternion curve. Thus, one should achieve exactly the same level of control over the angular velocity by manipulating the proper distance in intrinsic coordinates as

$$s(i, i+1) = \int_{ti}^{t_{i+1}} \sqrt{\frac{dx^j}{dt} g_{jk} \frac{dx^k}{dt}} \, dt, \tag{28.36}$$

or in extrinsic Euclidean coordinates using the quaternion values themselves:

$$s(i, i+1) = 2\cos^{-1}(q_i \cdot q_{i+1}). \tag{28.37}$$

Beyond Quaternions

These concluding chapters address briefly several ways in which it is known that the properties of quaternions can be generalized. Although quaternions are extremely unique, and are of particular relevance to the understanding and modeling of everyday life because of their relationship to 3D orientation frames, new insights into the nature of quaternions can be gained by investigating what happens when we attempt to generalize them.

These topics are definitely optional, and are intended for readers with advanced interests. However, many readers may find it very interesting to see how, in the Clifford algebra chapter, the true origin of the "square root" formula for 2D rotations

$$\mathbf{R}_2 = \begin{bmatrix} a^2 - b^2 & -2ab \\ 2ab & a^2 - b^2 \end{bmatrix},$$

written in terms of the half angles with $a = \cos(\theta/2)$ and $b = \sin(\theta/2)$, is shown to be an inevitable result of the 2D Clifford algebra and not an arbitrary invention.

The Relationship of 4D Rotations to Quaternions

29

29.1   WHAT HAPPENED IN THREE DIMENSIONS

In three dimensions, there were many ways to deduce the quadratic mapping from quaternions to the 3×3 rotation matrix belonging to the group $\mathbf{SO}(3)$ and implementing a rotation on ordinary 3D frames. The one most directly derived from the quaternion algebra conjugates "pure" quaternion three-vectors $v_i = (0, \mathbf{V}_i)$ and pulls out the elements of the rotation matrix in the following way:

$$\sum_{j=1}^{3} R(q)_{ij} v_j = q \star v_i \star q^{-1}.$$

We easily find that the quadratic relationship between $\mathbf{R}_3(q)$ and $q = (q_0, q_1, q_2, q_3)$ is

$$\mathbf{R}_3 = \begin{bmatrix} q_0^2 + q_1^2 - q_2^2 - q_3^2 & 2q_1q_2 - 2q_0q_3 & 2q_1q_3 + 2q_0q_2 \\ 2q_1q_2 + 2q_0q_3 & q_0^2 - q_1^2 + q_2^2 - q_3^2 & 2q_2q_3 - 2q_0q_1 \\ 2q_1q_3 - 2q_0q_2 & 2q_2q_3 + 2q_0q_1 & q_0^2 - q_1^2 - q_2^2 + q_3^2 \end{bmatrix}. \qquad [29.1]$$

377

29.2   QUATERNIONS AND FOUR DIMENSIONS

In the 4D case, which we should really regard as the more fundamental one because it includes the 3D transformation as a special case, we can find the induced $\mathbf{SO}(4)$ matrix by extending quaternion multiplication to act on full four-vector quaternions $v_\mu = (v_0, \mathbf{V})_\mu$ and not just three-vectors ("pure" quaternions) $v = (0, \mathbf{V})$ in the following way:

$$\sum_{\nu=0}^{3} R_{\mu\nu} v_\nu = p \star v_\mu \star q^{-1}.$$

Working out the algebra, we find that the 3D rotation matrix $\mathbf{R}_3$ is just the degenerate $p = q$ case of the following 4D rotation matrix:

$$\mathbf{R}_4 = \begin{bmatrix}
p_0 q_0 + p_1 q_1 + p_2 q_2 + p_3 q_3 & p_0 q_1 - p_1 q_0 - p_2 q_3 + p_3 q_2 \\
-p_0 q_1 + p_1 q_0 - p_2 q_3 + p_3 q_2 & p_0 q_0 + p_1 q_1 - p_2 q_2 - p_3 q_3 \\
-p_0 q_2 + p_2 q_0 + p_1 q_3 - p_3 q_1 & p_0 q_3 + p_3 q_0 + p_1 q_2 + p_2 q_1 \\
-p_0 q_3 + p_3 q_0 - p_1 q_2 + p_2 q_1 & -p_0 q_2 - p_2 q_0 + p_1 q_3 + p_3 q_1
\end{bmatrix}$$

$$\begin{bmatrix}
p_0 q_2 - p_2 q_0 + p_1 q_3 - p_3 q_1 & p_0 q_3 - p_3 q_0 - p_1 q_2 + p_2 q_1 \\
-p_0 q_3 - p_3 q_0 + p_1 q_2 + p_2 q_1 & p_0 q_2 + p_2 q_0 + p_1 q_3 + p_3 q_1 \\
p_0 q_0 - p_1 q_1 + p_2 q_2 - p_3 q_3 & -p_0 q_1 - p_1 q_0 + p_2 q_3 + p_3 q_2 \\
p_0 q_1 + p_1 q_0 + p_2 q_3 + p_3 q_2 & p_0 q_0 - p_1 q_1 - p_2 q_2 + p_3 q_3
\end{bmatrix} . \quad (29.2)$$

One may check that Equation 29.1 is just the lower right-hand corner of the degenerate $p = q$ case of Equation 29.2. An implementation of Equation 29.2 is presented in Table 29.1.

We may take this form and plug in

$$p_0 = \cos(\phi/2), \qquad \mathbf{p} = \hat{\mathbf{m}} \sin(\phi/2)$$

to get a new form of the 4D orthogonal rotation matrix *parameterized in terms of two separate three-sphere coordinates*:

$$\mathbf{R}_4 = [A_0 \quad A_1 \quad A_2 \quad A_3], \qquad (29.3)$$

```
QQTo4DRot[p_List,q_List] :=
  Module[{p0 = p[[1]], p1 = p[[2]], p2 = p[[3]], p3 = p[[4]],
          q0 = q[[1]], q1 = q[[2]], q2 = q[[3]], q3 = q[[4]]},
{{p0*q0 + p1*q1 + p2*q2 + p3*q3,
  p0*q1 - p1*q0 - p2*q3 + p3*q2,
  p0*q2 - p2*q0 + p1*q3 - p3*q1,
  p0*q3 - p3*q0 - p1*q2 + p2*q1},
 {-p0*q1 + p1*q0 - p2*q3 + p3*q2,
  p0*q0 + p1*q1 - p2*q2 - p3*q3,
  -p0*q3 - p3*q0 + p1*q2 + p2*q1,
  p0*q2 + p2*q0 + p1*q3 + p3*q1},
 {-p0*q2 + p2*q0 + p1*q3 - p3*q1,
  p0*q3 + p3*q0 + p1*q2 + p2*q1,
  p0*q0 - p1*q1 + p2*q2 - p3*q3,
  -p0*q1 - p1*q0 + p2*q3 + p3*q2},
 {-p0*q3 + p3*q0 - p1*q2 + p2*q1,
  -p0*q2 - p2*q0 + p1*q3 + p3*q1,
  p0*q1 + p1*q0 + p2*q3 + p3*q2,
  p0*q0 - p1*q1 - p2*q2 + p3*q3}} ]
```

TABLE 29.1 Mathematica code for the 4×4 orthogonal rotation matrix in terms of a double quaternion.

where

$$A_0 = \frac{1}{2} \begin{bmatrix} C_+ + C_- + \hat{\mathbf{m}} \cdot \hat{\mathbf{n}}(C_- - C_+) \\ -m_{23}^- C_- + m_{23}^- C_+ + m_1^+ S_- + m_1^- S_+ \\ -m_{31}^- C_- + m_{31}^- C_+ + m_2^+ S_- + m_2^- S_+ \\ -m_{12}^- C_- + m_{12}^- C_+ + m_3^+ S_- + m_3^- S_+ \end{bmatrix},$$

$$A_1 = \frac{1}{2} \begin{bmatrix} -m_{23}^- C_- + m_{23}^- C_+ - m_1^+ S_- - m_1^- S_+ \\ C_+ + C_- + (m_1 n_1 - m_2 n_2 - m_3 n_3)(C_- - C_+) \\ m_{12}^+ C_- - m_{12}^+ C_+ + m_3^- S_- + m_3^- S_+ \\ m_{31}^+ C_- - m_{31}^+ C_+ - m_2^- S_- - m_2^+ S_+ \end{bmatrix},$$

$$A_2 = \frac{1}{2} \begin{bmatrix} -m_{31}^- C_- + m_{31}^- C_+ - m_2^+ S_- - m_2^- S_+ \\ m_{12}^+ C_- - m_{12}^+ C_+ - m_3^- S_- - m_3^+ S_+ \\ C_+ + C_- + (-m_1 n_1 + m_2 n_2 - m_3 n_3)(C_- - C_+) \\ m_{23}^+ C_- - m_{23}^+ C_+ + m_1^- S_- + m_1^+ S_+ \end{bmatrix},$$

$$A_3 = \frac{1}{2} \begin{bmatrix} -m_{12}^- C_- + m_{12}^- C_+ - m_3^+ S_- - m_3^- S_+ \\ m_{31}^+ C_- - m_{31}^+ C_+ + m_2^- S_- + m_2^+ S_+ \\ m_{23}^+ C_- - m_{23}^+ C_+ - m_1^- S_- - m_1^+ S_+ \\ C_+ + C_- + (-m_1 n_1 - m_2 n_2 + m_3 n_3)(C_- - C_+) \end{bmatrix}.$$

Here, $C_\pm = \cos \frac{1}{2}(\phi \pm \theta)$, $S_\pm = \sin \frac{1}{2}(\phi \pm \theta)$, $m_i^\pm = (m_i \pm n_i)$, and $m_{ij}^\pm = (m_i n_j \pm m_j n_i)$.

Shoemake-style interpolation between two distinct 4D frames is now achieved by applying the desired SLERP-based interpolation method independently to a set of coordinates $p(t)$ on one three-sphere, and to a separate set of coordinates $q(t)$ on another. The resulting matrix $\mathbf{R}_4(t)$ gives geodesic interpolations for simple SLERPs, and smooth interpolations based on infinitesimal geodesic components when the spline methods of Chapter 25 are used in tandem on both quaternions of the pair at the same time.

Controls: A three-degree-of-freedom controller can in fact be used to generalize the two-degree-of-freedom rolling-ball controller [66] from 3D to 4D orientation control [34,72]. This 4D orientation control technique can be used with a 3D tracker or 3D haptic probe to carry out interactive view control or to specify keyframes for 4D double-quaternion interpolations. As pointed out by Shoemake [151], the Arcball controller can also be adapted with complete faithfulness of spirit to the 4D case, in that one can pick two points in a three-sphere to specify an initial 4D frame and then pick two more points in the three-sphere to define the current 4D frame. Note that Equation 29.2 gives the complete 4D rotation formula. Alternatively, one can replace the 4D rolling ball or virtual sphere controls described at the beginning by a pair (or more) of 3D controllers as noted by Hanson [66].

Quaternions and the Four Division Algebras

30

It is clear from the stories of William Rowan Hamilton that quaternions were discovered because he persisted in asking the question "What generalizes the complex numbers." Therefore, it is appropriate to continue our study of quaternions by asking the question "What do we discover when we attempt to generalize quaternions?" The answers fall into several categories. Some features, such as the algebraic properties of quaternions, have very limited generalizability to other dimensions. Other features, such as the constructability of double-valued parameterizations of Euclidean rotations (the *spin representations*) generalize to all dimensions. In this chapter we look at the properties that are very specific to quaternions, and which in fact show that there are only four possible algebraic systems of a certain type, of which quaternions are one.

30.1 DIVISION ALGEBRAS

We now place quaternions in their unique context as one of only four possible examples of an object known as a *division algebra* or a *division ring*. A *division algebra* is a system of numbers having the following properties:

- *Addition:* The first fundamental property of a division algebra is that there must exist a standard set of commutative and associative addition rules.
- *Multiplication:* There must in addition be a multiplication rule that is distributive with respect to addition. However, commutativity of multiplication is

relaxed for the quaternions, and associativity of multiplication is relaxed for the octonions.

- *Division:* The distinguishing, and most restrictive, property of a division algebra is that every nonzero element has a multiplicative inverse.

There is a remarkable relationship between the division algebras and some very old theorems about n-dimensional vectors. n-dimensional real vectors can be arranged to form algebras in many ways, but division (the multiplicative inverse for nonzero vectors) exists only for $n = 1, 2, 4$ for associative systems and only for $n = 8$ for nonassociative systems. This can be rephrased in yet another way. Hurwitz [101] examined algebras with norms and showed that requiring the norm of a product to be the product of the norms

$$|\mathbf{x} \star \mathbf{y}| = |\mathbf{x}||\mathbf{y}|$$

gives a constraint that can only by satisfied by real numbers, complex numbers, quaternions, and octonions. Further related interesting results were discovered by Bott and Milnor [122] using modern mathematical technology, again drawing the conclusion that only for dimensions 1, 2, 4, and 8 can there exist finite-dimensional real division algebras. These algebras of course correspond exactly to real numbers, complex numbers, quaternions, and octonions, whose properties we shall now systematically explore in turn.

30.1.1 THE NUMBER SYSTEMS WITH DIMENSIONS 1, 2, 4, AND 8

The four possible algebraic number systems are as follows.

- *Reals:* $\mathbb{R}$ denotes the real numbers with ordinary addition, subtraction, multiplication, and division, with the trivial multiplication algebra

$$(u) \star (v) = (uv),$$

where juxtaposition denotes multiplication and division is defined by u/v.
- *Complex numbers:* $\mathbb{C}$ denotes the complex numbers, with the commutative multiplication algebra

$$(u_0, u_1) \star (v_0, v_1) = (u_0 v_0 - u_1 v_1, u_0 v_1 + u_1 v_0).$$

A new operation, conjugation, is defined as $\overline{(u_0, u_1)} = (u_0, -u_1)$, and the norm-squared is the nonzero part remaining when a number is multiplied by its conjugate:

$$|u|^2 = (u_0)^2 + (u_1)^2.$$

Here, technically, we have extracted the scalar value of the norm from the algebra using $u \star \bar{u} = (|u|^2, 0)$. With these definitions, complex division is defined by the order-independent (commutative) relation

$$\frac{u}{v} = u \star \frac{\bar{v}}{(v_0)^2 + (v_1)^2} = \frac{u \star \bar{v}}{|v|^2}$$

$$= \frac{(u_0 v_0 + u_1 v_1, -u_0 v_1 + u_1 v_0)}{(v_0)^2 + (v_1)^2}.$$

- **Quaternions:** $\mathbb{Q}$, also often written as $\mathbb{H}$, formally denotes the quaternions with the noncommutative multiplication algebra for $u = (u_0, u_1, u_2, u_3) = (u_0, \mathbf{u})$ we have used throughout:

$$(u_0, u_1, u_2, u_3) \star (v_0, v_1, v_2, v_3) = \begin{bmatrix} u_0 v_0 - u_1 v_1 - u_2 v_2 - u_3 v_3 \\ u_1 v_0 + u_0 v_1 + u_2 v_3 - u_3 v_2 \\ u_2 v_0 + u_0 v_2 + u_3 v_1 - u_1 v_3 \\ u_3 v_0 + u_0 v_3 + u_1 v_2 - u_2 v_1 \end{bmatrix}$$

$$= (u_0 v_0 - \mathbf{u} \cdot \mathbf{v}, u_0 \mathbf{v} + v_0 \mathbf{u} + \mathbf{u} \times \mathbf{v}).$$

Conjugation is defined by $\overline{(u_0, u_1, u_2, u_3)} = (u_0, -u_1, -u_2, -u_3)$ or $\overline{(u_0, \mathbf{u})} = (u_0, -\mathbf{u})$, and the norm-squared by $|u|^2 = (u_0)^2 + (u_1)^2 + (u_2)^2 + (u_3)^2 = (u_0)^2 + \mathbf{u} \cdot \mathbf{u}$. Here, again, we implicitly extract the scalar value from $u \star \bar{u} = (|u|^2, \mathbf{0})$. Quaternion division is noncommutative, and is defined by the (order-dependent) relations

$$u/v = u \star \bar{v}/|v|^2$$

and

$$v \backslash u = \bar{v} \star u/|v|^2.$$

- **Octonions:** $\mathbb{O}$ denotes the octonions, which have eight components:

$$u = (u_0, u_1, u_2, u_3, u_4, u_5, u_6, u_7) = (u_0, \mathbf{u}).$$

Octonions possess a noncommutative, nonassociative multiplication algebra of the form

$$(u_0, u_1, u_2, u_3, u_4, u_5, u_6, u_7) \star (v_0, v_1, v_2, v_3, v_4, v_5, v_6, v_7)$$

$$= \begin{bmatrix} u_0 v_0 - u_1 v_1 - u_2 v_2 - u_3 v_3 - u_4 v_4 - u_5 v_5 - u_6 v_6 - u_7 v_7 \\ u_1 v_0 + u_0 v_1 + u_2 v_4 - u_4 v_2 + u_5 v_6 - u_6 v_5 + u_3 v_7 - u_7 v_3 \\ u_2 v_0 + u_0 v_2 + u_3 v_5 - u_5 v_3 + u_6 v_7 - u_7 v_6 + u_4 v_1 - u_1 v_4 \\ u_3 v_0 + u_0 v_3 + u_4 v_6 - u_6 v_4 + u_7 v_1 - u_1 v_7 + u_5 v_2 - u_2 v_5 \\ u_4 v_0 + u_0 v_4 + u_1 v_2 - u_2 v_1 + u_5 v_7 - u_7 v_5 + u_6 v_3 - u_3 v_6 \\ u_5 v_0 + u_0 v_5 + u_2 v_3 - u_3 v_2 + u_6 v_1 - u_1 v_6 + u_7 v_4 - u_4 v_7 \\ u_6 v_0 + u_0 v_6 + u_3 v_4 - u_4 v_3 + u_7 v_2 - u_2 v_7 + u_1 v_5 - u_5 v_1 \\ u_7 v_0 + u_0 v_7 + u_4 v_5 - u_5 v_4 + u_1 v_3 - u_3 v_1 + u_2 v_6 - u_6 v_2 \end{bmatrix}$$

$$= \big(u_0 v_0 - \mathbf{u} \cdot \mathbf{v}, \; u_0 \mathbf{v} + v_0 \mathbf{u} + \Omega(\mathbf{u}, \mathbf{v})\big),$$

where the mapping function Ω is defined by the explicit equation preceding. Conjugation can be abbreviated as $\overline{(u_0, \mathbf{u})} = (u_0, -\mathbf{u})$, and thus the norm-squared becomes $|u|^2 = (u_0)^2 + \mathbf{u} \cdot \mathbf{u}$. That this algebra is nonassociative can be verified by comparing (for example) products of octonions containing only single nonzero elements, such as

$$(u_1 \star v_2) \star w_3 = (uv)_4 \star w_3$$

$$= -(uvw)_6,$$

$$u_1 \star (v_2 \star w_3) = u_1 \star (vw)_5$$

$$= +(uvw)_6,$$

where we use the obvious notation $u_1 = (0, u, 0, 0, 0, 0, 0, 0)$, $(uv)_2 = (0, 0, uv, 0, 0, 0, 0, 0)$ (and so on) for octonions containing only a single nonzero element.

Interestingly, this apparently daunting algebra can be reconstructed without much trouble by drawing the circle shown in Figure 30.1, with the imaginary octonion variables $(1, 2, 3, 4, 5, 6, 7)$ forming a cyclic ring. Then each of the seven triples $(1, 2, 4)$, $(2, 3, 5)$, $(3, 4, 6)$, $(4, 5, 7)$, $(5, 6, 1)$, $(6, 7, 2)$, $(7, 1, 3)$—assembled by taking two successive elements and skipping one—forms a quaternion subalgebra. This of course has a close analogy

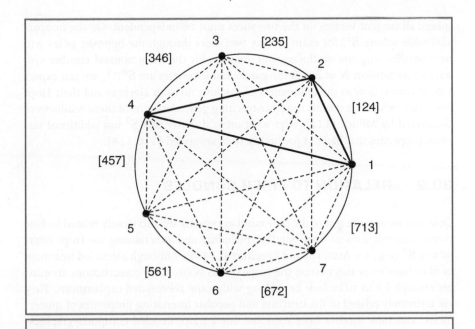

FIGURE 30.1 *Construction of the nonassociative octonion algebra from seven interleaved noncommutative quaternion algebras.*

to the three successive complex subalgebras $(1) = \mathbf{i}$, $(2) = \mathbf{j}$, $(3) = \mathbf{k}$ embedded within the quaternion algebra.

Disappointingly, as previously noted [101,122], this succession of constructions does not extend beyond octonions, and thus these four number systems are basically all we have to work with.

30.1.2 PARALLELIZABLE SPHERES

There are many peculiar properties of space that seem closely related to the uniqueness of the four number systems $(\mathbb{R}, \mathbb{C}, \mathbb{Q}, \mathbb{O})$. Classic results in geometry have shown, for example, that only *three spheres*—$\mathbf{S}^1$, $\mathbf{S}^3$, and $\mathbf{S}^7$—are parallelizable. Parallelizability means, roughly, that if you look at two slices through a diameter of the sphere all tangent vectors on the two slices must be independent. On the nonparallelizable sphere $\mathbf{S}^2$, for example, any two slices through the opposite poles will have parallel tangents on the equator. Because for the three nonreal number systems of dimension $N = 2, 4, 8$ the parallelizable spheres are $\mathbf{S}^{N-1}$, we can expect a deep connection to the uniqueness of the four division algebras and their Hopf fibrations, which we discuss next. Far-reaching generalizations of these results were discovered by Milnor [122], who went on to discover that $\mathbf{S}^7$ has additional unusual properties that had not previously been suspected [110,124].

30.2   RELATION TO FIBER BUNDLES

Quaternions and the rest of the four number systems are also closely related to four classic constructions of fiber bundles, including and generalizing the Hopf fibration of $\mathbf{S}^3$ (e.g., see Artin [6] and Eguchi et al. [45]). Although a detailed treatment of fiber bundles is also beyond the scope of this book, these constructions are simple enough for us to include here along with some abbreviated explanations. They are intimately related to the structure and peculiar interesting properties of quaternions, and thus merit a brief mention. For a more detailed computer-graphics-oriented treatment see, for example, Shoemake [152].

A fiber bundle is a geometric construction with two parts: a *base manifold* and a *fiber*. Think of any small piece of the base manifold as a shag rug, and the fibers as pieces of thread attached to specific points in the backing of the rug. If nothing much interesting happens, this is a *trivial bundle*, with each piece of the total bundle looking like a direct product of the space of the fiber with the space of the backing. However, interesting fiber bundles have more structure. They can have nontrivial base manifolds, and the fibers themselves can be sewn to their neighboring fibers in interesting ways, generating spaces you would not be able to identify as similar to a shag rug unless you looked only at a very small patch at a time.

We remark that there is a vast technology in classical differential geometry that includes carefully defining the global manifold properties of a fiber bundle by associating appropriate transition functions to local patches (which are often group elements), as well as involving the assignment of connections and associated curvatures. The latter is intimately related to Yang–Mills theory and Einstein's theory of gravitation (e.g., see Drechsler and Mayer [38], Eguchi et al. [45], Weinberg [168], and many other references in the physics literature). The generalized Hopf fibration builds a complete fiber bundle "by hand" out of the following fairly trivial ideas:

1   *Define the square length of the N real variables.* Let $u = (u_1, \ldots, u_N)$ be a generic
    vector. As noted, the square length of a vector in any of the division algebras
    is obtainable from the conjugate multiplication

$$|u|^2 = u \star \bar{u} = \sum_{i=1}^{N} (u_i)^2$$

    for any of the four number systems, with $N = (1, 2, 4, 8)$.

2   *Make a $(2N - 1)$-sphere from the constant-length sum of two squares.* Now let a and
    b be two *separate* N-vectors. Use these to create a $(2N - 1)$-sphere $\mathbf{S}^{2N-1}$
    embedded in $\mathbb{R}^{2N}$. By definition, the equation

$$|a|^2 + |b|^2 = 1$$

    describes this unit sphere, and we continue to allow only dimensions $N =
    (1, 2, 4, 8)$.

3   *Construct an $(N + 1)$-vector in $\mathbf{S}^N$ from quadratic forms.* One can combine the $2N$
    variables contained in a and b (constrained to lie on a point in $\mathbf{S}^{2N-1}$) into
    a list of $N + 1$ quadratic expressions $\mathbf{v}$ that, when considered as a Euclidean
    $(N + 1)$- vector, has unit length and thus describes a point in $\mathbf{S}^N$. This list of
    quadratic forms exists only because of a deep correspondence to each of the
    four algebras, as we shall illustrate in a moment.

4   *Identify total space, base space, and fiber.* The entire fiber bundle is the sphere $\mathbf{S}^{2N-1}$,
    and the quadratic form gives the base space $\mathbf{S}^N$. The fiber is basically what
    is left over; namely, the set of points in $\mathbf{S}^{2N-1}$ that all generate exactly the
    same point in $\mathbf{S}^N$ when the quadratic form is evaluated.

30.3   CONSTRUCTING THE HOPF FIBRATIONS

We now give the details for the construction of the "classic" Hopf fibrations for
each of the four division algebras in turn.

30.3.1   REAL: $\mathbf{S}^0$ fiber $+ \mathbf{S}^1$ base $= \mathbf{S}^1$ bundle

The construction of the simplest of all Hopf fibrations takes two numbers from $\mathbb{R}$
(namely, x_1 and x_2), puts them on a circle, and then constructs a two-vector of

quadratic forms (v_0, v_1) using the difference of the squares and the only other available quadratic object, the product of the two original numbers:

$$(x_1)^2 + (x_2)^2 = 1,$$

$\mathbf{S}^1$ Circle is full space,

$$v_0 = (x_1)^2 - (x_2)^2,$$
$$v_1 = 2x_1 x_2,$$

$$(v_0)^2 + (v_1)^2 = \left((x_1)^2 + (x_2)^2\right)^2 = 1,$$

A *different* $\mathbf{S}^1$, half the full space, is the base space.   (30.1)

Here it is essential to observe that although the circle equations for $\mathbf{v}$ and $\mathbf{x}$ appear identical there are two points on the circle $(x_1)^2 + (x_2)^2 = 1$ corresponding to each point in $(v_0)^2 + (v_1)^2 = 1$, the base space. The full space is a double covering of the base space, and that two-element double corresponds to the group of multiplication by -1. Figure 30.2 illustrates the full space, the base space, and the fiber's two points in the full space corresponding to a single point in the base. Note that although we have drawn the full space folded over to make the fibers line up that does not alter the fact that the full range of θ in the parameterization of x_1 and x_2 must still be used.

Details: The parametric solution $(x_1, x_2) = (\cos\theta, \sin\theta)$ of the circle equation we started with describes the full bundle. From this, the projection to the base space can immediately be written using the double-angle formulas as $(v_0, v_1) = (\cos 2\theta, \sin 2\theta)$. Thus, each point in the base space has two points (θ and $\theta + \pi$) in the bundle. The fiber is therefore two points (or $\mathbf{S}^0$) and the bundle is nontrivial because it is a single continuous curve (as shown in Figure 30.2), not two curves (as would result from the structure $\mathbf{S}^0 \times \mathbf{S}^1$). This is a bundle that is essentially the boundary of a Mobius band, and is the most perfect nontrivial example of a principal bundle imaginable.

Rigorous relation to Clifford algebra: The expression for (v_0, v_1) given previously is one column of the $\mathbf{SO}(2)$ rotation matrix that comes directly from the Clifford algebra of $\mathbb{R}^2$, as we shall see in Chapter 31. The entire matrix, interestingly, gives all possible choices of the vectors of quadratic forms that could be chosen for the real

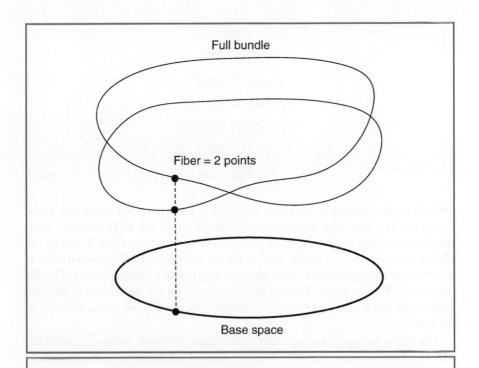

FIGURE 30.2 *Construction of the S^1 Hopf fibration, whose full bundle can be seen as the boundary of a Mobius band and whose fiber at each point of the base circle consists of the two points (S^0) at θ and $\theta + \pi$.*

Hopf fibration. This matrix should already be familiar to the reader:

$$\begin{bmatrix} (x_1)^2 - (x_2)^2 & -2x_1x_2 \\ 2x_1x_2 & (x_1)^2 - (x_2)^2 \end{bmatrix}.$$

30.3.2   COMPLEX: S^1 fiber $+ S^2$ base $= S^3$ bundle

The construction, as follows, takes two numbers from $\mathbb{C}$ (i.e., $z_1 = x_1 + iy_1$ and $z_2 = x_2 + iy_2$).

$$|z_1|^2 + |z_2|^2 = 1,$$
$$\mathbf{S}^3 \text{ is full space,}$$

$$v_0 = |z_1|^2 - |z_2|^2,$$
$$v_1 = 2(x_1 x_2 - y_1 y_2),$$
$$v_2 = 2(x_1 y_2 + x_2 y_1),$$

$$(v_0)^2 + (v_1)^2 + (v_2)^2 = \left(|z_1|^2 + |z_2|^2\right)^2 = 1,$$
$$\mathbf{S}^2 \text{ is base space.} \tag{30.2}$$

Notes: Instead of having single points projected to the same point of the base space in (v_0, v_1, v_2), this map has entire *circles*—the fiber of the $\mathbf{U}(1)$ Principal bundle characterizing the classic Hopf fibration. If one parameterizes $\mathbf{S}^3$ using the Clifford torus—$(|z_1|^2 = \cos^2\theta, |z_2|^2 = \sin^2\theta)$ embedded in $\mathbf{S}^3$ symmetrically at $\theta = \pi/4$—the diagonal lines on the torus corresponding to constant sums of angles map to the $\mathbf{S}^2$ base space. There is no dependence on the differences of toroidal angles, and that forms the required projection from the $\mathbf{S}^1$ fiber to single points in the base.

The vector (v_0, v_1, v_2) of quadratic forms given previously is one column of the $\mathbf{SO}(3)$ rotation matrix derivable from the Clifford algebra of $\mathbb{R}^3$, or equivalently from quaternions. The rows and columns of this matrix give all possible choices for the quadratic forms of the $\mathbf{S}^3$ Hopf fibration:

$$\begin{bmatrix} x_1^2 + y_1^2 - x_2^2 - y_2^2 & -2x_1 x_2 - 2y_1 y_2 & 2y_1 y_2 + 2x_1 x_2 \\ 2y_1 x_2 + 2x_1 y_2 & x_1^2 - y_1^2 + x_2^2 - y_2^2 & 2x_2 y_2 - 2x_1 y_1 \\ 2y_1 y_2 - 2x_1 x_2 & 2x_2 y_2 + 2x_1 y_1 & x_1^2 - y_1^2 - x_2^2 + y_2^2 \end{bmatrix}.$$

30.3.3 QUATERNION: $\mathbf{S}^3$ fiber $+ \mathbf{S}^4$ base $= \mathbf{S}^7$ bundle

The quaternion construction, as follows, takes two numbers from $\mathbb{Q}$ (i.e., $q = (q_0, q_1, q_2, q_3)$ and $p = (p_0, p_1, p_2, p_3)$) and builds an $\mathbf{S}^7$:

$$|q|^2 + |p|^2 = 1,$$
$$\mathbf{S}^7 \text{ is full space,}$$

$$v_0 = |q|^2 - |p|^2,$$

$$v_1 = q_0 p_0 - q_1 p_1 - q_2 p_2 - q_3 p_3,$$

$$v_2 = q_0 p_1 + q_1 p_0 + q_2 p_3 - q_3 p_2,$$

$$v_3 = q_0 p_2 + q_2 p_0 + q_3 p_1 - q_1 p_3,$$

$$v_4 = q_0 p_3 + q_3 p_0 + q_1 p_2 - q_2 p_1,$$

$$(v_0)^2 + (v_1)^2 + (v_2)^2 + (v_3)^2 + (v_4)^2 = \left(|q|^2 + |p|^2\right)^2 = 1,$$

$$\mathbf{S}^4 \text{ is base space.} \tag{30.3}$$

Notes: The v vector describes a point in $\mathbf{S}^4$, but projects the three parameters of an entire $\mathbf{S}^3$ to a single point, making the fiber $\mathbf{S}^3$ at each point. The column $(v_0, v_1, v_2, v_3, v_4)$ is essentially one column of an $\mathbf{SO}(5)$ rotation matrix, each row or column of which produces an alternative arrangement of the fibration's quadratic map.

30.3.4 OCTONION: $\mathbf{S}^7$ fiber $+ \, \mathbf{S}^8$ base $= \mathbf{S}^{15}$ bundle

The octonion construction takes from $\mathbb{O}$ the numbers

$$q = (q_0, q_1, q_2, q_3, q_4, q_5, q_6, q_7), \qquad p = (p_0, p_1, p_2, p_3, p_4, p_5, p_6, p_7),$$

constructs an $\mathbf{S}^{15}$, and assembles the nine-vector quadratic form:

$$|q|^2 + |p|^2 = 1,$$

$$\mathbf{S}^{15} \text{ is the full space,}$$

$$v_0 = |q|^2 - |p|^2,$$

$$v_1 = q_0 p_0 - q_1 p_1 - q_2 p_2 - q_3 p_3 - q_4 p_4 - q_5 p_5 - q_6 p_6 - q_7 p_7,$$

$$v_2 = q_1 p_0 + q_0 p_1 + q_2 p_4 - q_4 p_2 + q_5 p_6 - q_6 p_5 + q_3 p_7 - q_7 p_3,$$

$$v_3 = q_2 p_0 + q_0 p_2 + q_3 p_5 - q_5 p_3 + q_6 p_7 - q_7 p_6 + q_4 p_1 - q_1 p_4,$$

$$v_4 = q_3 p_0 + q_0 p_3 + q_4 p_6 - q_6 p_4 + q_7 p_1 - q_1 p_7 + q_5 p_2 - q_2 p_5,$$

$$v_5 = q_4 p_0 + q_0 p_4 + q_1 p_2 - q_2 p_1 + q_5 p_7 - q_7 p_5 + q_6 p_3 - q_3 p_6,$$

$$v_6 = q_5 p_0 + q_0 p_5 + q_2 p_3 - q_3 p_2 + q_6 p_1 - q_1 p_6 + q_7 p_4 - q_4 p_7,$$

$$v_7 = q_6 p_0 + q_0 p_6 + q_3 p_4 - q_4 p_3 + q_7 p_2 - q_2 p_7 + q_1 p_5 - q_5 p_1,$$

$$v_8 = q_7 p_0 + q_0 p_7 + q_4 p_5 - q_5 p_4 + q_1 p_3 - q_3 p_1 + q_2 p_6 - q_6 p_2,$$

$$(v_0)^2 + (v_1)^2 + (v_2)^2 + (v_3)^2 + (v_4)^2 + (v_5)^2 + (v_6)^2 + (v_7)^2 + (v_8)^2,$$
$$= \left(|q|^2 + |p|^2\right)^2 = 1,$$
$$\mathbf{S}^8 \text{ is the base space.} \tag{30.4}$$

Notes: The v vector describes a point in $\mathbf{S}^8$, but projects the seven parameters of an entire $\mathbf{S}^7$ to a single point, making the fiber $\mathbf{S}^7$ at each point. The elements of $(v_0, v_1, v_2, v_3, v_4, v_5, v_6, v_7, v_8)$ form one column of an $\mathbf{SO}(9)$ rotation matrix, each row or column of which describes an alternative form of the projection creating the Hopf fibration.

Remark: Octonions lead to another interesting fundamental algebra called the Jordan algebra, which closely resembles the algebra of the Pauli matrices used in quaternion constructions—except that it uses a 3×3 Hermitian matrix of octonions.

Summary: Table 30.1 summarizes the results of the Hopf fibrations following naturally from all four division algebras.

Domain	No. Vars.	Base Space	Fiber (Group)	Full Space
$\mathbb{R}$	$N = 1$	$\mathbf{S}^1$	$\mathbf{S}^0 = \mathbb{Z}_2$	Double cover of $\mathbf{S}^1$
$\mathbb{C}$	$N = 2$	$\mathbf{S}^2$	$\mathbf{S}^1 = \mathbf{U}(1)$	$\mathbf{S}^3$
$\mathbb{Q}$	$N = 4$	$\mathbf{S}^4$	$\mathbf{S}^3 = \mathbf{SU}(2)$	$\mathbf{S}^7$
$\mathbb{O}$	$N = 8$	$\mathbf{S}^8$	$\mathbf{S}^7$	$\mathbf{S}^{15}$

TABLE 30.1   Hopf fibrations of division algebras.

Clifford Algebras

31

The quaternion-based formalism for handling and visualizing rotations works well in dimensions 2, 3, and 4 because in these dimensions the Spin group (the double covering of the orthogonal group) has simple topology and geometry. It would be natural to expect that this simplicity continues to hold for rotations in any dimension, and that all of our 3D intuitions about labeling frames, interpolating frames, and simple frame-to-frame distance measures continue to be valid. Unfortunately, that is not the case: quaternions are quite unique to 3D, and only a serendipitous accident of topology allows an extension even to 4D.

On the other hand, there is a mathematical formalism that treats N-dimensional rotations in a very general way. *Clifford algebras* [8,117] are the basis of the standard mathematical approach to the group theory of orthogonal transformations in $\mathbb{R}^N$, the Euclidean space of N real dimensions. What happens is that Clifford algebras successfully *specialize* to the case of quaternions for $N = 3$, but the elegant features of the quaternion framework for 3D rotations *do not generalize* directly to higher dimensions.

Just as quaternions double cover the 3D rotations and provide a natural *square root* for all standard structures of the orthogonal group, we shall see that Clifford algebras embody double coverings for all N-dimensional orthogonal rotations, and that these coverings correspond to the *Spin groups* in arbitrary dimensions.

Furthermore, the Clifford-algebra approach provides additional depth to our understanding of dimensions 2, 3, and 4. We can get a better feel for which properties are accidents of the low dimension and which are general and extensible concepts. In fact, it is only by studying the Clifford algebra for two dimensions that we finally come full circle to a rigorous understanding of the heuristic argument that led us to consider the half-angle formulas for 2D rotations. From the Clifford algebra, we will see that the 2D half-angle formulas are not simply a superfluous algebraic manipulation but embody the mathematical core of all rotations.

Although there are many applications for Clifford algebras beyond those we will treat here (e.g., see Ablamowicz et al. [1] or Dorst et al. [37]), these are beyond our scope and we will focus exclusively on the exploitation of Clifford algebras for rotations.

31.1   INTRODUCTION TO CLIFFORD ALGEBRAS

The motivation to study Clifford algebras comes from the need to understand quaternions in a larger context. As we follow the path through complex numbers and 2D rotations to quaternions and 3D rotations, questions such as the following puzzle us.

- *Is 2D special?* Are there properties of 2D frames that cannot be found in other dimensions?
- *Is 3D special?* Are there properties of 3D frames that cannot be found in other dimensions?
- *What properties are only found in low dimensions?* If 2D and 3D *are* special for some reason, what properties do they have that are lost in higher dimensions, and why?
- *What properties are universal?* What properties do 2D and 3D rotations possess that in fact continue to have clear analogs in higher dimensions, and why?

What we shall see is that Clifford algebras do indeed provide higher-dimensional analogs of the quadratic forms we used in 2D and 3D to construct rotations in Euclidean space. Unhappily, the fact that complex numbers and quaternions are the only nontrivial associative division rings gives them unique and unduplicatable properties enabling the analysis and visualization of 2D and 3D rotations. (4D rotations, as noted in a previous chapter, also work fairly well due to the lucky coincidence that 4D rotations decompose into a pair of 3D rotations.)

The second major feature we will note about Clifford algebras is that the objects that are analogous to quaternions correspond to famous but nonintuitive mathematical structures, the Spin representations and their corresponding bases, the spinors. The square-root-like properties we have seen throughout our treatment of 2D and 3D rotations with half-angle formulas and quaternions are the precise properties that generalize to the Spin groups in higher dimensions. Curiously, whereas Clifford discovered the conceptual machinery for the Spin groups, the *really* interesting relationships to spin 1/2-elementary particles came much later (e.g., see Wigner [173]).

31.2 FOUNDATIONS

The fundamental Clifford algebra defined on an N-dimensional space is based on the properties of the chosen metric g_{ij} and a set of basis elements e_i, $i = 1, \ldots, N$. What this means informally is that points in the N-dimensional space are described abstractly by a set of basis vectors e_i, $i = 1, \ldots, N$, and real numbers v_i, and thus a vector looks like a sum of weights over the basis:

$$V = \sum_i v_i e_i.$$

The length of a vector is computed from the inner product with the metric:

$$\|V\|^2 = \langle V, V \rangle = \sum_{ij} v_i g_{ij} v_j.$$

The metric can in general be a function of the spatial position, but to handle the group theory of rotations in flat Euclidean space we will need only $g_{ij} = \delta_{ij}$, the identity matrix in Euclidean space. Technically, one must pay attention to the upper and lower indices more carefully than we do here, but this makes no difference for flat Euclidean spaces.

The *unique feature* Clifford discovered was that if the algebraic product of the basis vectors is assigned a particular nonintuitive multiplication rule, a vast number of other useful properties result. Clifford algebras are based on this multiplication rule:

Clifford Product

$$e_i e_j + e_j e_i = -2g_{ij}. \tag{31.1}$$

Because e_i for a Euclidean space obeys $(e_i)^2 = -1$, we need the concept of the *conjugate* variable,

$$(e_i)^\dagger = -e_i. \tag{31.2}$$

If there are multiple basis elements the conjugate reverses their order, for example,

$$(e_i e_j)^\dagger = (-e_j)(-e_i) = e_j e_i, \tag{31.3}$$

and so on.

Algebra = all possible products: In the context of Clifford algebras, the fundamental relation $e_i e_j + e_j e_i = -2g_{ij}$ is the *Clifford product*, whereas the *Clifford algebra* (or the *geometric algebra*) is the space $\text{Cl}(N)$ of all possible products. If the basis $\{e_i\}$ has dimension N, the dimension of the space of all possible products $\text{Cl}(N)$ is 2^N. A related important splitting is the space of all *even* products and the space of all *odd* products, each of which, including the identity and the volume element $\prod_i e_i$, has the same dimension 2^{N-1}.

   This is all there is to the abstract form of the Clifford algebra. Physicists should immediately recognize these formulas as those obeyed by the *Pauli matrices* and the *Dirac matrices*. That is, although the algebra is abstract it has concrete realizations using matrices. For example, the Pauli matrices $\boldsymbol{\sigma}$ times $-\sqrt{-1} = -i$ represent quaternions as $q = q_0 I_2 - i\mathbf{q} \cdot \boldsymbol{\sigma}$, where

$$\sigma_1 = \begin{bmatrix} 0 & 1 \\ 1 & 0 \end{bmatrix},$$

$$\sigma_2 = \begin{bmatrix} 0 & -i \\ i & 0 \end{bmatrix},$$

$$\sigma_3 = \begin{bmatrix} 1 & 0 \\ 0 & -1 \end{bmatrix},$$

$$I_2 = \begin{bmatrix} 1 & 0 \\ 0 & 1 \end{bmatrix}.$$

The matrices $\boldsymbol{\sigma}$ obey the algebra

$$\sigma_j \sigma_k = \delta_{jk} + i\epsilon_{jkl}\sigma_l, \tag{31.4}$$

where ϵ_{jkl} is the totally antisymmetric pseudotensor. (See Appendix F.) Thus, we reproduce the Clifford algebra by identifying e_k with $-i\sigma_k$:

$$e_j e_k + e_k e_j = i^2 \sigma_j \sigma_k + i^2 \sigma_k \sigma_j = -2\delta_{jk}. \qquad (31.5)$$

With the Pauli matrices producing the 3D algebra, the Euclidean Dirac matrices reproduce the 4D algebra. We see that the explicit choice of the 4×4 Euclidean Dirac matrices

$$\gamma_1 = \begin{bmatrix} 0 & \sigma_1 \\ \sigma_1 & 0 \end{bmatrix},$$

$$\gamma_2 = \begin{bmatrix} 0 & -i\sigma_2 \\ i\sigma_2 & 0 \end{bmatrix},$$

$$\gamma_3 = \begin{bmatrix} \sigma_3 & 0 \\ 0 & -\sigma_3 \end{bmatrix},$$

$$\gamma_4 = \begin{bmatrix} 1 & 0 & 0 & 0 \\ 0 & 1 & 0 & 0 \\ 0 & 0 & -1 & 0 \\ 0 & 0 & 0 & -1 \end{bmatrix}$$

gives the algebra

$$\gamma_j \gamma_k = \delta_{jk} + (i/2)\epsilon_{jkmn}[\gamma_m, \gamma_n]. \qquad (31.6)$$

Again, identifying e_k with $i\gamma_k$ we find that we have an explicit realization of the 4D Euclidean Clifford algebra:

$$e_j e_k + e_k e_j = i^2 \gamma_j \gamma_k + i^2 \gamma_k \gamma_j = -2\delta_{jk}. \qquad (31.7)$$

31.2.1   CLIFFORD ALGEBRAS AND ROTATIONS

The geometry of Clifford algebras has some interesting properties. As we shall see, it is almost like the "square root of a square root" in its relationship to rotations.

The Clifford algebra implements reflections: Our first step is to define a bare unit vector using the basis of the Clifford algebra, and then see what happens when we transform it by Clifford multiplication in ways analogous to those we used for quaternions. An

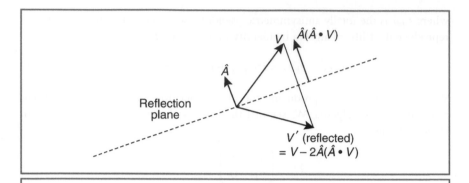

FIGURE 31.1 *Clifford algebra conjugation results in a reflection about* $\langle A, X \rangle = 0$.

initial bare unit vector is

$$A = \sum a_i e_i,$$
(31.8)

where

$$\|A\| = \sqrt{\langle A, A \rangle} = \sqrt{\sum (a_i)^2} = 1.$$

Introducing an arbitrary vector, $V = \sum v_i e_i$ (which does not have to be a unit vector) we define the Clifford conjugate operation as

$$A * V * A = V - 2A\langle A, V \rangle,$$

where $\|A\| = 1$ is critical. This is just a *reflection* of the component of V lying in the direction of A about the *plane*

$$\langle A, X \rangle = 0.$$

Thus, we can visualize the action of Clifford conjugation with a unit vector A as the transformation of the vector V into the vector V',

$$V' = A * V * A = V - 2A\langle A, V \rangle,$$

with the effect shown in Figure 31.1.

Pairs of Clifford reflections: Now let $B = \sum b_i e_i$ be another vector with $\|B\| = 1$. Repeating the Clifford multiplication rule,

$$
\begin{aligned}
V'' &= B * V' * B \\
&= A * B * V * B * A \\
&= V' - 2B\langle B, V' \rangle \\
&= V - 2A\langle A, V \rangle - 2B\big(B, V - 2A\langle A, V \rangle \big).
\end{aligned}
$$

This can be shown to be a *proper rotation* of the vector V as

$$
V'' = A * B * V * B * A = \sum_{ij} R_{ij} v_j e_i,
$$

where R_{ij} is an orthonormal matrix of unit determinant. One can see how this comes about from the representation of the double reflection in Figure 31.2.

Two Clifford reflections therefore result in a *proper rotation* of the vector V as

$$
V'' = A * B * V * B * A = R \cdot V, \tag{31.9}
$$

where R_{ij} is an orthonormal matrix of unit determinant. The full symbolic form of the matrix takes the form (remembering that we have substituted $\|A\| = 1$ and $\|B\| = 1$)

$$
R = \begin{bmatrix}
1 - 2A_1^2 - 2B_1^2 + 4A_1^2 B_1^2 + 4A_1 A_2 B_1 B_2 \\
-2A_1 A_2 + 4A_1 A_2 B_1^2 - 2B_1 B_2 + 4A_2^2 B_1 B_2 \\
-2A_1 A_2 + 4A_1^2 B_1 B_2 - 2B_1 B_2 + 4A_1 A_2 B_2^2 \\
1 - 2A_2^2 + 4A_1 A_2 B_1 B_2 - 2B_2^2 + 4A_2^2 B_2^2
\end{bmatrix} \tag{31.10}
$$

Because any rotation of this type can be transformed into a local 2D coordinate system, we can take $A = (\cos t_1, \sin t_1)$, $B = (\cos t_2, \sin t_2)$ and compute the amount of the rotation. Substituting the 2D expressions for A and B into the 2D form of the matrix, we find

$$
\begin{bmatrix}
\cos 2(t_2 - t_1) & -\sin 2(t_2 - t_1) \\
\sin 2(t_2 - t_1) & \cos 2(t_2 - t_1)
\end{bmatrix}. \tag{31.11}
$$

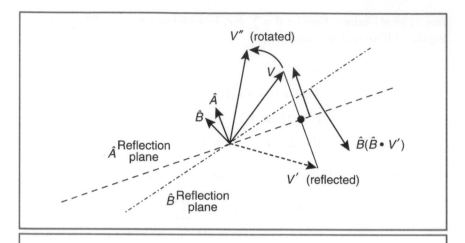

FIGURE 31.2 *Clifford algebra rotation graph.*

In other words, the rotation of the reflected vector is twice the angle between the pair of reflection directions,

$$A \cdot B = \cos(t_2 - t_1), \tag{31.12}$$

$$V'' \cdot V = \cos\big(2(t_1 - t_2)\big), \tag{31.13}$$

consistent with what we observe in Figure 31.2.

31.2.2 HIGHER–DIMENSIONAL CLIFFORD ALGEBRA ROTATIONS

We do not get anything given to us easily in the Clifford algebra world. Although the construction works perfectly in higher dimensions (in that any elementary rotation is ultimately a rotation in a 2D subplane), one may need *more* than a single (A, B) pair to exhaust all possible rotational degrees of freedom in the larger dimensions. One can work out a comparison between the necessary degrees of freedom and the available parameters in successive pairs of N-dimensional unit vectors (A_n, B_n) to generate the synopsis outlined in Table 31.1. In summary, the basic idea is as follows.

N	Pairs	Params.	Unit + Other Constraints	Freedom
2	1	$2*2=4$	$2+1=3$	1
3	1	$2*3=6$	$2+1=3$	3
4	1	$2*4=8$	$2+0=2$	6
5	2	$4*5=20$	$4+6=10$	10
6	2	$4*6=24$	$4+5=9$	15
7	2	$4*7=28$	$4+3=7$	21
8	2	$4*8=32$	$4+0=4$	28
9	3	$6*9=54$	$6+12=18$	36
10	3	$6*10=60$	$6+9=15$	45
11	3	$6*11=66$	$6+5=11$	55
12	3	$6*12=72$	$6+0=6$	66
13	4	$8*13=104$	$8+18=26$	78
14	4	$8*13=112$	$8+13=21$	91
.	.	.	.	.
N	$p=\lfloor\frac{N}{2}\rfloor$	$2p*N$	$p(2p+1)$	$\frac{N(N-1)}{2}$

TABLE 31.1 Number of conjugate unit vectors needed to construct a full $\frac{N(N-1)}{2}$-degree-of-freedom **SO**(N) rotation group element using the Clifford conjugate double-reflection construction.

- 2D: There are two 2D unit vectors (A, B), and hence two free parameters left due to $\|A\| = 1$, $\|B\| = 1$. However, 2D rotations have only one free parameter. From the explicit example given previously, the two apparent free parameters (t_1, t_2) do not appear independently in the resulting rotation matrix, but only the combination $(t_2 - t_1)$. Thus, only one independent parameter survives.
- 3D: Two 3D unit vectors (A, B) have four independent degrees of freedom, which could be parameterized using $A = (\cos t_1 \cos p_1, \sin t_1 \cos p_1, \sin p_1)$, $B = (\cos t_2 \cos p_2, \sin t_2 \cos p_2, \sin p_2)$, and thus (t_1, p_1, t_2, p_2) would be the parameters. Extracting the explicit linear dependence among the para-

meters is difficult, but the matrix is easily confirmed to be orthogonal using the dimension-independent algebraic calculation cited previously. An $\mathbf{SO}(3)$ matrix has only three independent parameters, and thus one of the four variables in (A, B) can be eliminated. Another approach would be to find an inverse map from the four parameters to the four quaternion variables. Then the $q \cdot q = 1$ constraint would alternatively eliminate the extra freedom.

- 4D: Four dimensions presents a new phenomenon. Because two 4D unit vectors (A, B) have six independent parameters, and $\mathbf{SO}(4)$ has six parameters, the parameterizations must be equivalent.

- 5D: Five and higher dimensions give another requirement. Because two 5D unit vectors (A, B) have *eight* independent parameters, and $\mathbf{SO}(5)$ has 10 degrees of freedom, the Clifford conjugate construction must be supplemented by two more variables—giving a total of four, (A_1, B_1, A_2, B_2), or 16 parameters. Six of these must be eliminatable, because only 10 degrees of freedom exist in the constructed matrix.

- 6D: Six dimensional orthogonal rotation matrices have 15 parametric degrees of freedom, whereas four Clifford reflection unit vectors (A_1, B_1, A_2, B_2) supply 20. Five are therefore eliminatable.

31.3   EXAMPLES OF CLIFFORD ALGEBRAS

Now that we see the basic process for generating rotations using the Clifford algebra, we will proceed to study several of the lower-dimensional cases in detail, exposing the truly remarkable fact that even our simplest 2D rotation example in classical geometry is truly only "natural" in some sense when we use the half-angle reformulation!

31.3.1 1D CLIFFORD ALGEBRA

We begin with the very simplest, but not entirely trivial, case of a single dimension. $N = 1$, and thus

$$e_1 e_1 = -1.$$

One might think this was the complex numbers, because we recall that the imaginary number i obeying $i^2 = -1$ was the fundamental quantity enabling us to define complex number systems. In fact, this is true, because the full basis includes the identity and therefore with $(1, e_1)$ and $(e_1)^2 = -1$ we can indeed represent all complex numbers.

The second thing one might think is that now we can use the $N = 1$ basis to represent 2D rotations, as we did in earlier chapters. However, be careful: there is only one dimension, and thus if we are to represent rotations as double reflections *the only possible reflection is $x \to -x$*. This is *not* enough to do 2D rotations! Complex numbers in fact have two equivalent bases, and the natural treatment of 2D rotations is given by the $N = 2$ algebra, which maps back to an equivalent $N = 1$ Clifford algebra, the complex numbers, more or less by accident.

31.3.2 2D CLIFFORD ALGEBRA

2D space in the Clifford algebra framework is described by the two elements (e_1, e_2) obeying the algebra

$$e_1 e_1 = -1,$$

$$e_1 e_2 + e_2 e_1 = 0,$$

$$e_2 e_2 = -1.$$

The full technology of the Clifford algebra now depends on producing an exhaustive list of *all possible* products until additional factors reduce to something already in the list. For $N = 1$, this was trivial—just the list containing the identity and the basis, $(1, e_1)$. When we start multiplying all possible $N = 2$ products, however, we find one new feature, a fourth component. The four-component basis

$$(1, e_1, e_2, e_1 e_2)$$

exhausts all possible Clifford products for $N = 2$.

Here again we are led into temptation. Because $e_1 e_2 e_1 e_2 = -1$, we could simply take

$$\mathbf{x}_i = e_1,$$

$$\mathbf{x}_j = e_2,$$

$$\mathbf{x}_k = e_3 \equiv e_1 e_2$$

and identify this basis with the *quaternions* $(1, \mathbf{x}_i, \mathbf{x}_j, \mathbf{x}_k)$. To check, we see that with the previous identification,

$$-1 = \mathbf{x}_i\mathbf{x}_i = \mathbf{x}_j\mathbf{x}_j = \mathbf{x}_k\mathbf{x}_k,$$

$$\mathbf{x}_i\mathbf{x}_j = \mathbf{x}_k,$$

$$\mathbf{x}_j\mathbf{x}_k = \mathbf{x}_i,$$

$$\mathbf{x}_k\mathbf{x}_i = \mathbf{x}_j,$$

which is precisely the quaternion basis algebra.

Where is the 2D rotation?: Again, we must be careful; there are only *two* dimensions, and thus there are insufficient degrees of freedom to do 3D rotations! The *true* basis of rotations in each dimension comes from the *even part* of the family of all Clifford products, which for $N = 2$ reduces to

$$(1, e_1e_2)$$

Complex Numbers Emerge

And what is e_1e_2? We can identify it exactly as the value of i, with $i^2 = -1$, which we used throughout our treatment of 2D rotations to condense the 2D matrix form into an algebra. Now everything fits into its proper place, as we see next.

31.3.3 2D ROTATIONS DONE RIGHT

To perform a 2D rotation using a Clifford double reflection, we take a general element of the *even basis* of the algebra:

$$R = a + be_1e_2, \tag{31.14}$$

$$R^\dagger = a - be_1e_2. \tag{31.15}$$

Note that although a standard rotation matrix is 2×2, this is an element of the algebra with only two elements (and in fact only one is independent, as we see in a moment). The action of the even Clifford algebra rotation on the (odd-dimensional) vector $V = v_1 e_1 + v_2 e_2$ can now be written as

$$R * V * R^\dagger = V' = v_1' e_1 + v_2' e_2. \tag{31.16}$$

Because V' is the result of the rotation acting on the 2D vector V, we can easily show that

$$\begin{bmatrix} v_1' \\ v_2' \end{bmatrix} = \begin{bmatrix} a^2 - b^2 & -2ab \\ 2ab & a^2 - b^2 \end{bmatrix} \begin{bmatrix} v_1 \\ v_2 \end{bmatrix}.$$

This is a rotation provided the determinant is unity, or

$$\det \begin{bmatrix} a^2 - b^2 & -2ab \\ 2ab & a^2 - b^2 \end{bmatrix} = (a^2 + b^2)^2 = 1, \tag{31.17}$$

and hence

$$a^2 + b^2 = 1, \tag{31.18}$$

and a and b correspond to a (cos, sin) pair.

What is i?: What we have previously called $i = \sqrt{-1}$ is really $i = e_1 e_2$, and with for example $a = \cos\alpha$, $b = \sin\alpha$ we have

$$R = e^{e_1 e_2 \alpha} \tag{31.19}$$

$$= e^{i\alpha} \tag{31.20}$$

$$= \cos\alpha + i\sin\alpha \tag{31.21}$$

as the Clifford algebra form of the standard 2D rotation.

Finally, we see that

$$a^2 - b^2 = \cos^2\alpha - \sin^2\alpha \equiv \cos(2\alpha), \tag{31.22}$$

$$2ab = 2\cos\alpha\sin\alpha \equiv \sin(2\alpha), \tag{31.23}$$

and thus the half-angle formula—which seemed an arbitrary manipulation when we introduced it—is *mandatory!* Our apparently artificial 2D transformation written in terms of half angles was not so silly after all; *no other procedure generalizes to N dimensions.* The Clifford algebra for $N = 2$ automatically produces

$$R_2(a, b) = \begin{bmatrix} a^2 - b^2 & -2ab \\ 2ab & a^2 - b^2 \end{bmatrix},$$

where $a^2 + b^2 = 1$, and we have the solution

$$a = \cos\alpha = \cos(\theta/2),$$
$$b = \sin\alpha = \sin(\theta/2)$$

or

$$R_2(a,b) = \begin{bmatrix} \cos 2\alpha & -\sin 2\alpha \\ \sin 2\alpha & \cos 2\alpha \end{bmatrix}$$
$$= \begin{bmatrix} \cos\theta & -\sin\theta \\ \sin\theta & \cos\theta \end{bmatrix}.$$

31.3.4 3D CLIFFORD ALGEBRA

The 3D case is of course a little trickier. Here, we start from the set (e_1, e_2, e_3) and the algebra

$$e_1 e_1 = -1,$$
$$e_1 e_2 + e_2 e_1 = 0, \qquad e_2 e_2 = -1,$$
$$e_1 e_3 + e_3 e_1 = 0, \qquad e_2 e_3 + e_3 e_2 = 0, \qquad e_3 e_3 = -1. \qquad (31.24)$$

Then the full list of all Clifford products starting from the set $\{e_i\}$ is 8-dimensional:

$$1$$
$$e_1 \qquad e_2 \qquad e_3$$
$$e_2 e_3 \qquad e_3 e_1 \qquad e_1 e_2 \qquad\qquad (31.25)$$
$$e_1 e_2 e_3$$

To relate this to the double reflections, which are the proper rotations, we again need to select the even part. (The odd part corresponds to improper rotations—those that contain a reflection and that cannot be reexpressed as a rotation.)

The basis of rotations in the Clifford algebra therefore consists of four components,

$$(1, e_2 e_3, e_3 e_1, e_1 e_2), \qquad\qquad (31.26)$$

which—having the benefit of hindsight from knowing the title of this book—we may deduce are *the quaternion basis of the 3D rotations*. We may identify these four objects either as simply the vector basis of the list of elements of the abstract quaternion algebra we have used throughout, or as the Pauli matrices

$$(I_2, -i\sigma_x, -i\sigma_y, -i\sigma_z),$$

or as the extended complex numbers used to define Hamilton's original quaternion basis:

$$(1, \mathbf{i}, \mathbf{j}, \mathbf{k}).$$

31.3.5 CLIFFORD IMPLEMENTATION OF 3D ROTATIONS

Finally, we are ready for the "right" way to do 3D Euclidean rotations—a method that generalizes to all dimensions and that allows us to obtain some insight into where our 3D quaternion methods in fact are generalizable and where they are too unique to generalize. Proper 3D rotations using the Clifford algebra framework are now characterized simply by writing down the most general four-vector in the even subset of the available 3D Clifford algebra components,

$$R = q_0 + q_1(e_2 e_3) + q_2(e_3 e_1) + q_3(e_1 e_2), \qquad (31.27)$$

and requiring that the corresponding 3×3 matrix be normalized with determinant one. The resulting action is simply

$$R * V * R^\dagger = \sum_{i=1}^{4} v_i' e_i, \qquad (31.28)$$

where the coefficients of v_i in v_i' are *precisely* our old friend, the quadratic quaternion map

$$R_3 = \begin{bmatrix} q_0^2 + q_1^2 - q_2^2 - q_3^2 & 2q_1 q_2 - 2q_0 q_3 & 2q_1 q_3 + 2q_0 q_2 \\ 2q_1 q_2 + 2q_0 q_3 & q_0^2 - q_1^2 + q_2^2 - q_3^2 & 2q_2 q_3 - 2q_0 q_1 \\ 2q_1 q_3 - 2q_0 q_2 & 2q_2 q_3 + 2q_0 q_1 & q_0^2 - q_1^2 - q_2^2 + q_3^2 \end{bmatrix}.$$

31.4   HIGHER DIMENSIONS

As one might expect, higher dimensions are much more complicated, and do not work out as neatly—except for a convenient accident in $N = 4$, which allows a double-quaternion form [69] (described in detail in Chapter 29). Note that Clifford algebras in dimensions greater than eight essentially repeat many fundamental properties. Clifford algebras obey a *periodicity theorem* that allows the construction of all algebras if you know the first eight very well.

The rank two Clifford subalgebras generate the rotational Lie algebra—the infinitesimal rotations. When exponentiated, these generate all other possible even elements, and from there we get the Clifford algebra form of the Lie group.

To get some idea of what is going on in general, we can do a little counting. First, with each basis $(e_1, e_2, \ldots, e_N)$ and its algebra $e_i e_j + e_j e_i = -2\delta_{ij}$ we see that there is essentially a Pascal's triangle of possible elements—for a total of $N + 1$ different lengths of elements from 0 to N—and that the first element is the identity, the second is the complete set of $N(N - 1)/2$ pairs, and the last is the volume element $\prod_i e_i$. The total number of even-element pairs is precisely the same as the number of independent values of an $N \times N$ orthogonal matrix, $N(N - 1)/2$. This is the number of elements in the Lie algebra of the rotation group, equivalent to the set of basis matrices from which all other matrices in the group can be constructed by exponentiation.

When all possible two-component elements are multiplied together, the result is a certain number of additional even elements. For **SO**(2) and **SO**(3), there are no other even elements besides the identity, and *this is why these rotation groups are very special*. For **SO**(4), the additional element is the volume, $e_1 e_2 e_3 e_4$. This plays the role of a "second identity" and allows the construction of two complete quaternion algebras that relate the eight even elements in 4D to a pair of unit quaternions. For **SO**(5), there are the usual 10, plus the identity, plus the five four-element forms that lack one element, for a total of 16. For **SO**(6), there are the usual 15 plus 17 more, for a total of 32. We can easily see that for **SO**(N) the elements include the identity, the usual $N(N - 1)/2$ pairs, and the remaining even elements, always giving the total number of even elements $= 2^{N-1}$.

It is interesting to compare two relations. First consider the $N \times N$ rotation matrices and how many independent degrees of freedom there are in the matrices, as summarized in Table 31.2. Then, in contrast, consider all even elements of each Clifford algebra compared to the number of possible free parameters representing a rotation by exponentiation of the two-element algebra (as summarized in Table 31.3).

Matrix Dim. N	Elements	Orthogonality Constraints	Elements $-$ Constraints $=$ Total DOF $= N(N-1)/2$
2	4	3	1
3	9	6	3
4	16	10	6
5	25	15	10
6	36	21	15
7	49	28	21
8	64	36	28
.	.	.	.
N	N^2	$\frac{N(N+1)}{2}$	$\frac{N(N-1)}{2}$

TABLE 31.2 *Relation between orthogonal $N \times N$ matrices and actual degrees of freedom.*

By analogy with our treatment of matrix logarithms in Chapter 18, we may consider the two-component Clifford algebra elements to be identified with the logarithms of rotations in N dimensions, now expressed abstractly as a Clifford algebra rather than concretely as a set of matrices:

$$\log \mathbf{R}_N = \sum_{i<j} q_{ij} e_i e_j. \tag{31.29}$$

The construction of the $\mathbf{SO}(N)$ rotation matrix $\mathbf{R}_N$ is basically the same, as

$$\mathbf{R} * V * \mathbf{R}^\dagger = \sum_{i=1}^{4} v_i' e_i, \tag{31.30}$$

where the matrix $\mathbf{R}_N$ is obtained from comparing the coefficients in

$$v' = \mathbf{R}_N \cdot v,$$

Dim.	Pairs	Total Even	Necessary Constraints	The Constraint
2	1	2	1	$\mathbf{S}^1$: $a^2 + b^2 = 1$
3	3	4	1	$\mathbf{S}^3$: $q_0^2 + q_1^2 + q_2^2 + q_3^2 = 1$
4	6	8	2	Double $\mathbf{S}^3$: $q \cdot q = 1$, $p \cdot p = 1$
5	10	16	6	??
6	15	32	17	??
7	21	64	43	??
8	28	128	100	??
.	.	.	.	.
N	$\frac{N(N-1)}{2}$	2^{N-1}	$2^{N-1} - \frac{N(N-1)}{2}$	??

TABLE 31.3   Relation between even Clifford algebra elements and the degrees of freedom of orthogonal $N \times N$ matrices. (We do not have neat forms for the constraints for $N > 4$.)

where we impose the requirement that the matrix have unit determinant.

Finally, we examine also the number of degrees of freedom versus required constraints that are realized in the *group* compared to the *algebra*. The even Clifford algebra has precisely the correct number of generators for the rotation group in question, but when iterated the number of Clifford elements becomes much larger and some knowledge of the constraints is useful to define the degrees of freedom. This information is summarized in Table 31.4.

31.5   PIN(N), SPIN(N), O(N), SO(N), AND ALL THAT...

The Spin representations of the orthogonal groups follow from the remaining elements of the Clifford algebra Cl(N). A top-level summary of the structure follows. The key concept is that transformations with no preceding $\mathbf{S}$ can have a negative

N	Dim. (even Clifford)	Dim. (rotations)	Constraints (the difference)
1	1	0	1
2	2	1	1
3	4	3	1
4	8	6	2
5	16	10	6
6	32	15	17
7	64	21	43
8	128	28	100
.	.	.	.
N	2^{N-1}	$\frac{N(N-1)}{2}$	$\frac{2^N - N^2 + N}{2}$

TABLE 31.4 N-dimensional degrees of freedom after iterating the even Clifford algebra, and the implied number of constraints on the resulting rotation coefficients.

determinant and be sign-reversing. If there is a preceding **S**, the group is "special," which means it has determinant plus one and cannot be sign-reversing.

- Pin(N): G is "Pin" if it is a general reflection. G includes *all* elements of Cl(N).
- Spin(N): G is "Spin" if it is a general rotation. G includes only *even* elements of Cl(N).
- **O**(N): $G * V * G^\dagger$ is "**O**" if G is in Pin and result is a vector reflection.
- **SO**(N): $G * V * G^\dagger$ is "**SO**" if G is in Spin and result is a vector rotation.

We will not pursue these concepts further here, but refer the interested reader to detailed treatments such as those of Atiyah et al. [8] and Lawson and Michelsohn [117].

Conclusions

We bring our mathematical composition to a close by repeating the theme introduced in Chapter 4, whose brevity and elegance we now see is deceptively simple. From the fundamental relations

$$p \star q = (p_0, p_1, p_2, p_3) \star (q_0, q_1, q_2, q_3)$$

$$= \begin{bmatrix} p_0q_0 - p_1q_1 - p_2q_2 - p_3q_3 \\ p_1q_0 + p_0q_1 + p_2q_3 - p_3q_2 \\ p_2q_0 + p_0q_2 + p_3q_1 - p_1q_3 \\ p_3q_0 + p_0q_3 + p_1q_2 - p_2q_1 \end{bmatrix}$$

$$= \begin{bmatrix} p_0 & -p_1 & -p_2 & -p_3 \\ p_1 & p_0 & -p_3 & p_2 \\ p_2 & p_3 & p_0 & -p_1 \\ p_3 & -p_2 & p_1 & p_0 \end{bmatrix} \begin{bmatrix} q_0 \\ q_1 \\ q_2 \\ q_3 \end{bmatrix}$$

$$= (p_0q_0 - \mathbf{p} \cdot \mathbf{q}, \; p_0\mathbf{q} + q_0\mathbf{p} + \mathbf{p} \times \mathbf{q}),$$

$$p \cdot q = (p_0, p_1, p_2, p_3) \cdot (q_0, q_1, q_2, q_3)$$

$$= p_0q_0 + p_1q_1 + p_2q_2 + p_3q_3$$

$$= p_0q_0 + \mathbf{p} \cdot \mathbf{q},$$

$$\bar{q} = (q_0, -q_1, -q_2, -q_3)$$

$$= (q_0, -\mathbf{q}),$$

$$q \star \bar{q} = (q \cdot q, \mathbf{0})$$

$$q \cdot q = (q_0)^2 + (q_1)^2 + (q_2)^2 + (q_3)^2$$
$$= (q_0)^2 + \mathbf{q} \cdot \mathbf{q} = 1,$$

we have found an enormous number of questions that can be asked and properties that can be computed by exploiting the power of the quaternion frame. We have found ways to interpolate, ways to optimize, ways to extract information from data, and new ways to reinterpret old concepts using quaternions. Many unusual technologies have been uncovered and applied, and we have been able to construct visualizations—namely, quantitative visual representations that assist our insights— for nearly all of them.

A handful of questions remain without completely satisfactory answers. For example, there is some uncertainty as to what a Bernstein polynomial basis might be on a sphere, or what the proper analogs of linear fractional transformations or NURBS (Nonuniform Rational B-Splines) might be. Another problem is that although certain joint systems of interest are described by just three degrees of freedom (and are therefore accessible to quaternion technology) many complex natural and synthetic joint systems are not. How do we find and exploit a topological space with a well-defined quaternion-like distance measure that extends to the degrees of freedom of an entire functional system instead of just one three-degree-of-freedom joint? Can further study of the geometry of spinors and spin representations lead to new pictures that assist our understanding in the way the belt trick helped us relate quaternions to real experience? There are many open problems that must be solved both to deeply understand quaternions themselves and to extend their features to more complete domains. We hope that what we have presented here can provide the framework upon which the solutions to problems such as these can be built.

Appendices

The following appendices serve to summarize a number of the principal mathematical approaches and formulas in the main text in a single, easy-to-locate place. They also provide a selection of useful supplementary material that did not have a clear context elsewhere.

Notation

A   If the reader is conversant with the conventions of 3D vector notation and has used complex variables, this appendix should be elementary and can be skipped. Nevertheless, for anyone who might benefit from a quick summary, or who might be accustomed to substantially different notational conventions from the author, the summary presented here may be essential in order to follow some of the notation in the main body of the book. In particular, the explanations of basic quaternion notation in Chapter 4 depend strongly on the notation given here.

A.1   VECTORS

A vector $\mathbf{x}$ is a set of real numbers we typically write in the form

$$\mathbf{x} = (x, y) \tag{A.1}$$

for two-dimensional (2D) vectors, and as

$$\mathbf{x} = (x, y, z) \tag{A.2}$$

for three-dimensional (3D) vectors. Technically, we should treat this notation as a shorthand for a column vector because we are thinking in the back of our minds of multiplying these vectors by rotation matrices to transform them to a new orientation. That is, in proper matrix notation

$$\mathbf{x} = x\hat{\mathbf{x}} + y\hat{\mathbf{y}}$$

$$= x\begin{bmatrix} 1 \\ 0 \end{bmatrix} + y\begin{bmatrix} 0 \\ 1 \end{bmatrix}$$

$$= \begin{bmatrix} x \\ y \end{bmatrix}$$

for 2D vectors and

$$\mathbf{x} = x\hat{\mathbf{x}} + y\hat{\mathbf{y}} + z\hat{\mathbf{z}}$$

$$= x\begin{bmatrix} 1 \\ 0 \\ 0 \end{bmatrix} + y\begin{bmatrix} 0 \\ 1 \\ 0 \end{bmatrix} + z\begin{bmatrix} 0 \\ 0 \\ 1 \end{bmatrix}$$

$$= \begin{bmatrix} x \\ y \\ z \end{bmatrix}$$

for 3D vectors.

A.2   LENGTH OF A VECTOR

The length of a Euclidean vector is computed from the Pythagorean theorem, generalized to higher dimensions. Several equivalent notations are common for the squared length of a Euclidean vector. Examples are

$$\|\mathbf{u}\|^2 = \mathbf{u}^2 = \mathbf{u} \cdot \mathbf{u} = x^2 + y^2$$

in 2D and

$$\|\mathbf{u}\|^2 = \mathbf{u}^2 = \mathbf{u} \cdot \mathbf{u} = x^2 + y^2 + z^2$$

in 3D. The inner product (or dot product) notation is generalized to arbitrary pairs of vectors in the following. The length of the vector $\mathbf{u}$ is then the square root of its squared length, and is typically written using double vertical bars as

$$\|\mathbf{u}\| = \sqrt{x^2 + y^2}$$

in 2D and

$$\|\mathbf{u}\| = \sqrt{x^2 + y^2 + z^2}$$

in 3D.

A.3   UNIT VECTORS

A unit vector $\hat{\mathbf{u}}$ is the vector that results when we divide a (nonzero) vector $\mathbf{u} = (x, y)$ or $\mathbf{u} = (x, y, z)$ by its Euclidean length, $\|\mathbf{u}\|$. That is,

$$\hat{\mathbf{u}} = \frac{\mathbf{u}}{\|\mathbf{u}\|},$$

and we see that obviously the Euclidean length of $\hat{\mathbf{u}}$ is one, or "unity," and hence the terminology *unit vector*.

A.4   POLAR COORDINATES

As a basis for some of the concepts we use to understand both complex variables and quaternions, we note that the Cartesian coordinates defined in Equations A.1 and A.2 have alternative forms based on their *magnitude* $\|\mathbf{u}\|$, which we write as a radius $r = \|\mathbf{u}\|$. Then the 2D polar form of a Euclidean vector uses elementary trigonometry to express the components in terms of r and the angle θ between the vector $\mathbf{u}$ and the $\hat{\mathbf{x}}$ axis; namely,

$$\mathbf{u} = (x, y) = (r \cos \theta, r \sin \theta).$$

This is a point on a circle of radius r. A 3D polar coordinate then becomes a point on a sphere of radius r, expressible in terms of trigonometric functions as

$$\mathbf{u} = (x, y, z) = (r \cos \theta \cos \phi, r \sin \theta \cos \phi, r \sin \phi).$$

Beyond 2D, there are various alternatives for polar coordinates, and quaternions will be expressed in several equivalent 4D polar coordinate systems.

A.5   SPHERES

Spheres are labeled by the dimension of the space that results if you cut out a bit of the sphere in every direction around the North Pole and flatten it out. A circle then has dimension one, and a balloon dimension two (an exploded balloon can be flattened like a sheet of paper). Thus, a circle is the one-sphere $\mathbf{S}^1$, a balloon is a two-sphere $\mathbf{S}^2$, and so on. Now that we know some sample equations for polar coordinates, we can start from there and define a sphere mathematically as a set of points that all lie at a fixed radius from the center. The circle, or one-sphere $\mathbf{S}^1$, embedded in two dimensions then obeys the equation

$$\|\mathbf{u}\|^2 = x^2 + y^2 = r^2 \text{ constant} \qquad \text{(A.3)}$$

and the "ordinary sphere" (or two-sphere $\mathbf{S}^2$) embedded in three dimensions obeys the equation

$$\|\mathbf{u}\|^2 = x^2 + y^2 + z^2 = r^2 \text{ constant.} \qquad \text{(A.4)}$$

We saw in Chapter 4 that quaternions obey the equation of the hypersphere—the three-sphere, written formally as $\mathbf{S}^3$.

A.6   MATRIX TRANSFORMATIONS

In our context, matrices are rectangular arrays of real numbers. We will typically deal with square matrices. A square matrix $\mathbf{R}$ acting by right-multiplication on a vector $\mathbf{x}$ produces a new vector $\mathbf{x}'$ as follows:

$$\mathbf{x}' = \mathbf{R} \cdot \mathbf{x}$$
$$= \begin{bmatrix} r_{11} & r_{12} \\ r_{21} & r_{22} \end{bmatrix} \begin{bmatrix} x \\ y \end{bmatrix} = \begin{bmatrix} xr_{11} + yr_{12} \\ xr_{21} + yr_{22} \end{bmatrix}$$
$$= \begin{bmatrix} x' \\ y' \end{bmatrix}$$

for the 2D case and

$$\mathbf{x}' = \mathbf{R} \cdot \mathbf{x}$$

$$= \begin{bmatrix} r_{11} & r_{12} & r_{13} \\ r_{21} & r_{22} & r_{23} \\ r_{31} & r_{32} & r_{33} \end{bmatrix} \begin{bmatrix} x \\ y \\ z \end{bmatrix} = \begin{bmatrix} xr_{11} + yr_{12} + zr_{13} \\ xr_{21} + yr_{22} + zr_{23} \\ xr_{31} + yr_{32} + zr_{33} \end{bmatrix}$$

$$= \begin{bmatrix} x' \\ y' \\ z' \end{bmatrix}$$

for the 3D case.

A.7   FEATURES OF SQUARE MATRICES

Matrices in general are rectangular arrays of numbers. Square matrices are used to transform vectors into similar vectors, and thus have unique features. The two features we will need to use on occasion are the *trace*, which is the sum of the diagonal elements,

$$\text{Trace } \mathbf{R} \equiv \text{tr} \, \mathbf{R} = \sum_{i=1}^{n} r_{ii},$$

so that in 2D

$$\text{Trace } \mathbf{R} \equiv \text{tr} \, \mathbf{R} = r_{11} + r_{22}$$

and in 3D

$$\text{Trace } \mathbf{R} \equiv \text{tr} \, \mathbf{R} = r_{11} + r_{22} + r_{33}.$$

The other frequently encountered property is the *determinant*. Determinants in general have elegant expressions in terms of totally antisymmetric products, for which we refer the interested reader to Appendix F. For the special cases of square 2D and 3D matrices, the determinant can be given explicitly. In 2D we write

$$\text{Determinant } \mathbf{R} \equiv \det \mathbf{R} = r_{11}r_{22} - r_{12}r_{21}$$

and in 3D we write

$$\text{Determinant } \mathbf{R} \equiv \det \mathbf{R} = r_{11}r_{22}r_{33} - r_{11}r_{23}r_{32}$$

$$- r_{12}r_{21}r_{33} + r_{12}r_{23}r_{31}$$

$$+ r_{13}r_{21}r_{32} - r_{13}r_{22}r_{31}.$$

A.8   ORTHOGONAL MATRICES

An orthogonal matrix is a square matrix whose transpose is its own inverse. What this means is that if we let I_2 and I_3 be the 2×2 and 3×3 identity matrices, respectively, where

$$I_2 = \begin{bmatrix} 1 & 0 \\ 0 & 1 \end{bmatrix}$$

and

$$I_3 = \begin{bmatrix} 1 & 0 & 0 \\ 0 & 1 & 0 \\ 0 & 0 & 1 \end{bmatrix},$$

if the superscript T denotes the transposed matrix and $\mathbf{R}_N$ is an orthogonal $N \times N$ matrix, then

$$\mathbf{R}_2 \cdot (\mathbf{R}_2)^T = I_2$$

and

$$\mathbf{R}_3 \cdot (\mathbf{R}_3)^T = I_3.$$

A.9   VECTOR PRODUCTS

There are two products of vectors that concern us because they have particularly useful properties when we transform the component vectors by applying an orthogonal matrix $\mathbf{R}$. As before, we use the notation $\mathbf{x}_i = (x_i, y_i)$ for 2D vectors and $\mathbf{x}_i = (x_i, y_i, z_i)$ for 3D vectors.

A.9.1 2D DOT PRODUCT

In 2D, the first of these is the inner product (or *dot product*),

$$\mathbf{x}_1 \cdot \mathbf{x}_2 = x_1 x_2 + y_1 y_2,$$

which is *invariant* under the action of orthogonal matrix multiplication:

$$\mathbf{x}_1' \cdot \mathbf{x}_2' = \mathbf{R}\mathbf{x}_1 \cdot \mathbf{R}\mathbf{x}_2$$
$$= \mathbf{x}_1 \mathbf{R}^T \mathbf{R}\mathbf{x}_2$$
$$= \mathbf{x}_1 I \mathbf{x}_2$$
$$= \mathbf{x}_1 \cdot \mathbf{x}_2.$$

The dot product may be thought of as the generalization of the squared Euclidean length to a pair of vectors. It can be shown that in any dimension the dot product is invariant under the action of orthogonal matrices and is proportional to the *cosine* of the angle between the two vectors:

$$\mathbf{x}_1 \cdot \mathbf{x}_2 = \|\mathbf{x}_1\| \|\mathbf{x}_2\| \cos\theta_{12}.$$

A.9.2 2D CROSS PRODUCT

The concept of the 2D cross product is unconventional, but in retrospect extremely natural from the point of view of N-dimensional geometry (e.g., see Hanson [67]). Here, we can simply define the *2D cross product* as the procedure that generates a new vector $\mathbf{c}$ that is *perpendicular* to a given 2D vector $\mathbf{x}$:

$$\mathbf{c} = \times\mathbf{x}$$
$$= \det \begin{bmatrix} x & \hat{\mathbf{x}} \\ y & \hat{\mathbf{y}} \end{bmatrix}$$
$$= (-y, x).$$

One can easily verify that $\mathbf{c} \cdot \mathbf{x} = 0$.

A.9.3 3D DOT PRODUCT

In 3D, a construction analogous to the 2D case yields the invariant dot product, again proportional to the cosine:

$$\mathbf{x}_1 \cdot \mathbf{x}_2 = x_1 x_2 + y_1 y_2 + z_1 z_2 = \|\mathbf{x}_1\| \|\mathbf{x}_2\| \cos\theta_{12}.$$

A.9.4 3D CROSS PRODUCT

The cross product in 3D is defined as

$$\mathbf{c} = \mathbf{x}_1 \times \mathbf{x}_2$$

$$= \det \begin{bmatrix} x_1 & x_2 & \hat{\mathbf{x}} \\ y_1 & y_2 & \hat{\mathbf{y}} \\ z_1 & z_2 & \hat{\mathbf{z}} \end{bmatrix}$$

$$= (y_1 z_2 - y_2 z_1, z_1 x_2 - z_2 x_1, x_1 y_2 - x_2 y_1),$$

where now we find that $\mathbf{c}$ is perpendicular to *each* of the component vectors:

$$\mathbf{c} \cdot \mathbf{x}_1 = \mathbf{c} \cdot \mathbf{x}_2 = 0.$$

The *magnitude* of the 3D cross product is proportional to the *sine* of the angle between the vectors and is thus equal to the *area* of the parallelogram defined by the two vectors in the plane they span:

$$\text{Area of Parallelogram} = \|\mathbf{c}\| = \|\mathbf{x}_1\|\|\mathbf{x}_2\| \sin\theta_{12}.$$

A.10 COMPLEX VARIABLES

Complex variables were once thought to be so unnatural that they became known as *imaginary numbers*—because, presumably, there was no way they could be connected with reality. In fact, nothing could be further from the truth. Quantum mechanics—the best theoretical description yet developed to predict wide ranges of important, very real, physical processes—depends in essential and inescapable ways on complex numbers. Perhaps a more traditional concrete example is the fact that quadratic algebraic equations cannot be solved for all ranges of their parameters unless complex numbers are allowed. The essence of complex numbers thus comes directly from examining the two equations

$$z^2 = +1,$$
$$z^2 = -1$$

and attempting to solve them. The only way we can solve both of these innocent-looking equations is to invent a new object, the pure "imaginary" number i, which is endowed with the remarkable property that

$$i^2 = -1 \quad \Rightarrow \quad i = \sqrt{-1}.$$

The solutions of the pair of equations previously introduced are thus

$$z = \pm 1,$$

$$z = \pm i,$$

respectively. With some effort, one can show that by including i in our world we can miraculously express the solution of any algebraic equation of one variable. If we allow only real variables, this is no longer possible.

We can represent a general complex number z—which can be an arbitrary combination of a real part x and an imaginary part proportional to $i = \sqrt{-1}$—in the following alternative ways (note that one of these choices is actually a 2×2 matrix):

$$z = x + iy,$$

$$z = r\cos\theta + ir\sin\theta,$$

$$z = \begin{bmatrix} x & -y \\ y & x \end{bmatrix} = r \begin{bmatrix} \cos\theta & -\sin\theta \\ \sin\theta & \cos\theta \end{bmatrix}. \qquad (A.5)$$

We note that the complex identity

$$e^{i\theta} = \cos\theta + i\sin\theta$$

is one of the most fundamental equations in all of mathematics, and is often referred to as "Euler's identity." With this identity, we take the polar coordinate expression just given and reexpress it as

$$z = r\cos\theta + ir\sin\theta = re^{i\theta}.$$

The algebra of complex multiplication follows from any of the formulas shown in Equation A.5, either by explicitly using $i^2 = -1$ or from matrix multiplication. Thus, for example,

$$z_1 z_2 = (x_1 + iy_1)(x_2 + iy_2)$$

$$= (x_1 x_2 - y_1 y_2) + i(x_1 y_2 + x_2 y_1),$$

$$z_1 z_2 = r_1 e^{i\theta_1} r_2 e^{i\theta_2}$$

$$= r_1 r_2 e^{i(\theta_1 + \theta_2)}$$

$$= r_1 r_2 \cos(\theta_1 + \theta_2) + i r_1 r_2 \sin(\theta_1 + \theta_2),$$

$$z_1 z_2 = \begin{bmatrix} x_1 & -y_1 \\ y_1 & x_1 \end{bmatrix} \cdot \begin{bmatrix} x_2 & -y_2 \\ y_2 & x_2 \end{bmatrix}$$

$$= \begin{bmatrix} x_1 x_2 - y_1 y_2 & -(x_1 y_2 + x_2 y_1) \\ x_1 y_2 + x_2 y_1 & x_1 x_2 - y_1 y_2 \end{bmatrix},$$

$$z_1 z_2 = \begin{bmatrix} x_1 & -y_1 \\ y_1 & x_1 \end{bmatrix} \begin{bmatrix} x_2 \\ y_2 \end{bmatrix}$$

$$= \begin{bmatrix} x_1 x_2 - y_1 y_2 \\ x_1 y_2 + x_2 y_1 \end{bmatrix}.$$

Thus, we see that all of these forms are exactly equivalent to the more abstract statement that the *complex multiplication algebra* is defined by associating the complex product of two pairs of real numbers with a *third pair* constructed from their elements as follows:

$$z_1 \star z_2 = (x_1, y_1) \star (x_2, y_2) = \begin{bmatrix} x_1 \\ y_1 \end{bmatrix} \star \begin{bmatrix} x_2 \\ y_2 \end{bmatrix}$$

$$= (x_1 x_2 - y_1 y_2, x_1 y_2 + x_2 y_1)$$

$$= \begin{bmatrix} x_1 x_2 - y_1 y_2 \\ x_1 y_2 + x_2 y_1 \end{bmatrix}.$$

2D Complex Frames

B

The basic 2D rotation matrix changing the axes by an angle θ and its interpolation in terms of a local frame on a curve, with tangent vector $\mathbf{T}$ and normal vector $\mathbf{N}$, may be defined as

$$R_2 = \begin{bmatrix} \cos\theta & -\sin\theta \\ \sin\theta & \cos\theta \end{bmatrix} = [\hat{\mathbf{T}} \quad \hat{\mathbf{N}}]. \tag{B.1}$$

A double-valued quaternion-like parameterization of the 2D frame may be written in the form

$$R_2 = \begin{bmatrix} a^2 - b^2 & -2ab \\ 2ab & a^2 - b^2 \end{bmatrix}. \tag{B.2}$$

We can easily verify that if (a, b) is a point on $\mathbf{S}^1$ embedded in $\mathbb{R}^2$ (i.e., $a^2 + b^2 = 1$) this is an orthonormal parameterization of the frame, and furthermore that θ is related to (a, b) by the half-angle formulas:

$$a = \cos\frac{\theta}{2}, \qquad b = \sin\frac{\theta}{2}.$$

If desired, the redundant parameter can be eliminated locally by using projective coordinates such as $c = b/a = \tan(\theta/2)$ to get the unconstrained single-parameter form

$$R_2 = [\hat{\mathbf{T}} \quad \hat{\mathbf{N}}] = \frac{1}{1 + c^2} \begin{bmatrix} 1 - c^2 & -2c \\ 2c & 1 - c^2 \end{bmatrix}. \tag{B.3}$$

If we now define $W_1 = \begin{bmatrix} a & -b \\ b & a \end{bmatrix}$ and $W_2 = \begin{bmatrix} -b & -a \\ a & -b \end{bmatrix}$, and take derivatives with respect to θ, we may write

$$\hat{\mathbf{T}}' = 2W_1 \cdot \begin{bmatrix} a' \\ b' \end{bmatrix}, \qquad\qquad (\text{B.4})$$

$$\hat{\mathbf{N}}' = 2W_2 \cdot \begin{bmatrix} a' \\ b' \end{bmatrix}. \qquad\qquad (\text{B.5})$$

Taking explicit derivatives of the trigonometric forms, we see that

$$a' = -\frac{b}{2}\frac{d\theta}{dt}, \qquad\qquad (\text{B.6})$$

$$b' = +\frac{a}{2}\frac{d\theta}{dt}. \qquad\qquad (\text{B.7})$$

If we allow proper-distance rescaling (e.g., to convert the time scale t to an arc-length or proper-distance scale $\tau(t)$), we need to introduce the following two notions.

- *Geometric curvature*: This is the intrinsic, reparameterization-independent curvature describing the bending of the curve in space, which is the same at any point in space regardless of any locally chosen parameters or coordinate system:

$$\kappa(t) = \frac{d\theta}{dt}.$$

- *Parametric velocity*: This is the intrinsic velocity or rate of change of the spatial distance with respect to a particular parameterization:

$$v(t) = \frac{d\tau(t)}{dt}.$$

We may now express the right-hand side of the 2D frame equations as

$$\hat{\mathbf{T}}' = W_1 \cdot \begin{bmatrix} 0 & -\kappa \\ +\kappa & 0 \end{bmatrix} \cdot \begin{bmatrix} a \\ b \end{bmatrix} = +v\kappa\hat{\mathbf{N}}$$

and

$$\hat{\mathbf{N}}' = W_2 \cdot \begin{bmatrix} 0 & -\kappa \\ +\kappa & 0 \end{bmatrix} \cdot \begin{bmatrix} a \\ b \end{bmatrix} = -v\kappa\hat{\mathbf{T}}.$$

Matching terms and multiplying by $W_i^T = W_i^{-1}$, we find that the equation

$$\begin{bmatrix} a' \\ b' \end{bmatrix} = \frac{1}{2}v \begin{bmatrix} 0 & -\kappa \\ +\kappa & 0 \end{bmatrix} \cdot \begin{bmatrix} a \\ b \end{bmatrix} \qquad (\text{B.8})$$

contains both the frame equations $\hat{\mathbf{T}}' = +\kappa\hat{\mathbf{N}}$ and $\hat{\mathbf{N}}' = -\kappa\hat{\mathbf{T}}$, but is an intrinsically simpler system with two variables and one constraint, replacing the classic system with four variables and three constraints.

If we take the angular range $0 \to 4\pi$ instead of $0 \to 2\pi$, we have a 2:1 quadratic mapping from (a, b) to $(\hat{\mathbf{T}}, \hat{\mathbf{N}})$ because $(a, b) \sim (-a, -b)$ (see Equation B.2). Equation B.8 is the *square root* of the frame equations (note the factor of $(1/2)$). The curvature matrix is basically $g^{-1}dg$, with $g = W_1$, and is an element of the Lie algebra for the 2D rotation spin group with the explicit form

$$\begin{bmatrix} a & b \\ -b & a \end{bmatrix}\begin{bmatrix} a' & -b' \\ b' & a' \end{bmatrix} = \begin{bmatrix} aa' + bb' & -ab' + ba' \\ ab' - ba' & aa' + bb' \end{bmatrix}$$

$$= \begin{bmatrix} 0 & -\frac{\theta'}{2} \\ +\frac{\theta'}{2} & 0 \end{bmatrix}.$$

Here, $aa' + bb' = 0$ due to the constraint $a^2 + b^2 = 1$, and

$$ab' - ba' = \cos\frac{\theta}{2}\left[\frac{\theta'}{2}\cos\frac{\theta}{2}\right] - \sin\frac{\theta}{2}\left[-\frac{\theta'}{2}\sin\frac{\theta}{2}\right]$$

$$= \frac{\theta'}{2},$$

giving the identification $v\kappa = \theta'$ when we pull out the factor of $1/2$ (as in Equation B.8).

Alternatively, we may write

$$\frac{d\theta}{dt} = \frac{d\tau}{dt}\frac{d\theta}{d\tau}$$

and

$$v(t) = \left| \frac{d\mathbf{x}(t)}{dt} \right|,$$

$$v^2 = \frac{ds^2}{dt^2} = \frac{d\mathbf{x}(t) \cdot d\mathbf{x}(t)}{dt^2}.$$

The actual group properties in (a, b) space follow from the multiplication rule (easily deduced from the formulas for the trigonometric functions of sums of angles)

$$(a, b) \star (\tilde{a}, \tilde{b}) = (a\tilde{a} - b\tilde{b}, a\tilde{b} + b\tilde{a}),$$

which is in turn isomorphic to *complex multiplication* with $(a, b) = a + ib = e^{i\theta/2}$. This is no surprise, in that $\mathbf{SO}(2)$ and its double-covering spin group are subgroups of the corresponding 3D rotation groups, and complex numbers are a subset of the quaternions.

An alternative approach to the entire 2D orientation system is in fact to represent a particular frame by the unit-length complex variable z, where

$$z = \left(\cos(\theta/2), \sin(\theta/2) \right) = e^{i\theta/2},$$

and the action of a rotation on the frame by

$$R(\theta') \cdot z = e^{i\theta'/2} e^{i\theta/2} = e^{i(\theta' + \theta)/2}.$$

The frame derivatives then become

$$\dot{R}(\theta) = i \frac{\dot{\theta}}{2} R(\theta)$$

$$= i v \frac{\kappa}{2} R(\theta).$$

Replacing commutative complex arithmetic by noncommutative quaternion arithmetic leads almost immediately to hypotheses for the corresponding expressions for 3D frames and rotations in terms of quaternions.

3D Quaternion Frames

We next outline the basic features for 3D orientation and quaternion frames, following the pattern now established in Appendix B for 2D orientation and complex numbers. A quaternion frame is a unit four-vector $q = (q_0, q_1, q_2, q_3) = (q_0, \mathbf{q})$ with the following features.

C.1   UNIT NORM

If we define the inner product of two quaternions as

$$q \cdot p = q_0 p_0 + q_1 p_1 + q_2 p_2 + q_3 p_3, \tag{C.1}$$

the components of a quaternion frame obey the constraint

$$q \cdot q = (q_0)^2 + (q_1)^2 + (q_2)^2 + (q_3)^2 = 1 \tag{C.2}$$

and therefore lie on $\mathbf{S}^3$, the three-sphere embedded in 4D Euclidean space $\mathbb{R}^4$.

C.2   MULTIPLICATION RULE

The quaternion product of two quaternions p and q is defined to give a positive cross product in the vector part, and may be written as

$$p \star q = (p_0 q_0 - \mathbf{p} \cdot \mathbf{q}, \ p_0 \mathbf{q} + q_0 \mathbf{p} + \mathbf{p} \times \mathbf{q}),$$

or more explicitly in component form as

$$p \star q = \begin{bmatrix} [p \star q]_0 \\ [p \star q]_1 \\ [p \star q]_2 \\ [p \star q]_3 \end{bmatrix} = \begin{bmatrix} p_0 q_0 - p_1 q_1 - p_2 q_2 - p_3 q_3 \\ p_1 q_0 + p_0 q_1 + p_2 q_3 - p_3 q_2 \\ p_2 q_0 + p_0 q_2 + p_3 q_1 - p_1 q_3 \\ p_3 q_0 + p_0 q_3 + p_1 q_2 - p_2 q_1 \end{bmatrix}. \tag{C.3}$$

This rule is isomorphic to left-multiplication in the group $\mathbf{SU}(2)$, the double covering of the ordinary 3D rotation group $\mathbf{SO}(3)$. What is more useful for our purposes is the fact that it is also isomorphic to multiplication by a member of the group of orthogonal transformations in $\mathbb{R}^4$, given by

$$p \star q = \mathbf{P}q = \begin{bmatrix} p_0 & -p_1 & -p_2 & -p_3 \\ p_1 & p_0 & -p_3 & p_2 \\ p_2 & p_3 & p_0 & -p_1 \\ p_3 & -p_2 & p_1 & p_0 \end{bmatrix} \begin{bmatrix} q_0 \\ q_1 \\ q_2 \\ q_3 \end{bmatrix}, \tag{C.4}$$

where $\mathbf{P}$ is an orthogonal matrix, $\mathbf{P}^T \cdot \mathbf{P} = I_4$, and $\det \mathbf{P} = (p \cdot p)^2 = 1$. Because $\mathbf{P}$ has only three free parameters, it does not itself include all 4D rotations. However, we may recover the remaining three parameters by considering transformation by right-multiplication to be an independent operation, resulting in a similar matrix but with the signs in the lower right-hand off-diagonal 3×3 section reversed:

$$q \star p = \bar{\mathbf{P}}q = \begin{bmatrix} p_0 & -p_1 & -p_2 & -p_3 \\ p_1 & p_0 & p_3 & -p_2 \\ p_2 & -p_3 & p_0 & p_1 \\ p_3 & p_2 & -p_1 & p_0 \end{bmatrix} \begin{bmatrix} q_0 \\ q_1 \\ q_2 \\ q_3 \end{bmatrix}. \tag{C.5}$$

This observation reflects the well-known decomposition of the 4D rotation group into two 3D rotations (e.g., see Hanson [66]).

If two quaternions a and b are transformed by multiplying them by the same quaternion p, their inner product $a \cdot b$ transforms as

$$(p \star a) \cdot (p \star b) = (a \cdot b)(p \cdot p) \tag{C.6}$$

and is therefore invariant if p is a unit-quaternion frame representing a rotation. This also follows trivially from the fact that the matrix $\mathbf{P}$ is orthogonal.

The *inverse* of a unit quaternion satisfies $q \star q^{-1} = (1, \mathbf{0})$ and is easily shown to take the form $q^{-1} = (q_0, -\mathbf{q})$. Hence, like complex numbers quaternions permit the inverse to be constructed by *conjugation* of the "imaginary" part:

$$\bar{q} = q^{-1} = (q_0, -\mathbf{q}).$$

The relative quaternion rotation t transforming between two quaternions may be represented using the product

$$t = p \star q^{-1} = (p_0 q_0 + \mathbf{p} \cdot \mathbf{q}, q_0 \mathbf{p} - p_0 \mathbf{q} - \mathbf{p} \times \mathbf{q}).$$

This has the convenient property that the zeroth component is the invariant 4D inner product $p \cdot q = \cos(\theta/2)$, where θ is the angle of the rotation in 3D space needed to rotate along a geodesic from the frame denoted by q to that given by p. In fact, the 4D inner product reduces to

$$p \star q^{-1} + q \star p^{-1} = (2q \cdot p, \mathbf{0}),$$

whereas the 3D dot product and cross product arise from the symmetric and antisymmetric sums of quaternions containing only a three-vector part:

$$\mathbf{p} \star \mathbf{q} \equiv (0, \mathbf{p}) \star (0, \mathbf{q}) = (-\mathbf{p} \cdot \mathbf{q}, \mathbf{p} \times \mathbf{q}),$$

$$\mathbf{p} \star \mathbf{q} + \mathbf{q} \star \mathbf{p} = -2\mathbf{p} \cdot \mathbf{q} = (-2\mathbf{p} \cdot \mathbf{q}, \mathbf{0}),$$

$$\mathbf{p} \star \mathbf{q} - \mathbf{q} \star \mathbf{p} = 2\mathbf{p} \times \mathbf{q} = (0, 2\mathbf{p} \times \mathbf{q}).$$

C.3   MAPPING TO 3D ROTATIONS

Every possible 3D rotation R (a 3×3 orthogonal matrix) can be constructed from either of two related quaternions, $q = (q_0, q_1, q_2, q_3)$ or $-q = (-q_0, -q_1, -q_2, -q_3)$, using the quadratic relationship $R_q(\mathbf{V}) = q \star (0, \mathbf{V}) \star q^{-1}$, written explicitly as

$$R = \begin{bmatrix} q_0^2 + q_1^2 - q_2^2 - q_3^2 & 2q_1 q_2 - 2q_0 q_3 & 2q_1 q_3 + 2q_0 q_2 \\ 2q_1 q_2 + 2q_0 q_3 & q_0^2 - q_1^2 + q_2^2 - q_3^2 & 2q_2 q_3 - 2q_0 q_1 \\ 2q_1 q_3 - 2q_0 q_2 & 2q_2 q_3 + 2q_0 q_1 & q_0^2 - q_1^2 - q_2^2 + q_3^2 \end{bmatrix}. \quad \text{[C.7]}$$

The signs here result from choosing the left-multiplication convention $R_p R_q(\mathbf{V}) = R_{pq}(\mathbf{V}) = (p \star q) \star (0, \mathbf{V}) \star (p \star q)^{-1}$. Algorithms for the inverse mapping from R to q require careful singularity checking, and are detailed (for example) in Nielson [132], Shoemake [148], Shuster [153], and Shuster and Natanson [154].

The basic algorithm for finding the vector part of a quaternion follows from examining

$$
R - R^T = \begin{bmatrix} 0 & -4q_0 q_3 & 4q_0 q_2 \\ +4q_0 q_3 & 0 & -4q_0 q_1 \\ -4q_0 q_2 & 4q_0 q_1 & 0 \end{bmatrix}
$$

and searching for the largest value, and then normalizing to find $\hat{\mathbf{n}}$ from $\mathbf{q}$. If $\theta \approx 0$, one extracts the value of $\hat{\mathbf{n}}$ from the largest difference of the off-diagonal terms. The rotation angle, even if zero or close to zero, follows reliably from the trace:

$$
\begin{aligned}
\operatorname{tr} R &= 1 + 2\cos\theta \\
&= 1 + 2\big(\cos^2(\theta/2) - \sin^2(\theta/2)\big) \\
&= 4\cos^2(\theta/2) - 1, \\
(q_0)^2 &= \cos^2(\theta/2) = \frac{1}{4}(\operatorname{tr} R + 1).
\end{aligned}
$$

The analog of Equation C.7 of the projective coordinates for 2D rotations noted in Equation B.3 is obtained by converting to the projective variable $\mathbf{c} = \mathbf{q}/q_0 = \tan(\theta/2)\hat{\mathbf{n}}$ and observing that

$$
(q_0)^2 = \frac{(q_0)^2}{(q_0)^2 + \mathbf{q} \cdot \mathbf{q}} = \frac{1}{1 + \mathbf{q} \cdot \mathbf{q}/(q_0)^2} = \frac{1}{1 + \|\mathbf{c}\|^2}.
$$

We then find the three-parameter form of the rotation matrix with no constraints:

$$
R = \frac{1}{1 + \|\mathbf{c}\|^2} \begin{bmatrix} 1 + c_1^2 - c_2^2 - c_3^2 & 2c_1 c_2 - 2c_3 & 2c_1 c_3 + 2c_2 \\ 2c_1 c_2 + 2c_3 & 1 - c_1^2 + c_2^2 - c_3^2 & 2c_2 c_3 - 2c_1 \\ 2c_1 c_3 - 2c_2 & 2c_2 c_3 + 2c_1 & 1 - c_1^2 - c_2^2 + c_3^2 \end{bmatrix}. \quad \text{(C.8)}
$$

C.4   ROTATION CORRESPONDENCE

When we substitute $q(\theta, \hat{\mathbf{n}}) = (\cos\frac{\theta}{2}, \hat{\mathbf{n}}\sin\frac{\theta}{2})$ into Equation C.7, where $\hat{\mathbf{n}} \cdot \hat{\mathbf{n}} = 1$ is a unit three-vector lying on the two-sphere $\mathbf{S}^2$, $\mathbf{R}(\theta, \hat{\mathbf{n}})$ becomes the standard matrix for a rotation by θ in the plane perpendicular to $\hat{\mathbf{n}}$. The quadratic form ensures that the two distinct unit quaternions q and $-q$ in $\mathbf{S}^3$ correspond to the *same* $\mathbf{SO}(3)$ rotation. The explicit form of $\mathbf{R}(\theta, \hat{\mathbf{n}})$ is

$$\mathbf{R}(\theta, \hat{\mathbf{n}}) = \begin{bmatrix} c + (n_1)^2(1-c) & n_1n_2(1-c) - sn_3 & n_3n_1(1-c) + sn_2 \\ n_1n_2(1-c) + sn_3 & c + (n_2)^2(1-c) & n_3n_2(1-c) - sn_1 \\ n_1n_3(1-c) - sn_2 & n_2n_3(1-c) + sn_1 & c + (n_3)^2(1-c) \end{bmatrix},$$

$$\text{(C.9)}$$

where $c = \cos\theta$, $s = \sin\theta$, and $\hat{\mathbf{n}} \cdot \hat{\mathbf{n}} = 1$. For example, choosing the quaternion $q = (\cos\frac{\theta}{2}, 0, 0, \sin\frac{\theta}{2})$ yields the rotation matrix

$$\mathbf{R} = \begin{bmatrix} \cos\theta & -\sin\theta & 0 \\ \sin\theta & \cos\theta & 0 \\ 0 & 0 & 1 \end{bmatrix},$$

producing a right-handed rotation of the basis vectors $\hat{\mathbf{x}} = (1, 0, 0)$ and $\hat{\mathbf{y}} = (0, 1, 0)$ around the $\hat{\mathbf{z}}$ axis.

C.5   QUATERNION EXPONENTIAL FORM

Just as the 2D rotation implemented in terms of complex variables has an exponential form, there is a way to write the 3D rotations in terms of exponentials of quaternions. We can write the following exponential and expand it in a power series to find a quaternion expression corresponding exactly to the complex expression for 2D rotations:

$$e^{\mathbf{i}\cdot\hat{\mathbf{n}}\theta/2} = \cos(\theta/2) + \mathbf{i}\cdot\hat{\mathbf{n}}\sin(\theta/2)$$

$$= q_0 + \mathbf{i}\cdot\mathbf{q}$$

$$= q_0 + iq_1 + jq_2 + kq_3.$$

Here, the three components of $\mathbf{i} = (i, j, k)$ are the quaternion imaginaries of Hamilton's original notation (Chapter 1), obeying $i^2 = j^2 = k^2 = ijk = -1$. As a consequence, for each individual component—say just the i component with $q_2 = q_3 = 0$—we recover precisely $e^{i\theta/2}e^{i\theta/2} = e^{i\theta}$, and we see again that $e^{i\theta/2}$ is literally the square root of the original complex representation of this 2D subset of the 3D rotations.

Frame and Surface Evolution

D

D.1 QUATERNION FRAME EVOLUTION

Using Equation C.7, we can express all 3D coordinate frames in the form of quaternions. If we assume that the columns of Equation C.7 are the vectors $(\hat{\mathbf{T}}, \hat{\mathbf{N}}_1, \hat{\mathbf{N}}_2)$, respectively, one can explicitly express each vector in terms of the matrices

$$[W_1] = \begin{bmatrix} q_0 & q_1 & -q_2 & -q_3 \\ q_3 & q_2 & q_1 & q_0 \\ -q_2 & q_3 & -q_0 & q_1 \end{bmatrix}, \tag{D.1}$$

$$[W_2] = \begin{bmatrix} -q_3 & q_2 & q_1 & -q_0 \\ q_0 & -q_1 & q_2 & -q_3 \\ q_1 & q_0 & q_3 & q_2 \end{bmatrix}, \tag{D.2}$$

$$[W_3] = \begin{bmatrix} q_2 & q_3 & q_0 & q_1 \\ -q_1 & -q_0 & q_3 & q_2 \\ q_0 & -q_1 & -q_2 & q_3 \end{bmatrix}, \tag{D.3}$$

with the result that $\hat{\mathbf{T}} = [W_3] \cdot [q]$, $\hat{\mathbf{N}}_1 = [W_1] \cdot [q]$, and $\hat{\mathbf{N}}_2 = [W_2] \cdot [q]$, where $[q]$ is the column vector with components (q_0, q_1, q_2, q_3). Differentiating each of these expressions and substituting

$$\begin{bmatrix} \hat{\mathbf{T}}'(t) & \hat{\mathbf{N}}_1'(t) & \hat{\mathbf{N}}_2'(t) \end{bmatrix}$$

$$= \begin{bmatrix} \hat{\mathbf{T}}(t) & \hat{\mathbf{N}}_1(t) & \hat{\mathbf{N}}_2(t) \end{bmatrix} v(t) \begin{bmatrix} 0 & -k_y(t) & +k_x(t) \\ +k_y(t) & 0 & -k_z(t) \\ -k_x(t) & +k_z(t) & 0 \end{bmatrix}, \qquad \text{(D.4)}$$

one finds that factors of the matrices $[W_i]$ can be pulled out and a single universal equation linear in the quaternions remains:

$$\begin{bmatrix} q_0' \\ q_1' \\ q_2' \\ q_3' \end{bmatrix} = \frac{v}{2} \begin{bmatrix} 0 & -k_x & -k_y & -k_z \\ +k_x & 0 & -k_z & +k_y \\ +k_y & +k_z & 0 & -k_x \\ +k_z & -k_y & +k_x & 0 \end{bmatrix} \cdot \begin{bmatrix} q_0 \\ q_1 \\ q_2 \\ q_3 \end{bmatrix}. \qquad \text{(D.5)}$$

The first occurrence of this equation we are aware of is found in the work of Tait [162]. Here, $v(t) = \|\mathbf{x}'(t)\|$ is the scalar magnitude of the curve derivative if a unit-speed parameterization is not being used for the curve. One may consider Equation D.5 in some sense to be the *square root* of the 3D frame equations. Alternatively, we can deduce directly from $R_q(\mathbf{V}) = q \star (0, \mathbf{V}) \star q^{-1}$, $dq = q \star (q^{-1} \star dq)$ and $(q^{-1} \star dq) = -(dq^{-1} \star q)$ that the 3D vector equations are equivalent to the quaternion form

$$q' = \frac{1}{2} vq \star (0, k_x, k_y, k_z) = \frac{1}{2} vq \star (0, \mathbf{k}), \qquad \text{(D.6)}$$

$$(q^{-1})' = -\frac{1}{2} v(0, \mathbf{k}) \star q^{-1}, \qquad \text{(D.7)}$$

where $\mathbf{k} = 2(q_0 \, d\mathbf{q} - \mathbf{q} \, dq_0 - \mathbf{q} \times d\mathbf{q})$, or, explicitly,

$$k_0 = 2(dq_x q_x + dq_y q_y + dq_z q_z + dq_0 q_0) = 0,$$

$$k_x = 2(q_0 dq_x - q_x dq_0 - q_y dq_z + q_z dq_y),$$

$$k_y = 2(q_0 dq_y - q_y dq_0 - q_z dq_x + q_x dq_z),$$

$$k_z = 2(q_0 dq_z - q_z dq_0 - q_x dq_y + q_y dq_x).$$

Here, $k_0 = 0$ is the diagonal value in Equation D.5. The quaternion approach to the frame equations exemplified by Equation D.5 (or Equation D.6) has the following key properties.

- $q(t) \cdot q'(t) = 0$ by construction. Thus, all unit quaternions remain unit quaternions as they evolve by this equation.
- The number of equations has been reduced from nine coupled equations with six orthonormality constraints to four coupled equations incorporating a single constraint that keeps the solution vector confined to the three-sphere.

D.2   QUATERNION SURFACE EVOLUTION

The same set of equations can be considered to work on curves that are paths in a surface, thus also permitting a quaternion equivalent to the Weingarten equations for the classical differential geometry of surfaces. Explicit forms permitting the recovery of the classical equations follow from reexpressing those equations,

$$D_{\mathbf{U}}\hat{\mathbf{N}} \times D_{\mathbf{V}}\hat{\mathbf{N}} = K(\mathbf{U} \times \mathbf{V}), \tag{D.8}$$

$$D_{\mathbf{U}}\hat{\mathbf{N}} \times \mathbf{V} + \mathbf{U} \times D_{\mathbf{V}}\hat{\mathbf{N}} = 2H(\mathbf{U} \times \mathbf{V}), \tag{D.9}$$

with $\mathbf{U} = \mathbf{x}_u \cdot \nabla$ and $\mathbf{V} = \mathbf{x}_v \cdot \nabla$, in quaternion form.

The Weingarten curvature equation is essentially the cross product of two derivatives of the form of Equations D.6 and D.7. Because the cross product is obtainable from a quaternion multiplication of two pure vectors, we can find our way to the formula by replacing the parameter t on a curve by a pair of parameters (u, v) corresponding to the motion of a frame on a surface along two separate curves. When the frame equations for the two curves are multiplied together in the following way, the cross product needed to derive the Weingarten equations appears immediately.

$$q_u \star q_v^{-1} = -\frac{1}{4} q \star (0, \mathbf{a}) \star (0, \mathbf{b}) \star q^{-1}$$

$$= -\frac{1}{4} q \star (-\mathbf{a} \cdot \mathbf{b}, \mathbf{a} \times \mathbf{b}) \star q^{-1}$$

$$= -\frac{1}{4}\left[-\mathbf{a}\cdot\mathbf{b}\hat{\mathbf{I}} + (\mathbf{a}\times\mathbf{b})_x\hat{\mathbf{T}}_1 + (\mathbf{a}\times\mathbf{b})_y\hat{\mathbf{T}}_2 + (\mathbf{a}\times\mathbf{b})_z\hat{\mathbf{N}}\right]. \quad \text{(D.10)}$$

Here, Equation 21.12 defines the quaternion frame vectors—$\hat{\mathbf{N}} \equiv (0, \hat{\mathbf{N}}) = q \star (0, \hat{\mathbf{z}}) \star q^{-1}$, and so on—and we have introduced the identity element $\hat{\mathbf{I}} = (1, \mathbf{0}) = q \star (1, \mathbf{0}) \star q^{-1}$ as a fourth quaternion basis vector. The mean curvature equation has only one derivative and a free vector field. An expression producing the right combination of terms is

$$\hat{\mathbf{T}}_1 \star q \star q_u^{-1} + \hat{\mathbf{T}}_2 \star q \star q_v^{-1}$$

$$= -\frac{1}{2}q \star (-\hat{\mathbf{x}}\cdot\mathbf{a} - \hat{\mathbf{y}}\cdot\mathbf{b}, \hat{\mathbf{x}}\times\mathbf{a} + \hat{\mathbf{y}}\times\mathbf{b}) \star q^{-1}$$

$$= -\frac{1}{2}\left[-(a_x + b_y)\hat{\mathbf{I}} + b_z\hat{\mathbf{T}}_1 - a_z\hat{\mathbf{T}}_2 + (a_y - b_x)\hat{\mathbf{N}}\right]. \quad \text{(D.11)}$$

Projecting out the $\hat{\mathbf{N}}$ component of these equations recovers the scalar and mean curvatures:

$$K = \det\begin{bmatrix} +a_y(u,v) & -a_x(u,v) \\ +b_y(u,v) & -b_x(u,v) \end{bmatrix} = a_xb_y - a_yb_x,$$

$$H = \frac{1}{2}\text{tr}\begin{bmatrix} +a_y(u,v) & -a_x(u,v) \\ +b_y(u,v) & -b_x(u,v) \end{bmatrix} = \frac{1}{2}(a_y - b_x).$$

Quaternion Survival Kit

This appendix summarizes the essential quaternion utilities (Tables E.1 through E.7) needed to implement many of the concepts presented in the book. Selected programs are duplicated in the text as applicable.

```
double MIN_NORM = 1.0e-7;

void
QuaternionProduct
(double p0, double p1, double p2, double p3,
 double q0, double q1, double q2, double q3,
 double *Q0, double *Q1, double *Q2, double *Q3)
{ *Q0 = p0*q0 - p1*q1 - p2*q2 - p3*q3;
  *Q1 = p1*q0 + p0*q1 + p2*q3 - p3*q2;
  *Q2 = p2*q0 + p0*q2 + p3*q1 - p1*q3;
  *Q3 = p3*q0 + p0*q3 + p1*q2 - p2*q1;
}

double
QuaternionDot
(double p0, double p1, double p2, double p3,
 double q0, double q1, double q2, double q3)
{return(p0*q0 + p1*q1 + p2*q2 + p3*q3); }

void
QuaternionConjugate
(double q0, double q1, double q2, double q3,
 double *Q0, double *Q1, double *Q2, double *Q3)
{ *Q0 =  q0;
  *Q1 = -q1;
  *Q2 = -q2;
  *Q3 = -q3;
}

void
NormalizeQuaternion
(double *q0, double *q1, double *q2, double *q3)
{double denom;
 denom =
   sqrt((*q0)*(*q0) + (*q1)*(*q1) + (*q2)*(*q2) + (*q3)*(*q3));
 if(denom > MIN_NORM) { *q0 = (*q0)/denom;
                        *q1 = (*q1)/denom;
                        *q2 = (*q2)/denom;
                        *q3 = (*q3)/denom; }
}
```

TABLE E.1   *Elementary C code implementing the quaternion operations of Equations 4.1 through 4.3, and forcing unit magnitude as required by Equation 4.4. In this straight C-coding method, we return multiple values as results only through pointers such as* double *Q0.

```
typedef struct tag_Quat { double w, x, y, z; } Quat;

typedef struct tag_Point { double x, y, z; } Point;

void QuatMult(Quat *q1, Quat *q2, Quat *res)
{res->w =
  q1->w * q2->w - q1->x * q2->x - q1->y * q2->y - q1->z * q2->z;
 res->x =
  q1->w * q2->x + q1->x * q2->w + q1->y * q2->z - q1->z * q2->y;
 res->y =
  q1->w * q2->y + q1->y * q2->w + q1->z * q2->x - q1->x * q2->z;
 res->z =
  q1->w * q2->z + q1->z * q2->w + q1->x * q2->y - q1->y * q2->x;
}

double  QuatDot(Quat *q1, Quat *q2)
{return(
 q1->w * q2->w + q1->x * q2->x + q1->y * q2->y + q1->z * q2->z);
}

void  ConjQuat(Quat *q, Quat *qc)
{   qc->w =   q->w;
    qc->x = - q->x; qc->y = - q->y; qc->z = - q->z; }

double MIN_NORM = 1.0e-7;

void UnitQuat(Quat *v)
{double denom =
        sqrt( v->w*v->w + v->x*v->x + v->y*v->y + v->z*v->z );
 if(denom > MIN_NORM){ v->x /= denom;
                       v->y /= denom;
                       v->z /= denom;
                       v->w /= denom; }  }

double  ScalarQuat(Quat *q) { return (q->w); }

void VectorQuat(Quat *q, Point *v) {
    v->x = q->x;
    v->y = q->y;
    v->z = q->z; }
```

TABLE E.2 *An alternative quaternion code kit corresponding to Equations 4.1 through 4.4. The use of structures and pointers reduces the overhead for transmission of argument values compared to the version in Table E.1.*

```
MatToQuat(double m[4][4], Quat * quat)
{
  double  tr, s, q[4];
  int     i, j, k;

  int nxt[3] = {1, 2, 0};

  tr = m[0][0] + m[1][1] + m[2][2];

  /* check the diagonal */
  if (tr > 0.0) {
    s = sqrt (tr + 1.0);
    quat->w = s / 2.0;
    s = 0.5 / s;
    quat->x = (m[2][1] - m[1][2]) * s;
    quat->y = (m[0][2] - m[2][0]) * s;
    quat->z = (m[1][0] - m[0][1]) * s;
  } else {
    /* diagonal is negative */
    i = 0;
    if (m[1][1] > m[0][0]) i = 1;
    if (m[2][2] > m[i][i]) i = 2;
    j = nxt[i];
    k = nxt[j];

    s = sqrt ((m[i][i] - (m[j][j] + m[k][k])) + 1.0);

    q[i] = s * 0.5;

    if (s != 0.0) s = 0.5 / s;

    q[3] = (m[k][j] - m[j][k]) * s;
    q[j] = (m[i][j] + m[j][i]) * s;
    q[k] = (m[i][k] + m[k][i]) * s;

    quat->x = q[0];
    quat->y = q[1];
    quat->z = q[2];
    quat->w = q[3];
  }
}
```

TABLE E.3   *Basic C programs for manipulating quaternions, part I.*

```
QuatToMatrix(Quat * quat, double m[4][4])
{
  double wx, wy, wz, xx, yy, yz, xy, xz, zz, x2, y2, z2;

  /* calculate coefficients */

  x2 = quat->x + quat->x;
  y2 = quat->y + quat->y;
  z2 = quat->z + quat->z;
  xx = quat->x * x2;    xy = quat->x * y2;    xz = quat->x * z2;
  yy = quat->y * y2;    yz = quat->y * z2;    zz = quat->z * z2;
  wx = quat->w * x2;    wy = quat->w * y2;    wz = quat->w * z2;

  m[0][0] = 1.0 - (yy + zz);    m[0][1] = xy - wz;
  m[0][2] = xz + wy;            m[0][3] = 0.0;

  m[1][0] = xy + wz;            m[1][1] = 1.0 - (xx + zz);
  m[1][2] = yz - wx;            m[1][3] = 0.0;

  m[2][0] = xz - wy;            m[2][1] = yz + wx;
  m[2][2] = 1.0 - (xx + yy);    m[2][3] = 0.0;

  m[3][0] = 0;                  m[3][1] = 0;
  m[3][2] = 0;                  m[3][3] = 1;
}
```

TABLE E.4   *Basic C programs for manipulating quaternions, part II.*

```
/* Adapted from Graphics Gems Module:        GLquat.c */
void
slerp(Quat *q1, Quat *q2, double alpha, int spin, Quat *qOut)
{
    double beta;                // complementary interp parameter
    double theta;              // angle between A and B
    double sin_t, cos_t;       // sine, cosine of theta
    double phi;                // theta plus spins
    int     bflip;             // use negation of B?

    // cosine theta = dot product of A and B
    cos_t = QuatDot(q1,q2);

    // if B is on opposite hemisphere from A, use -B instead
    if (cos_t < 0.0) {
    cos_t = -cos_t;
    bflip = 1;
    } else
    bflip = 0;

  /* if B is (within precision limits) the same as A,
   * just linear interpolate between A and B.
   * Can't do spins, since we don't know what direction to spin.
   */
    if (1.0 - cos_t < 1e-7) {
    beta = 1.0 - alpha;
    } else {                    /* normal case */
    theta = acos(cos_t);
    phi   = theta + spin * M_PI;
    sin_t = sin(theta);
    beta  = sin(theta - alpha*phi) / sin_t;
    alpha = sin(alpha*phi) / sin_t;
    }

    if (bflip)
    alpha = -alpha;

    /* interpolate */
    qOut->x = beta*q1->x + alpha*q2->x;
    qOut->y = beta*q1->y + alpha*q2->y;
    qOut->z = beta*q1->z + alpha*q2->z;
    qOut->w = beta*q1->w + alpha*q2->w;

    UnitQuat(qOut);
}
```

TABLE E.5   *Basic C programs for quaternion SLERPs, part I.*

```
void
CatmullQuat(Quat *q00, Quat *q01, Quat *q02, Quat *q03,
            double t, Quat *qOut)
{   Quat q10,q11,q12,q20,q21;

    slerp(  q00,   q01,   t+1,    0, &q10 );
    slerp(  q01,   q02,   t,      0, &q11 );
    slerp(  q02,   q03,   t-1,    0, &q12 );

    slerp( &q10,  &q11,(t+1)/2, 0, &q20 );
    slerp( &q11,  &q12,   t/2,   0, &q21 );

    slerp( &q20,  &q21,   t,     0, qOut );
}

void
BezierQuat(Quat *q00, Quat *q01, Quat *q02, Quat *q03,
           double t, Quat *qOut)
{   Quat q10,q11,q12,q20,q21;

    slerp(  q00,   q01, t, 0, &q10 );
    slerp(  q01,   q02, t, 0, &q11 );
    slerp(  q02,   q03, t, 0, &q12 );

    slerp( &q10,  &q11, t, 0, &q20 );
    slerp( &q11,  &q12, t, 0, &q21 );

    slerp( &q20,  &q21, t, 0, qOut );
}

void
UniformBSplineQuat(Quat *q00, Quat *q01, Quat *q02, Quat *q03,
                   double t, Quat *qOut)
{   Quat q10,q11,q12,q20,q21;

    slerp(  q00,   q01, (t+2.0)/3.0, 0, &q10 );
    slerp(  q01,   q02, (t+1.0)/3.0, 0, &q11 );
    slerp(  q02,   q03,   t/3.0,     0, &q12 );

    slerp( &q10,  &q11, (t+1.0)/2.0, 0, &q20 );
    slerp( &q11,  &q12,   t/2.0,     0, &q21 );

    slerp( &q20,  &q21, t, 0,qOut);
}
```

TABLE E.6   *Basic C programs for quaternion SLERPs, part II.*

```
qprod[q_List,p_List] :=
{q[[1]]*p[[1]] - q[[2]]*p[[2]] - q[[3]]*p[[3]] - q[[4]]*p[[4]],
 q[[1]]*p[[2]] + q[[2]]*p[[1]] + q[[3]]*p[[4]] - q[[4]]*p[[3]],
 q[[1]]*p[[3]] + q[[3]]*p[[1]] + q[[4]]*p[[2]] - q[[2]]*p[[4]],
 q[[1]]*p[[4]] + q[[4]]*p[[1]] + q[[2]]*p[[3]] - q[[3]]*p[[2]]}
  /; Length[q] == Length[p] == 4

QuatToRot[q_List] :=
  Module[{q0= q[[1]], q1=q[[2]], q2 = q[[3]], q3 = q[[4]]},
   Module[{d23 = 2 q2 q3, d1 = 2 q0 q1,
           d31 = 2 q3 q1, d2 = 2 q0 q2,
           d12 = 2 q1 q2, d3 = 2 q0 q3,
           q0sq = q0^2, q1sq = q1^2, q2sq = q2^2, q3sq = q3^2},
          {{q0sq + q1sq - q2sq - q3sq, d12 - d3, d31 + d2},
           {d12 + d3, q0sq - q1sq + q2sq - q3sq, d23 - d1},
           {d31 - d2, d23 + d1, q0sq - q1sq - q2sq + q3sq}} ]]

makeQRot[angle_:0, n_List:{0,0,1}] :=
      Module[{c = Cos[angle/2], s = Sin[angle/2]},
      {c, n[[1]]*s, n[[2]]*s, n[[3]]*s}//N]

RotToQuat[mat_List] := Module[{q0,q1,q2,q3,trace,s,t1,t2,t3},
  trace = Sum[mat[[i,i]],{i,1,3}];
  If[trace > 0,    s = Sqrt[trace + 1.0]; q0 = s/2; s = 1/(2 s);
     q1 = (mat[[3,2]] - mat[[2,3]])*s;
     q2 = (mat[[1,3]] - mat[[3,1]])*s;
     q3 = (mat[[2,1]] - mat[[1,2]])*s,
   If[(mat[[1,1]] >= mat[[2,2]]) && (mat[[1,1]] >= mat[[3,3]]),
        (* i=0,  j = 1, k = 2 *)
      s = Sqrt[mat[[1,1]] - mat[[2,2]] - mat[[3,3]] + 1.0];
      q1 = s/2;  s = 1/(2 s);
      q0 = (mat[[3,2]] - mat[[2,3]])*s;
      q2 = (mat[[2,1]] + mat[[1,2]])*s;
      q3 = (mat[[1,3]] + mat[[3,1]])*s,
   If[(mat[[1,1]] < mat[[2,2]]) && (mat[[1,1]] >= mat[[3,3]]),
        (* i=1,  j = 2, k = 0 *)
      s = Sqrt[mat[[2,2]] - mat[[3,3]] - mat[[1,1]] + 1.0];
      q2 = s/2;  s = 1/(2 s);
      q0 = (mat[[1,3]] - mat[[3,1]])*s;
      q3 = (mat[[3,2]] + mat[[2,3]])*s;
      q1 = (mat[[2,1]] + mat[[1,2]])*s,
  (* Else:   i=2,  j = 0, k = 1
     (mat[[1,1]] < mat[[2,2]]) && (mat[[1,1]] < mat[[3,3]]) *)
      s = Sqrt[mat[[3,3]] - mat[[1,1]] - mat[[2,2]] + 1.0];
      q3 = s/2;  s = 1/(2 s);
      q0 = (mat[[2,1]] - mat[[1,2]])*s;
      q1 = (mat[[1,3]] + mat[[3,1]])*s;
      q2 = (mat[[3,2]] + mat[[2,3]])*s]]];
  {q0,q1,q2,q3}/Norm[{q0,q1,q2,q3}]]
```

TABLE E.7   *Basic Mathematica programs for manipulating quaternions.*

Quaternion Methods

F

This appendix presents a family of miscellaneous algorithms and methods we have found useful from time to time.

F.1 QUATERNION LOGARITHMS AND EXPONENTIALS

If we parametcrize a quaternion in a simplified way (with $0 \leqslant \theta < 2\pi$) to cover the entire $\mathbf{S}^3$ as $q(\theta, \hat{\mathbf{n}}) = (\cos\theta, \sin\theta\hat{\mathbf{n}})$, we can show using the quaternion algebra that this parameterization of $\mathbf{S}^3$ follows from the exponential series

$$\exp(0, \theta\hat{\mathbf{n}}) = (\cos\theta, \sin\theta\hat{\mathbf{n}}).$$

The logarithm of a structure is the object that, when exponentiated, produces the generic structure. Hence, we must have

$$\log q = \log(\cos\theta, \sin\theta\hat{\mathbf{n}}) = (0, \theta\hat{\mathbf{n}}).$$

Raising a quaternion to a power is defined via the exponential power series as well:

$$q^t = e^{t \log q} = e^{(0, t\theta\hat{\mathbf{n}})}$$

$$= (\cos t\theta, \hat{\mathbf{n}} \sin t\theta).$$

451

F.2   THE QUATERNION SQUARE ROOT TRICK

Quaternion square roots are often needed in various calculations, and there are several approaches of varying elegance. The most obvious is simply to express the quaternion in eigenvector coordinates and take half the total rotation:

$$q(\theta, \hat{\mathbf{n}}) = \left(\cos \tfrac{\theta}{2}, \hat{\mathbf{n}} \sin \tfrac{\theta}{2} \right),$$

$$p = \sqrt{q} \quad \text{iff} \quad p \star p = q,$$

$$p = \left(\cos \tfrac{\theta}{4}, \hat{\mathbf{n}} \sin \tfrac{\theta}{4} \right).$$

However, there is also a more elegant algebraic approach, motivated by the observation that if x is the cosine of an angle, the cosine of $1/2$ the angle is

$$y = \frac{\sqrt{1+x}}{\sqrt{2}}$$

$$= \frac{1+x}{\sqrt{2(1+x)}}.$$

Letting $q = (q_0, \mathbf{q})$, and noting the relation

$$(q_0)^2 = 1 - \mathbf{q} \cdot \mathbf{q},$$

we can use the half-angle formula to motivate an algebraic solution to $p \star p = q$ of the form

$$p = \frac{1+q}{\sqrt{2(1+q_0)}},$$

where $1 + q = (1 + q_0, q_x, q_y, q_z) = (1 + q_0, \mathbf{q})$. Because

$$(1+q) \star (1+q) = \left((1+q_0)^2 - \mathbf{q} \cdot \mathbf{q}, 2(1+q_0)\mathbf{q} \right)$$
$$= \left(2q_0(1+q_0), 2(1+q_0)\mathbf{q} \right)$$
$$= 2(1+q_0)q,$$

the identity $p \star p = q$ follows at once.

F.3   THE $\hat{a} \rightarrow \hat{b}$ FORMULA SIMPLIFIED

We frequently use the knowledge that a unit vector (direction) $\hat{a}$ can be aligned with another unit vector (direction) $\hat{b}$ by applying the rotation matrix $R(\theta, \hat{n})$ with

$$\cos\theta = \hat{a} \cdot \hat{b},$$

$$\sin\theta = \|\hat{a} \times \hat{b}\|,$$

$$\hat{n} = \frac{\hat{a} \times \hat{b}}{\|\hat{a} \times \hat{b}\|} = \frac{\hat{a} \times \hat{b}}{\sin\theta}.$$

(We can normally restrict $0 < \theta < \pi$ so $\sin\theta > 0$.)

The corresponding quaternion

$$r(\hat{a}, \hat{b}) = \left(\cos\frac{\theta}{2}, \hat{n}\sin\frac{\theta}{2} \right)$$

can be simplified if we let $z = \cos\theta = \hat{a} \cdot \hat{b}$ and recall that

$$\cos\frac{\theta}{2} = \sqrt{\frac{1+z}{2}},$$

$$\sin\frac{\theta}{2} = \sqrt{\frac{1-z}{2}},$$

$$\sin\theta = 2\cos\frac{\theta}{2}\sin\frac{\theta}{2}.$$

Then

$$r(\hat{a}, \hat{b}) = \left(\cos\frac{\theta}{2}, \hat{a} \times \hat{b}\frac{\sin\frac{\theta}{2}}{2\cos\frac{\theta}{2}\sin\frac{\theta}{2}} \right)$$

$$= \left(\cos\frac{\theta}{2}, \hat{a} \times \hat{b}\frac{1}{2\cos\frac{\theta}{2}} \right)$$

$$= \left(\sqrt{\frac{1+z}{2}}, \sqrt{\frac{1}{2(1+z)}}\,\hat{a} \times \hat{b} \right).$$

F.4   GRAM–SCHMIDT SPHERICAL INTERPOLATION

To find an interpolated unit vector that is guaranteed to remain on the sphere, and thus preserve its length of unity, we first assume that one vector (say, $\mathbf{q}_0$) is the starting point of the interpolation. To apply a standard rotation formula using a rigid length-preserving orthogonal transformation in the plane of $\mathbf{q}_0$ and $\mathbf{q}_1$, we need a unit vector *orthogonal* to $\mathbf{q}_0$ that contains some portion of the direction $\mathbf{q}_1$. Defining

$$\mathbf{q}_1' = \frac{\mathbf{q}_1 - \mathbf{q}_0(\mathbf{q}_0 \cdot \mathbf{q}_1)}{|\mathbf{q}_1 - \mathbf{q}_0(\mathbf{q}_0 \cdot \mathbf{q}_1)|},$$

we see that by construction

$$\mathbf{q}_1' \cdot \mathbf{q}_0 = 0,$$

$$\mathbf{q}_1' \cdot \mathbf{q}_1' = 1,$$

where we of course continue to require $\mathbf{q}_0 \cdot \mathbf{q}_0 = 1$. The denominator of this expression has the curious property that

$$\left| \mathbf{q}_1 - \mathbf{q}_0(\mathbf{q}_0 \cdot \mathbf{q}_1) \right|^2 = 1 - 2\cos^2 \phi + \cos^2 \phi$$

$$= \sin^2 \phi,$$

where we define $\cos \phi = \mathbf{q}_0 \cdot \mathbf{q}_1$. Note that because we have imposed $0 \leqslant \phi < \pi$, the sine is always nonnegative, and we can replace $|\sin \phi|$ by $\sin \phi$, which we will find convenient in the following. When $\phi = 0$, there is no interpolation to be done in any event.

Referring to the graphical construction shown in Figure 10.6, we next rephrase the unit-length-preserving rotation using the angle $t\phi$—where $0 \leqslant t \leqslant 1$ takes us from a unit vector aligned with $\mathbf{q}_0$ at $t = 0$ to one aligned with $\mathbf{q}_1$ at $t = 1$. Our new orthonormal basis can now be used to express the interpolation as:

$$\mathbf{q}(t) = \mathbf{q}_0 \cos t\phi + \mathbf{q}_1' \sin t\phi$$

$$= \mathbf{q}_0 \cos t\phi + (\mathbf{q}_1 - \mathbf{q}_0 \cos \phi) \frac{\sin t\phi}{\sin \phi}$$

$$= \mathbf{q}_0 \frac{\cos t\phi \sin \phi - \sin t\phi \cos \phi}{\sin \phi} + \mathbf{q}_1 \frac{\sin t\phi}{\sin \phi}$$

$$= \mathbf{q}_0 \frac{\sin(1-t)\phi}{\sin\phi} + \mathbf{q}_1 \frac{\sin t\phi}{\sin\phi}. \qquad \text{(F.1)}$$

This is the *SLERP formula*, which guarantees that

$$\mathbf{q}(t) \cdot \mathbf{q}(t) \equiv 1$$

by construction. The SLERP interpolator rotates one unit vector into another, keeping the intermediate vector in the mutual plane of the two limiting vectors while guaranteeing that the interpolated vector preserves its unit length throughout and therefore always remains on the sphere. The formula is true in *any dimension whatsoever* because it depends only on the local 2D plane determined by the two limiting vectors.

F.5   DIRECT SOLUTION FOR SPHERICAL INTERPOLATION

Let q be a unit vector on a sphere. Assume that q is located partway between two other unit vectors q_0 and q_1, with the location defined by some constants c_0 and c_1:

$$q = c_0 q_0 + c_1 q_1. \qquad \text{(F.2)}$$

As shown in Figure 10.7, q must partition the angle ϕ between q_0 and q_1 (where $\cos\phi = q_0 \cdot q_1$) into two subangles, ϕ_0 and ϕ_1, where $\cos\phi_0 = q \cdot q_1$ and $\cos\phi_1 = q \cdot q_0$ and $\phi = \phi_0 + \phi_1$. The labeling is chosen so that $\phi_0 = \phi$ makes $q = q_0$, and $\phi_1 = \phi$ makes $q = q_1$. No matter what the dimension of the unit-length qs, taking two dot products reduces this to a solvable linear system:

$$q \cdot q_0 = \cos\phi_1 = c_0 + c_1 \cos\phi,$$

$$q \cdot q_1 = \cos\phi_0 = c_0 \cos\phi + c_1.$$

Using Cramer's rule, we immediately find c_0 and c_1:

$$c_0 = \frac{\det \begin{bmatrix} \cos\phi_1 & \cos\phi \\ \cos\phi_0 & 1 \end{bmatrix}}{\det \begin{bmatrix} 1 & \cos\phi \\ \cos\phi & 1 \end{bmatrix}}$$

$$= \frac{\cos\phi_1 - \cos\phi_0 \cos\phi}{1 - \cos^2\phi},$$

$$c_1 = \frac{\det\begin{bmatrix} 1 & \cos\phi_1 \\ \cos\phi & \cos\phi_0 \end{bmatrix}}{\det\begin{bmatrix} 1 & \cos\phi \\ \cos\phi & 1 \end{bmatrix}}$$

$$= \frac{\cos\phi_0 - \cos\phi_1 \cos\phi}{1 - \cos^2\phi}.$$

Making the substitution $\phi = \phi_0 + \phi_1$, we find

$$c_0 = \frac{(\cos\phi_1 \sin\phi_0 + \cos\phi_0 \sin\phi_1)\sin\phi_0}{\sin^2\phi}$$

$$= \frac{\sin\phi_0}{\sin\phi}$$

$$= \frac{\sin t_0\phi}{\sin\phi},$$

$$c_1 = \frac{(\cos\phi_1 \sin\phi_0 + \cos\phi_0 \sin\phi_1)\sin\phi_1}{\sin^2\phi}$$

$$= \frac{\sin\phi_1}{\sin\phi}$$

$$= \frac{\sin t_1\phi}{\sin\phi},$$

where we have defined $t_0 = \phi_0/\phi$ and $t_1 = \phi_1/\phi$ to obtain a partition of unity, $t_0 + t_1 = 1$. Choosing, for example, $t_0 = 1 - t$ and $t_1 = t$, we recover the standard SLERP formula:

$$\mathbf{q}(t) = \mathbf{q}_0 \frac{\sin(1-t)\phi}{\sin\phi} + \mathbf{q}_1 \frac{\sin t\phi}{\sin\phi}. \tag{F.3}$$

Thus, the SLERP formula can be verified in a number of ways.

F.6   CONVERTING LINEAR ALGEBRA TO QUATERNION ALGEBRA

The equation describing a linear Euclidean interpolation

$$x(t) = a + t(b - a) \qquad \text{(F.4)}$$

transforms into the quaternion relation

$$q(t) = a\left(a^{-1}b\right)^t \qquad \text{(F.5)}$$

when we apply the following simple rules.

- Convert sums to products.
- Convert minus signs to products with the inverse.
- Convert multiplicative scale factors to exponents.

Because quaternions are not necessarily commutative, however, one must be careful about the order of multiplication. The order given in Equation F.5 will generally provide consistent results.

F.7   USEFUL TENSOR METHODS AND IDENTITIES

At various points in the text, we make use of some elegant formulas from linear algebra that are commonly used in mathematics and physics—for example, for group theoretical calculations, Maxwell's equations, and so on—but less often seen in the computer science literature. Here, we tabulate a few of these objects and some useful properties.

F.7.1   EINSTEIN SUMMATION CONVENTION

A shortcut attributed to Albert Einstein eliminates the need for large numbers of awkward summation symbols in complex tensor equations. It is rumored that Einstein was encouraged by his printer's typesetter, who wanted to simplify his job for Einstein's extremely complex mathematical formulas. The convention is simple: *unless otherwise noted, any repeated index is assumed to be summed over its allowed values.* Thus, we have

$$A_i B^i \equiv \sum_{i=0}^{3} A_i B^i,$$

$$A_{ij} A_{ik} \equiv \sum_{i=0}^{3} A_{ij} A_{ik},$$

$$A_{ijk} B^{imn} \equiv \sum_{i=0}^{3} A_{ijk} B^{imn},$$

$$A_{ijk} B^{ijn} \equiv \sum_{i=0}^{3} \sum_{j=0}^{3} A_{ijk} B^{ijn},$$

$$A_{ijk} B^{ijk} \equiv \sum_{i=0}^{3} \sum_{j=0}^{3} \sum_{k=0}^{3} A_{ijk} B^{ijk},$$

and so on.

F.7.2 KRONECKER DELTA

The Kronecker delta, δ_{ij}, is essentially a fancy way of writing the *identity matrix* as

$$\delta_{ij} = 0 \quad \text{if } i \neq j,$$
$$\delta_{ij} = 1 \quad \text{if } i = j$$

or

$$\text{Identity Matrix}_{ij} = \delta_{ij}.$$

F.7.3 LEVI-CIVITA SYMBOL

The Levi-Civita symbol—ϵ_{ij}, ϵ_{ijk}, ϵ_{ijkl}, and so on—is totally antisymmetric in its indices, and although it appears to be a Euclidean tensor in general it behaves differently under reflections than a tensor does. It is in fact a *pseudotensor*, and thus its alternate name is "the totally antisymmetric pseudotensor." The Levi-Civita symbol is defined as follows.

$$\epsilon_{ijk...} = 0 \quad \text{[If any two indices are equal]},$$

$$\epsilon_{ijk...} = +1 \quad \text{[If the indices are in cyclic order]},$$

$$\epsilon_{ijk...} = -1 \quad \text{[If the indices are in anticyclic order]}.$$

One of the many useful properties of the Levi-Civita symbol is that it can be used to write down an algebraic equation for a determinant. For example, using the Einstein summation convention we see that

$$\det \begin{bmatrix} a_{11} & a_{12} \\ a_{21} & a_{22} \end{bmatrix} = \frac{1}{2}\epsilon_{il}\epsilon_{jm}a_{ij}a_{lm}$$

$$= a_{11}a_{22} - a_{12}a_{21}$$

for any 2D matrix. In general, the Levi-Civita symbol can be used the same way for any dimension, and thus

$$\det[A] = \frac{1}{N!}\epsilon_{i_1 i_2 ... i_N}\epsilon_{j_1 j_2 ... j_N} \cdots \epsilon_{n_1 n_2 ... n_N} a_{i_1 j_1 ... n_1} a_{i_2 j_2 ... n_2} \cdots a_{i_N j_N ... n_N},$$

where there are N factors of the matrix elements $a_{...}$ and of the Levi-Civita symbols.

Finally, partial sums of Levi-Civita symbols result in determinants of Kronecker deltas, giving rise to a generalization of some familiar identities from 3D linear algebra such as

$$(\mathbf{A} \times \mathbf{B}) \cdot (\mathbf{C} \times \mathbf{D}) = \mathbf{A} \cdot \mathbf{C} - \mathbf{B} \cdot \mathbf{D},$$

which follows from the two definitions

$$(\mathbf{A} \times \mathbf{B})_i = \epsilon_{ijk}\mathbf{A}_j\mathbf{B}_k,$$

$$\epsilon_{ijk}\epsilon_{ilm} = \det \begin{bmatrix} \delta_{jl} & \delta_{jm} \\ \delta_{kl} & \delta_{km} \end{bmatrix}.$$

All possible relations of this type follow from the generic form

$$\epsilon_{i_1 i_2 ... i_N}\epsilon_{j_1 j_2 ... j_N} = \det \begin{bmatrix} \delta_{i_1 j_1} & \delta_{i_1 j_2} & \cdots & \delta_{i_1 j_N} \\ \delta_{i_2 j_1} & \delta_{i_2 j_2} & \cdots & \delta_{i_2 j_N} \\ \vdots & \vdots & \ddots & \vdots \\ \delta_{i_N j_1} & \delta_{i_N j_2} & \cdots & \delta_{i_N j_N} \end{bmatrix}.$$

Quaternion Path Optimization Using Surface Evolver

The Surface Evolver system (Brakke [22]) is a very sophisticated optimization package that has been in active use by the scientific and mathematics community for well over a decade and is still actively maintained by its developer, Ken Brakke, at Susquehanna University. The Surface Evolver can handle a number of issues specifically related to quaternion optimization. The symmetry group option

```
symmetry_group "central_symmetry"
```

identifies the quaternion q with $-q$ if desired during the variation to prevent reflected double traversals (such as those depicted in Figure 22.5) from varying independently. The system also supports metrics such as the pull-back metric on the sphere

$$ds^2 = \sum_{i,j} dx_i\, dx_j\, r^{-4}\left(r^2\delta_{ij} - x_i x_j\right)$$

to compute distances directly on the three-sphere. Computation using this metric, however, is very slow. Thus, in practice we have used the Euclidean $\mathbb{R}^4$ chord approximation, which works quite well for closely spaced samples and is much faster. Yet another alternative proposed by Brakke (private communication) is to use periodic coordinates on $\mathbf{S}^3$ of the form

$$(x_1 = \sin r \cos s,\, x_2 = \sin r \sin s,\, x_3 = \cos r \cos t,\, x_4 = \cos r \sin t)$$

and to vary directly on an $\mathbb{R}^3$ space with $(x = r, y = s, z = t)$ and the metric (see Equation 28.7):

$$\begin{bmatrix} 1 & 0 & 0 \\ 0 & \sin^2 x & 0 \\ 0 & 0 & \cos^2 x \end{bmatrix}.$$

To use the Evolver for large numbers of "sliding ring" sample points on a curve, it may be necessary to locate the parameter #define BDRYMAX 20 in skeleton.h, set it to the desired (large) value, and recompile. One must also remember to set space_dimension 4 when working with the quaternion hypersphere $\mathbf{S}^3$ in $\mathbb{R}^4$. To set up the constraint equations, it is necessary to write a piece of code (similar to the following Mathematica fragment) to define the boundary constraints for each fixed vector (tangent or normal) and the chosen initial quaternion frame.

```
Do[ringeqn = Qprod[makeQfromVec[veclist[[i]],P1],
                qOlist[[i]]]//Chop;
  Write[file, "boundary ",i," parameters 1"];
  Write[file, "x1: ",  CForm[ ringeqn[[2]]]];
  Write[file, "x2: ",  CForm[ ringeqn[[3]]]];
  Write[file, "x3: ",  CForm[ ringeqn[[4]]]];
  Write[file, "x4: ",  CForm[ ringeqn[[1]]]],
  {i,1,Length[veclist]}]
```

Note that because the Surface Evolver displays only the first three coordinates we have moved the scalar quaternion to the end. The Surface Evolver will then display our preferred projection automatically.

Quaternion Frame Integration

HClassical differential geometry texts present a time-honored method of finding the shape of a curve given a specification for its curvature and torsion throughout the curve along with its initial conditions. For example, an excellent modern treatment that is very compatible with the philosophy of this book, and especially the Mathematica-friendly computational philosophy we often exploit for visualization, is that of Alfred Gray [61].

Table H.1 shows the code for the traditional method of solving the differential equation, as implemented by Gray. Using the quaternion frame methods of Chapter 20, we are led to the much more compact program (shown in Table H.2), which is also slightly more general in that we accommodate a wider class of possible frame transport definitions. Accounting for a change-of-sign convention between the two approaches, the results shown in Figure H.1 are exactly identical to the results published in Gray's book. The following are calls to the quaternion frame program shown in Table H.2.

```
qplot3dx[{-Abs[#1] &, 0.3 &, 0 &},
{0, {0, 0, 0}, {1, 0, 0, 0}}, {-10,  10},
Axes -> None, AxesLabel -> {x, y, z}, PlotPoints
-> 500]

qplot3dx[{-1.3 &, .5 Sin[#1] &, 0 &},
{0, {0, 0, 0}, {1, 0, 0, 0}}, {0, 4 Pi},
Axes -> None, AxesLabel -> {x, y, z}, PlotPoints ->
200].
```

```
plotintrinsic3d[{kk_,tt_},
  {a_:0,{p1_:0,p2_:0,p3_:0},
        {q1_:1,q2_:0,q3_:0},
        {r1_:0,r2_:1,r3_:0}},
        {smin_:10,smax_:10},opts___] :=

 ParametricPlot3D[Module[
   {x1,x2,x3,t1,t2,t3,n1,n2,n3,b1,b2,b3},
   {x1[s],x2[s],x3[s]} /.
NDSolve[{x1'[ss] == t1[ss],
         x2'[ss] == t2[ss],
         x3'[ss] == t3[ss],
         t1'[ss] == kk[ss] n1[ss],
         t2'[ss] == kk[ss] n2[ss],
         t3'[ss] == kk[ss] n3[ss],
         n1'[ss] == -kk[ss] t1[ss] + tt[ss] b1[ss],
         n2'[ss] == -kk[ss] t2[ss] + tt[ss] b2[ss],
         n3'[ss] == -kk[ss] t3[ss] + tt[ss] b3[ss],
         b1'[ss] == -tt[ss] n1[ss],
         b2'[ss] == -tt[ss] n2[ss],
         b3'[ss] == -tt[ss] n3[ss],
         x1[a] == p1, x2[a] == p2, x3[a] == p3,
         t1[a] == q1, t2[a] == q2, t3[a] == q3,
         n1[a] == r1, n2[a] == r2, n3[a] == r3,
         b1[a] == q2*r3 - q3*r2,
         b2[a] == q3*r1 - q1*r3,
         b3[a] == q1*r2 - q2*r1},
        {x1,x2,x3,t1,t2,t3,n1,n2,n3,b1,b2,b3},
        {ss,smin,smax}]]//Evaluate,
        {s,smin,smax},opts];
```

TABLE H.1   *The Mathematica program* plotintrinsic3d[curvature,
torsion, initial conditions] *used in Gray [61] to plot a curve given its
initial conditions and curvatures as input parameters.*

```
qplot3dx[{k1_, k2_, k3_},
         {sz_:0, {xz1_:0, xz2_:0, xz3_:0},
         {qz0_:1, qz1_:0, qz2_:0, qz3_:0}},
         {smin_:0, smax_:1},
         opts___] :=

ParametricPlot3D[
  Evaluate[
     Module[{x1, x2, x3, q0, q1, q2, q3},
       {x1[s], x2[s], x3[s]} /.
NDSolve[
 { q0'[ss] ==
     (1/2)*(-k2[ss] q1[ss] - k3[ss] q2[ss] + k1[ss] q3[ss]),
   q1'[ss] ==
     (1/2)*(k2[ss] q0[ss] - k1[ss] q2[ss] + k3[ss] q3[ss]),
   q2'[ss] ==
     (1/2)*(k3[ss] q0[ss] + k1[ss] q1[ss] + k2[ss] q3[ss]),
   q3'[ss] ==
     (1/2)*(-k1[ss] q0[ss] - k3[ss] q1[ss] - k2[ss] q2[ss]),
   x1'[ss] == q0[ss]^2 + q1[ss]^2 - q2[ss]^2 - q3[ss]^2,
   x2'[ss] == 2 q1[ss] q2[ss] + 2 q0[ss] q3[ss],
   x3'[ss] == 2 q3[ss] q1[ss] - 2 q0[ss] q2[ss],
   x1[sz] == xz1, x2[sz] == xz2, x3[sz] == xz3,
   q0[sz] == qz0, q1[sz] == qz1, q2[sz] == qz2, q3[sz] == qz3},
   {x1, x2, x3, q0, q1, q2, q3},
   {ss, smin, smax} ]  ]  ],
   {s, smin, smax}, opts]
```

TABLE H.2   *The Mathematica quaternion frame program that is exactly equivalent to the much more complex standard frame method implemented in Table H.1.*

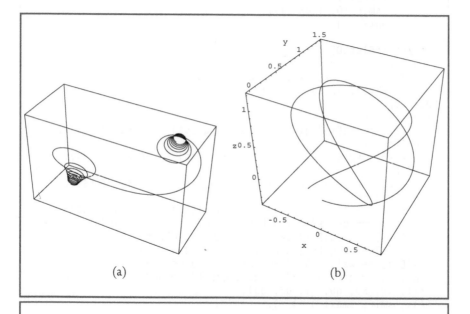

(a)   (b)

FIGURE H.1 The results of the calls to the quaternion frame program in Table H.2 generate exactly the same results as those shown in Gray [61], using the program in Table H.1.

Hyperspherical Geometry

Our quaternion analysis has made essential use of the properties of the three-sphere $\mathbf{S}^3$ and its spherical geometry. Spheres themselves exist in all dimensions, and have a list of similar properties that may help to answer some general questions about the special case $\mathbf{S}^3$ that has been so important to us, and to place it in a useful general context.

I.1   DEFINITIONS

A sphere $\mathbf{S}^N$ is most naturally defined as the embedding of the constant-radius equation in $(N+1)$-dimensional Euclidean space $\mathbb{R}^{N+1}$:

$$(x_1)^2 + \cdots + (x_N)^2 + (x_{N+1})^2 = r^2.$$

A hypersurface element, or solid-angle integral, of a hypersphere $\mathbf{S}^N$ is [47,166]

$$\Omega_N = \frac{2\pi^{(N+1)/2}}{\Gamma(\frac{1}{2}(N+1))}.$$

The volume of a ball B^N of radius r is found by integrating shells of the surface element of the sphere $\mathbf{S}^N$ out to the desired radius to give the volume:

$$V\left(B^{N+1}\right) = \int_0^r \Omega_N r^N \, dr = \Omega_N \frac{r^{N+1}}{N+1}.$$

A table of these values follows. (*Note:* $\Gamma(1) = 1$, $\Gamma(1/2) = \sqrt{\pi}$, $\Gamma(3/2) = (1/2)\sqrt{\pi}$, and in general $\alpha\Gamma(\alpha) = \Gamma(\alpha + 1)$. Thus, for example, $n! = \Gamma(n + 1)$ for nonnegative integers n.)

N-**Sphere Dim.**	Ω_N-**Solid Angle**
0	2
1	2π
2	4π
3	$2\pi^2$
4	$\frac{8\pi^2}{3}$
5	π^3
6	$\frac{16\pi^3}{15}$
7	$\frac{\pi^4}{3}$
8	$\frac{32\pi^4}{105}$
9	$\frac{\pi^5}{12}$
10	$\frac{64\pi^5}{945}$
$\vdots$	$\vdots$
N	$\dfrac{2\pi^{(N+1)/2}}{\Gamma(\frac{1}{2}(N+1))}$

Ball dimension	**Volume of ball**
B^1	$2r$
B^2	πr^2
B^3	$\frac{4\pi r^3}{3}$
B^4	$\frac{\pi^2 r^4}{2}$
B^5	$\frac{8\pi^2 r^5}{15}$
B^6	$\frac{\pi^3 r^6}{6}$
B^7	$\frac{16\pi^3 r^7}{105}$
B^8	$\frac{\pi^4 r^8}{24}$
B^9	$\frac{32\pi^4 r^9}{945}$
B^{10}	$\frac{\pi^5 r^{10}}{120}$
B^{11}	$\frac{64\pi^5 r^{11}}{10395}$
$\vdots$	$\vdots$
B^{N+1}	$\dfrac{r^{(N+1)}}{N+1}\dfrac{2\pi^{(N+1)/2}}{\Gamma((N+1)/2)}$

I.2   METRIC PROPERTIES

The metric on a sphere can be defined in a number of forms, depending on the choice of embedding coordinates and the resulting induced metric. For each embedding, one obtains volume element, a line element, a connection, and an equation for geodesic paths within the spherical manifold.

- *Polar coordinates:* $0 \leqslant x_n < 2\pi$, whereas all others are $0 \leqslant x_k \leqslant \pi$. If you switch $\cos(x_k)$ and $\sin(x_k)$, the range is $-\pi/2 \leqslant x_k \leqslant \pi/2$.

$$y_1 = r \cos x_1,$$

$$y_2 = r \sin x_1 \cos x_2,$$

$$y_3 = r \sin x_1 \sin x_2 \cos x_3,$$

$$\vdots$$

$$y_{n-1} = r \sin x_1 \sin x_2 \sin x_3 \sin x_4 \ldots \cos x_{n-1},$$

$$y_n = r \sin x_1 \sin x_2 \sin x_3 \sin x_4 \ldots \sin x_{n-1} \cos x_n,$$

$$y_{n+1} = r \sin x_1 \sin x_2 \sin x_3 \sin x_4 \ldots \sin x_{n-1} \sin x_n.$$

- Diagonal elements of induced metrics:

$$g_{11} = r^2,$$

$$g_{22} = r^2 \sin^2 x_1,$$

$$g_{33} = r^2 \sin^2 x_1 \sin^2 x_2,$$

$$\vdots$$

$$g_{nn} = r^2 \sin^2 x_1 \sin^2 x_2 \ldots \sin^2 x_{n-1}.$$

- Line elements from induced metrics:

$$ds^2 = r^2(dx_1)^2,$$

$$ds^2 = r^2(dx_1)^2 + \sin^2 x_1 (dx_2)^2,$$

$$ds^2 = r^2(dx_1)^2 + \sin^2 x_1 \big((dx_2)^2 + \sin^2 x_2 (dx_3)^2\big),$$

$$\vdots$$

$$(ds_N)^2 = r^2(dx_1)^2 + \sin^2 x_1 \big((dx_2)^2 + \sin^2 x_2 \big((dx_3)^2$$

$$+ \cdots + \sin^2 x_{N-1}(dx_N)^2\big)\ldots\big).$$

- Line elements from hemisphere-embedding coordinates:

$$ds^2 = d\mathbf{x} \cdot d\mathbf{x} + \frac{(\mathbf{x} \cdot d\mathbf{x})(\mathbf{x} \cdot d\mathbf{x})}{1 - \mathbf{x} \cdot \mathbf{x}}.$$

- *Stereographic projection:* The $\mathbf{S}^N$ metric induced from the stereographic projection in $\mathbb{R}^{N+1}$

$$\mathbf{x}' = \frac{(2\mathbf{x}, 1 - |\mathbf{x}|^2)}{1 + |\mathbf{x}|^2}$$

  is

$$g_{ij} = \frac{4\delta_{ij}}{(|\mathbf{x}|^2 + 1)^2}$$

  and the oriented volume element is

$$d\Omega_N = \left(\frac{2}{|\mathbf{x}|^2 + 1}\right)^N dx_1 \wedge dx_2 \wedge \cdots \wedge dx_N,$$

  where we use standard differential form notation (e.g., see Flanders [52]).

References

[1]   R. Ablamowicz, P. Lounesto, and J. M. Parra. *Clifford Algebras with Numeric and Symbolic Computations*. Boston, MA: Birkhäuser, 1996.

[2]   P. Alfeld, M. Neamtu, and L. L. Schumaker. "Bernstein–Bézier Polynomials on Spherical and Sphere-like Surfaces," *CAGD Journal* 13:333–349, 1996.

[3]   B. Alpern, L. Carter, M. Grayson, and C. Pelkie. "Orientation Maps: Techniques for Visualizing Rotations (A Consumer's Guide)," in *Proceedings of Visualization '93*, pp. 183–188. IEEE Computer Society Press, 1993.

[4]   S. L. Altmann. *Rotations, Quaternions, and Double Groups*. Oxford: Clarendon, 1986.

[5]   S. L. Altmann. "Hamilton, Rodrigues, and the Quaternion Scandal," *Mathematics Magazine* 62(5):291–308, December 1989.

[6]   M. Artin. *Algebra*. Englewood Cliffs, NJ: Prentice Hall, 1991.

[7]   B. Artmann. *The Concept of Number: From Quaternions to Monads to Topological Fields*. New York: Halsted Press, 1988.

[8]   M. F. Atiyah, R. Bott, and A. Shapiro. "Clifford Modules," *Topology* 3, Suppl. 1:3–38, 1986.

[9]   J. Baez. "The Octonions," *Bull. Amer. Math. Soc.* 39:145–205, 2002. Corrected version available at *www.arxiv.org/abs/math.RA/0105155* and *http://math.ucr.edu/home/baez/Octonions/*.

471

[10]   T. Banchoff and J. Werner. *Linear Algebra Through Geometry*. New York:
         Springer-Verlag, 1983.

[11]   T. F. Banchoff. "Visualizing Two-dimensional Phenomena in Four-
         dimensional Space: A Computer Graphics Approach," in E. Wegman and
         D. Priest (eds.), *Statistical Image Processing and Computer Graphics*, pp. 187–202.
         New York: Marcel Dekker, 1986.

[12]   T. F. Banchoff. *Beyond the Third Dimension: Geometry, Computer Graphics, and Higher
         Dimensions*. New York: Scientific American Library, 1990.

[13]   D. Banks. "Interactive Manipulation and Display of Two-dimensional Sur-
         faces in Four-dimensional Space," in D. Zeltzer (ed.), *Computer Graphics
         (1992 Symposium on Interactive 3D Graphics)*, vol. 25, pp. 197–207. New York:
         ACM Press, 1992.

[14]   D. C. Banks. "Illumination in Diverse Codimensions," in *Computer Graphics
         (SIGGRAPH '94 Proceedings, Annual Conference Series)*, pp. 327–334. New York:
         ACM Press, 1994.

[15]   A. H. Barr, B. Currin, S. Gabriel, and J. F. Hughes. "Smooth Interpolation
         of Orientations with Angular Velocity Constraints Using Quaternions,"
         in E. E. Catmull (ed.), *Computer Graphics (SIGGRAPH '92 Proceedings)*, vol. 26,
         pp. 313–320, July 1992.

[16]   M. Berger. *Geometry I, II*. Berlin: Springer-Verlag, 1987.

[17]   L. C. Biedenharn and J. D. Louck. "Angular Momentum in Quantum
         Physics," in G.-C. Rota (ed.), *Encyclopedia of Mathematics and Its Applications*,
         vol. 8. New York: Cambridge University Press, 1984.

[18]   L. C. Biedenharn and J. D. Louck. "The Racah–Wigner Algebra in Quan-
         tum Theory," in G.-C. Rota (ed.), *Encyclopedia of Mathematics and Its Applications*,
         vol. 9. New York: Cambridge University Press, 1984.

[19]   R. L. Bishop. "There Is More Than One Way to Frame a Curve," *Amer. Math.
         Monthly* 82(3):246–251, March 1975.

[20]   W. Blaschke. *Kinematik und Quaternionen*. Berlin: VEB Deutscher Verlag der Wissenschaften, 1960.

[21]   J. Bloomenthal. "Calculation of Reference Frames Along a Space Curve," in A. Glassner (ed.), *Graphics Gems*, pp. 567–571. Cambridge, MA: Academic Press, 1990.

[22]   K. A. Brakke. "The Surface Evolver," *Experimental Mathematics* 1(2):141–165, 1992. The "Evolver" system, manual, and sample data files are available by anonymous ftp from *geom.umn.edu*, The Geometry Center, Minneapolis, MN.

[23]   D. W. Brisson (ed.). *Hypergraphics: Visualizing Complex Relationships in Art, Science and Technology*, vol. 24. Boulder: Westview Press, 1978.

[24]   P. Brou. "Using the Gaussian Image to Find the Orientation of Objects," *International Journal of Robotics Research* 3(4):89–125, 1984.

[25]   J. L. Brown and A. J. Worsey. "Problems with Defining Barycentric Coordinates for the Sphere," *Mathematical Modeling and Numerical Analysis* 26:37–49, 1992.

[26]   S. R. Buss and J. P. Fillmore. "Spherical Averages and Applications to Spherical Splines and Interpolation," *ACM Transactions on Graphics (TOG)* 20(2):95–126, 2001.

[27]   S. A. Carey, R. P. Burton, and D. M. Campbell. "Shades of a Higher Dimension," *Computer Graphics World*, pp. 93–94, October 1987.

[28]   A. Cayley. "On Certain Results Relating to Quaternions," *Phil. Mag.* 26:141–145, 1845.

[29]   A. Cayley. "On Jacobi's Elliptic Functions, in Reply to the Rev. B. Bronwin; and on Quaternions," *Phil. Mag.* 26:208–211, 1845.

[30]   M. Chen, S. J. Mountford, and A. Sellen. "A Study in Interactive 3-D Rotation Using 2-D Control Devices," in *Computer Graphics (SIGGRAPH '88 Proceedings)*, vol. 22, pp. 121–130, 1988.

[31]   B. Cipra. "Mathematicians Gather to Play the Numbers Game," *Science* 259:894–895, 1993. Description of Alfred Gray's knots colored to represent variation in curvature and torsion.

[32]   W. K. Clifford. *Mathematical Papers*. New York: Chelsea Publishing Co., 1968. Reprinted from the 1882 edition.

[33]   H. S. M. Coxeter. *Regular Complex Polytopes* (2nd ed.). New York: Cambridge University Press, 1991.

[34]   R. A. Cross and A. J. Hanson. "Virtual Reality Performance for Virtual Geometry," in *Proceedings of Visualization '94*, pp. 156–163. Los Alamitos, CA: IEEE Computer Society Press, 1994.

[35]   M. J. Crowe. *A History of Vector Analysis*. Mineola, NY: Dover Press, 1964.

[36]   R. W. R. Darling. *Differential Forms and Connections*. Cambridge: Cambridge University Press, 1994.

[37]   C. Dorst, L. Doran, and J. Lasenby. *Applications of Geometric Algebra in Computer Science and Engineering*. Boston, MA: Birkhäuser, 2002.

[38]   W. Drechsler and M. E. Mayer. *Fiber Bundle Techniques in Gauge Theories*. Lecture Notes in Physics, No. 67. New York: Springer-Verlag, 1977.

[39]   P. DuVal. *Homographies, Quaternions and Rotations*. Oxford, 1964.

[40]   D. Eberly. "A Linear Algebraic Approach to Quaternions," 2002. www.geometrictools.com/Documentation/LinearAlgebraicQuaternions.pdf.

[41]   D. Eberly. "Quaternion Algebra and Calculus," 2002. www.geometrictools.com/Documentation/Quaternions.pdf.

[42]   D. Eberly. "Rotation Representations and Performance Issues," 2002. www.geometrictools.com/Documentation/RotationIssues.pdf.

[43]   A. R. Edmonds. *Angular Momentum in Quantum Mechanics*. Princeton, NJ: Princeton University Press, 1957.

[44] N. V. Efimov and E. R. Rozendorn. *Linear Algebra and Multi-Dimensional Geometry*. Moscow: Mir Publishers, 1975.

[45] T. Eguchi, P. B. Gilkey, and A. J. Hanson. "Gravitation, Gauge Theories and Differential Geometry." *Physics Reports* 66(6):213–393, December 1980.

[46] L. P. Eisenhart. *A Treatise on the Differential Geometry of Curves and Surfaces*. New York: Dover, 1960.

[47] A. Erdélyi, W. Magnus, M. F. Oberhettinger, and F. G. Tricomi. *Higher Transcendental Functions*, vol. 1. New York: McGraw–Hill, 1954. Based, in part, on notes left by Harry Bateman and compiled by the staff of the Bateman Manuscript Project.

[48] A. Erdélyi, W. Magnus, M. F. Oberhettinger, and F. G. Tricomi. *Higher Transcendental Functions*, vol. 2. New York: McGraw–Hill, 1954. Based, in part, on notes left by Harry Bateman and compiled by the staff of the Bateman Manuscript Project.

[49] A. Erdélyi, W. Magnus, M. F. Oberhettinger, and F. G. Tricomi. *Higher Transcendental Functions*, vol. 3. New York: McGraw–Hill, 1954. Based, in part, on notes left by Harry Bateman and compiled by the staff of the Bateman Manuscript Project.

[50] A. Erdélyi, M. F. Oberhettinger, and F. G. Tricomi. *Tables of Integral Transforms*. New York: McGraw–Hill, 1954. 2 volumes based, in part, on notes left by Harry Bateman and compiled by the staff of the Bateman Manuscript Project.

[51] G. Fischer. *Mathematische Modelle*, vols. I and II. Friedr. Braunschweig/Wiesbaden: Vieweg & Sohn, 1986.

[52] H. Flanders. *Differential Forms*. New York: Academic Press, 1963.

[53] J. D. Foley, A. van Dam, S. K. Feiner, and J. F. Hughes. *Computer Graphics, Principles and Practice* (2nd ed.). Reading, MA: Addison–Wesley, 1990.

[54]   A. R. Forsyth. *Geometry of Four Dimensions*. Cambridge: Cambridge University Press, 1930.

[55]   G. K. Francis. *A Topological Picturebook*. New York: Springer-Verlag, 1987.

[56]   R. Gilmore. *Lie Groups, Lie Algebras and Some of Their Applications*. New York: Wiley, 1974.

[57]   A. S. Glassner (ed.). *Graphics Gems*. Cambridge, MA: Academic Press, 1990.

[58]   H. Goldstein. *Classical Mechanics* (2nd ed.). Reading, MA: Addison–Wesley, 1981.

[59]   F. S. Grassia. "Practical Parameterization of Rotations Using the Exponential Map," *Journal of Graphics Tools* 3(3):29–48, 1998.

[60]   A. Gray. *Modern Differential Geometry of Curves and Surfaces*. Boca Raton, FL: CRC Press, 1993.

[61]   A. Gray. *Modern Differential Geometry of Curves and Surfaces* (2nd ed.). Boca Raton, FL: CRC Press, 1998.

[62]   J. J. Gray. "Olinde Rodrigues' Paper of 1840 on Transformation Groups," *Archive for the History of the Exact Sciences* 21:376–385, 1980.

[63]   C. M. Grimm and J. F. Hughes. "Modeling Surfaces with Arbitrary Topology Using Manifolds," in *Computer Graphics (SIGGRAPH '95 Proceedings, Annual Conference Series)*, pp. 359–368, 1995.

[64]   W. R. Hamilton. *Lectures on Quaternions*. Cambridge: Cambridge University Press, 1853.

[65]   W. R. Hamilton. *Elements of Quaternions*. Cambridge: Cambridge University Press, 1866. Republished by Chelsea, 1969.

[66]   A. J. Hanson. "The Rolling Ball," in D. Kirk (ed.), *Graphics Gems III*, pp. 51–60. Cambridge, MA: Academic Press, 1992.

[67]   A. J. Hanson. "Geometry for n-dimensional Graphics," in P. Heckbert (ed.), *Graphics Gems IV*, pp. 149–170. Cambridge, MA: Academic Press, 1994.

[68]   A. J. Hanson. "Quaternion Frenet Frames," Technical Report 407, Indiana University Computer Science Department, 1994.

[69]   A. J. Hanson. "Rotations for n-dimensional Graphics," in A. Paeth (ed.), *Graphics Gems V*, pp. 55–64. Cambridge, MA: Academic Press, 1995.

[70]   A. J. Hanson. "Constrained Optimal Framings of Curves and Surfaces Using Quaternion Gauss Maps," in *Proceedings of Visualization '98*, pp. 375–382. IEEE Computer Society Press, 1998. Missing pages 2 and 4 in original Proceedings. Correct version in second printing of Proceedings, on conference CD-ROM, or *ftp://ftp.cs.indiana.edu/pub/hanson/Vis98/vis98.quat.pdf*.

[71]   A. J. Hanson. "Quaternion Gauss Maps and Optimal Framings of Curves and Surfaces," Technical Report 518, Indiana University Computer Science Department, October 1998.

[72]   A. J. Hanson and R. A. Cross. "Interactive Visualization Methods for Four Dimensions," in *Proceedings of Visualization '93*, pp. 196–203. IEEE Computer Society Press, 1993.

[73]   A. J. Hanson and P. A. Heng. "Visualizing the Fourth Dimension Using Geometry and Light," in *Proceedings of Visualization '91*, pp. 321–328. IEEE Computer Society Press, 1991.

[74]   A. J. Hanson and P. A. Heng. "Four-dimensional Views of 3D Scalar Fields," in *Proceedings of Visualization '92*, pp. 84–91. IEEE Computer Society Press, 1992.

[75]   A. J. Hanson and P. A. Heng. "Foursight," in *Siggraph Video Review*, vol. 85. ACM Siggraph, 1992. Scene 11, presented in the Animation Screening Room at SIGGRAPH '92, Chicago, Illinois, July 28–31, 1992.

[76]   A. J. Hanson and P. A. Heng. "Illuminating the Fourth Dimension," *Computer Graphics and Applications* 12(4):54–62, July 1992.

[77]   A. J. Hanson, S. Hughes, and E. A. Wernert. "Constrained Navigation Interfaces," in H. Hagen and H.-C. Rodrian (eds.), *Scientific Visualization.* Los Alamitos, CA: IEEE Computer Society Press, Includes papers from the Dagstuhl June 1997 Workshop on Scientific Visualization.

[78]   A. J. Hanson and H. Ma. "Visualizing Flow with Quaternion Frames," in *Proceedings of Visualization '94*, pp. 108–115. Los Alamitos, CA: IEEE Computer Society Press, 1994.

[79]   A. J. Hanson and H. Ma. "Parallel Transport Approach to Curve Framing," Technical Report 425, Indiana University Computer Science Department, 1995.

[80]   A. J. Hanson and H. Ma. "Quaternion Frame Approach to Streamline Visualization," *IEEE Transactions on Visualization and Computer Graphics* 1(2):164–174, June 1995.

[81]   A. J. Hanson and H. Ma. "Space Walking," in *Proceedings of Visualization '95*, pp. 126–133. Los Alamitos, CA: IEEE Computer Society Press, 1995.

[82]   A. J. Hanson, T. Munzner, and G. K. Francis. "Interactive Methods for Visualizable Geometry," *IEEE Computer* 27(7):73–83, July 1994.

[83]   A. J. Hanson and E. A. Wernert. "Constrained 3D Navigation with 2D Controllers," in *Proceedings of Visualization '97*, pp. 175–182. Los Alamitos, CA: IEEE Computer Society Press, 1997.

[84]   A. J. Hanson, K. Ishkov, and J. Ma. "Meshview," a portable 4D geometry viewer written in OpenGL/Motif, available by anonymous ftp from *ftp.cs. indiana.edu:pub/hanson.*

[85]   A. J. Hanson, K. Ishkov, and J. Ma. "Meshview: Visualizing the Fourth Dimension," Overview of the Meshview 4D geometry viewer.

[86]   G. H. Hardy and E. M. Wright. *An Introduction to the Theory of Numbers.* Oxford: Oxford, 1954.

[87]   J. C. Hart, G. K. Francis, and L. H. Kauffman. "Visualizing Quaternion Rotation," *ACM Transactions on Graphics* 13(3):256–276, 1994.

[88]   E. J. Haug. *Computer-Aided Kinematics and Dynamics of Mechanical Systems Volume I: Basic Methods.* Boston: Allyn and Bacon, 1989.

[89]   L. Herda, R. Urtasun, and P. Fua. "Hierarchical Implicit Surface Joint Limits to Constrain Video-Based Motion Capture," in *ECCV*, Prague, Czech Republic, May 2004.

[90]   L. Herda, R. Urtasun, and P. Fua. "Hierarchical Implicit Surface Joint Limits to Constrain Video-based Motion Capture," in *European Conference on Computer Vision*, May 2004.

[91]   L. Herda, R. Urtasun, A. Hanson, and P. Fua. "An Automatic Method for Determining Quaternion Field Boundaries for Ball-and-Socket Joint Limits," in *Proceedings of the 5th International Conference on Automated Face and Gesture Recognition (FGR)*, pp. 95–100. Washington, DC, May 2002. IEEE Computer Society.

[92]   L. Herda, R. Urtasun, A. Hanson, and P. Fua. "Automatic Determination of Shoulder Joint Limits Using Experimentally Determined Quaternion Field Boundaries," *International Journal of Robotics Research* 22(6):419–434, June 2003.

[93]   D. Hilbert and S. Cohn-Vossen. *Geometry and the Imagination.* New York: Chelsea, 1952.

[94]   J. G. Hocking and G. S. Young. *Topology.* Reading, MA: Addison–Wesley, 1961.

[95]   M. Hofer and H. Pottmann. "Energy-minimizing Splines in Manifolds," *Transactions on Graphics* 23(3):284–293, 2004.

[96]   C. Hoffmann and J. Zhou. "Some Techniques for Visualizing Surfaces in Four-dimensional Space," *Computer-Aided Design* 23:83–91, 1991.

[97] S. Hollasch. "Four-space Visualization of 4D Objects," Master's thesis, Arizona State University, August 1991.

[98] P. Hughes. *Spacecraft Attitude Dynamics*. New York: Wiley, 1986.

[99] J. P. M. Hultquist. "Constructing Stream Surfaces in Steady 3D Vector Fields," in *Proceedings of Visualization '92*, pp. 171–178. Los Alamitos, CA: IEEE Computer Society Press, 1992.

[100] T. W. Hungerford. *Algebra*. New York: Springer-Verlag, 1974.

[101] A. Hurwitz. "Über die Composition der Quadratischen Formen von Beliebig Vielen Variabeln," *Nachr. Königl. Gesell. Wiss. Göttingen, Math.-Phys. Klasse*, pp. 309–316, 1898.

[102] M. Johnson. "Exploiting Quaternions," Ph.D. thesis, MIT, 2002.

[103] J. Junkins and J. Turner. *Optimal Spacecraft Rotational Maneuvers*. Elsevier, 1986.

[104] B. Jüttler. "Visualization of Moving Objects Using Dual Quaternion Curves," *Computers and Graphics* 18(3):315–326, 1994.

[105] B. Jüttler and M. G. Wagner. "Computer-aided Design with Spatial Rational B-spline Motions," *Journal of Mechanical Design* 118:193–201, June 1996.

[106] J. T. Kajiya. "Anisotropic Reflection Models," in *Computer Graphics (SIGGRAPH '85 Proceedings)*, vol. 19, pp. 15–21, 1985.

[107] J. T. Kajiya and S. Gabriel. "Spline Interpolation in Curved Space, 1985," course notes, SIGGRAPH '85.

[108] J. T. Kajiya and T. L. Kay. "Rendering FUR with Three-dimensional Textures," in J. Lane (ed.), *Computer Graphics (SIGGRAPH '89 Proceedings)*, vol. 23, pp. 271–280, July 1989.

[109] I. L. Kantor and A. S. Solodovnikov. *Hypercomplex Numbers: An Elementary Introduction to Algebras*. New York: Springer-Verlag, 1989.

[110]   M. A. Kervaire and J. W. Milnor. "Groups of Homotopy Spheres: I," *Annals of Mathematics* 77:504–537, 1963.

[111]   M.-J. Kim, M.-S. Kim, and S. Y. Shin. "A General Construction Scheme for Unit Quaternion Curves with Simple High Order Derivatives," in *Computer Graphics (SIGGRAPH '95 Proceedings, Annual Conference Series)*, pp. 369–376, 1995.

[112]   F. Klein. *The Icosahedron*. New York: Dover, 1956. Republication of 1913 English translation by B. B. Morrice of the original German edition of 1884.

[113]   F. Klein and A. Sommerfeld. *Über die Theorie des Kreisels*. Leipzig: Teubner, 1910.

[114]   F. Klock. "Two Moving Coordinate Frames for Sweeping Along a 3D Trajectory," *Computer Aided Geometric Design* 3:217–229, 1986.

[115]   J. B. Kuipers. *Quaternions and Rotation Sequences*. Princeton, NJ: Princeton University Press, 1999.

[116]   A. Kyrala. *Theoretical Physics: Applications of Vectors, Matrices, Tensors and Quaternions*. W. B. Saunders, 1967.

[117]   H. B. Lawson and M.-L. Michelsohn. *Spin Geometry*. Princeton, NJ: Princeton University Press, 1989.

[118]   J. Lee. *Riemannian Manifolds: An Introduction to Curvature*. New York: Springer, 1997.

[119]   H. Ma. "Curve and Surface Framing for Scientific Visualization and Domain Dependent Navigation," Ph.D. thesis, Indiana University, February 1996.

[120]   N. L. Max. "Computer Representation of Molecular Surfaces," *IEEE Computer Graphics and Applications* 3(5):21–29, August 1983.

[121]   N. L. Max. "DNA Animation, from Atom to Chromosome," *Journal of Molecular Graphics* 3(2):69–71, 1985.

[122]   J. Milnor. "Some Consequences of a Theorem of Bott," *Annals of Mathematics* 68:444–449, 1958.

[123]   J. Milnor. *Topology from the Differentiable Viewpoint.* Charlottesville, VA: The University Press of Virginia, 1965.

[124]   J. W. Milnor. "Topological Manifolds and Smooth Manifolds," in *Proc. Internat. Congr. Mathematicians (Stockholm, 1962)*, pp. 132–138. Djursholm: Inst. Mittag-Leffler, 1963.

[125]   C. W. Misner, K. S. Thorne, and J. A. Wheeler. *Gravitation.* New York: W. H. Freeman, 1973.

[126]   C. W. Misner, K. S. Thorne, and J. A. Wheeler. *Gravitation.* New York: W. H. Freeman, 1973. See "Orientation Entanglement Relation," p. 1149.

[127]   C. Moller. *The Theory of Relativity.* Oxford University Press, 1983.

[128]   H. S. Morton, Jr. "A Formulation of Rigid-body Rotational Dynamics Based on Generalized Angular Momentum Variables Corresponding to Euler Parameters," in *Proceedings of AIAA/Astrodynamics Conference*, 1984.

[129]   H. S. Morton, Jr. "Hamiltonian and Lagrangian Formulations of Rigid-body Rotational Dynamics Based on the Euler Parameters," *Journal of Astronautical Sciences* 41:569–592, 1994.

[130]   H. R. Müller. *Sphärische Kinematik.* Berlin: VEB Deutscher Verlag der Wissenschaften, 1962.

[131]   B. A. Murtagh and M. A. Saunder. "MINOS 5.0 User's Guide," Technical Report SOL 83-20, Dept. of Operations Research, Stanford University, 1983.

[132]   G. M. Nielson. "Smooth Interpolation of Orientations," in N. M. Thal-
         man and D. Thalman (eds.), *Computer Animation '93*, pp. 75–93, New York:
         Springer-Verlag, 1993.

[133]   M. A. Noll. "A Computer Technique for Displaying n-dimensional Hy-
         perobjects," *Communications of the ACM* 10(8):469–473, August 1967.

[134]   School of Mathematics and Scotland Statistics, University of St. Andrews.
         "History of Mathematics" web page. This web site contains an exhaustive
         history of mathematical developments, including many details on Hamil-
         ton and the history of quaternions.

[135]   F. C. Park and B. Ravani. "Smooth Invariant Interpolation of Rotations,"
         ACM Transactions on Graphics 16(3):277–295, July 1997.

[136]   R. Penrose and W. Rindler. *Spinors and Space–Time*. Cambridge: Cambridge
         University Press, 1984.

[137]   M. Phillips, S. Levy, and T. Munzner. "Geomview: An Interactive Geome-
         try Viewer," *Notices of the American Mathematical Society* 40(8):985–988, Octo-
         ber 1993. Available by anonymous ftp from *geom.umn.edu*, The Geometry
         Center, Minneapolis, MN.

[138]   J. C. Platt and A. H. Barr. "Constraint Methods for Flexible Models," in
         J. Dill (ed.), *Computer Graphics (SIGGRAPH '88 Proceedings)*, vol. 22, pp. 279–
         288, August 1988.

[139]   D. Pletincks. "Quaternion Calculus as a Basic Tool in Computer Graphics,"
         The Visual Computer 5(1):2–13, 1989.

[140]   R. Ramamoorthi and A. H. Barr. "Fast Construction of Accurate Quater-
         nion Splines," in T. Whitted (ed.), *SIGGRAPH '97 Proceedings, Annual Conference
         Series*, pp. 287–292. New York: ACM Press, 1997.

[141]   D. W. Ritchie and G. J. L. Kemp. "Fast Computation, Rotation, and Com-
         parison of Low Resolution Spherical Harmonic Molecular Surfaces," *Jour-
         nal of Computational Chemistry* 20:383–395, 1999. For details, see *www.math.
         chalmers.se/~kemp/publications/*.

[142]   O. Rodrigues. "Des Lois Géométriques Qui Régissent les Déplacements d'un Système Solide Dans L'espace, et la Variation des Coordonnées Provenant de Ses Déplacments Consideérés Indépendamment des Causes Qui Peuvent les Produire," *Journal de Mathématiques Pure at Appliquées* 5:380–440, 1840.

[143]   J. C. Russ. *Image Processing Handbook* (2nd ed.). Boca Raton, FL: CRC Press, 1995.

[144]   D. J. Sandin, L. H. Kauffman, and G. K. Francis. "Air on the Dirac Strings," Siggraph Video Review, vol. 93, 1993.

[145]   J. Schlag. "Using Geometric Constructions to Interpolate Orientation with Quaternions," in J. Arvo (ed.), *Graphics Gems II*, pp. 377–380. Reading, MA: Academic Press, 1991.

[146]   P. J. Schneider and D. H. Eberly. *Geometric Tools for Computer Graphics*. San Francisco: Morgan Kaufman, 2003.

[147]   U. Shani and D. H. Ballard. "Splines as Embeddings for Generalized Cylinders," *Computer Vision, Graphics, and Image Processing* 27:129–156, 1984.

[148]   K. Shoemake. "Quaternion Calculus for Animation," Siggraph '89 Course 23: Math for SIGGRAPH, 1987.

[149]   K. Shoemake. "Animating Rotation with Quaternion Curves," in *Computer Graphics (SIGGRAPH '85 Proceedings)*, vol. 19, pp. 245–254, 1985.

[150]   K. Shoemake. "Animation with Quaternions," Siggraph Course Lecture Notes, 1987.

[151]   K. Shoemake. "Arcball Rotation Control," in P. Heckbert (ed.), *Graphics Gems IV*, pp. 175–192. Cambridge, MA: Academic Press, 1994.

[152]   K. Shoemake. "Fiber Bundle Twist Reduction," in P. Heckbert (ed.), *Graphics Gems IV*, pp. 230–236. Cambridge, MA: Academic Press, 1994.

[153]   M. D. Shuster. "A Survey of Attitude Representations," *Journal of Astronautical Sciences* 41(4):439–517, 1993.

[154]   M. D. Shuster and G. A. Natanson. "Quaternion Computation from a Geometric Point of View," *Journal of Astronautical Sciences* 41(4):545–556, 1993.

[155]   D. M. Y. Sommerville. *An Introduction to the Geometry of N Dimensions*. New York: Dover Press, 1958.

[156]   N. Steenrod. *The Topology of Fibre Bundles*. Princeton Mathematical Series 14. Princeton, NJ: Princeton University Press, 1951.

[157]   K. V. Steiner and R. P. Burton. "Hidden Volumes: The 4th Dimension," *Computer Graphics World*, pp. 71–74, February 1987.

[158]   D. J. Struik. *Lectures on Classical Differential Geometry*. Reading, MA: Addison–Wesley, 1961.

[159]   R. Szeliski and D. Tonnesen. "Surface Modeling with Oriented Particle Systems," in *Computer Graphics (SIGGRAPH '92 Proceedings)*, vol. 26, pp. 185–194, 1992.

[160]   P. G. Tait. *An Elementary Treatise on Quaternions*. Cambridge: Cambridge University Press, 1867.

[161]   P. G. Tait. *Introduction to Quaternions*. Cambridge: Cambridge University Press, 1873.

[162]   P. G. Tait. *An Elementary Treatise on Quaternions* (rev. ed.). Cambridge: Cambridge University Press, 1890.

[163]   B. L. van der Waerden. *Group Theory and Quantum Mechanics*. Berlin: Springer-Verlag, 1974.

[164]   B. L. van der Waerden. "Hamilton's Discovery of Quaternions," *Mathematics Magazine*, 49:227–234, 1976.

[165]   M. Wade. "Apollo 10," *www.astronautix.com/flights/apollo10.htm*.

[166]   S. Waner. *Introduction to Differential Geometry and General Relativity*. Department of Mathematics, Hofstra University, 2002.

[167]   J. R. Weeks. *The Shape of Space* (2nd ed.). New York: Marcel Dekker, 2002.

[168]   S. Weinberg. *Gravitation and Cosmology: Principles and Applications of General Relativity*. New York: Wiley, 1972.

[169]   J. Wertz. *Spacecraft Attitude Determination and Control*. D. Reidel, 1985.

[170]   H. Whitney. "The Self-intersections of a Smooth n-manifold in $2n$-space," *Annals of Mathematics* 45(2):230–246, 1944.

[171]   H. Whitney. "The Singularities of a Smooth n-manifold in $(2n - 1)$-space," *Annals of Mathematics* 45(2):247–293, 1944.

[172]   E. T. Whittaker. *A Treatise on the Analytical Dynamics of Particles and Rigid Bodies*. New York: Dover, 1944.

[173]   E. Wigner. "Unitary Representations of the Inhomogeneous Lorentz Group," *Annals of Mathematics* 40:149, 1939.

[174]   A. Witkin and D. Baraff. Siggraph 2001 course notes on "Physically Based Modeling."

[175]   H. J. Woltring. "3-D Attitude Representation of Human Joints: A Standardization Proposal," *Journal of Biomechanics* 27:1399–1414, 1994.

[176]   W. D. Woods and F. O'Brien. "Apollo 15: Solo Orbital Operations—3," *www.hq.nasa.gov/office/pao/History/ap15fj/15solo_ops3.htm*.

Index

Printed and bound by CPI Group (UK) Ltd, Croydon, CR0 4YY

03/10/2024

01040312-0002